Plant Biopolymer Science
Food and Non-food Applications

Plant Biopolymer Science
Food and Non-food Applications

Edited by

D. Renard
Unité de Physico-Chimie des Macromolécules, INRA, Nantes, France

G. Della Valle
Unité de Recherches sur les Polysaccharides, leurs Organisations et leurs Interactions, INRA, Nantes, France

Y. Popineau
Unité de Biochimie et Technologie des Protéines, INRA, Nantes, France

ROYAL SOCIETY OF CHEMISTRY

The proceedings of the Workshop on Plant Biopolymer Science: Food and Non-food Applications held in Nantes, France on 25–27 June 2001.

Special Publication No. 276

ISBN 0-85404-856-1

A catalogue record for this book is available from the British Library

Published by The Royal Society of Chemistry
Thomas Graham House, Science Park, Milton Road, Cambridge CB4 0WF, UK

Registered Charity Number 207890

For further information see our web site at www.rsc.org

Typeset by Vision Typesetting, Manchester, UK
Printed and bound by Athenaeum Press Ltd, Gateshead, Tyne and Wear, UK

Preface

This book presents the proceedings of the Workshop on Plant Biopolymer Science: Food and Non-food Applications, which was held in Nantes at the Congress Centre on 24–27th June 2001. This meeting followed in the footsteps of the one organised in September 1998 in Montpellier at ENSAM, in collaboration with Professor Stephane Guilbert. The main idea behind this first meeting was that scientific approaches dealing with the structural or functional properties of macromolecules are rather similar whatever the type of biopolymer – protein, starch or cell wall polysaccharide. As this concept is now widely accepted, and since this first meeting was highly successful, it was consequently decided to organise this second 'Plant biopolymer workshop' here in Nantes.

Nantes was chosen as the location because of its INRA Research Centre, which is renowned for its basic and applied research on plant biopolymers, starch, proteins and cell wall polysaccharides. The main objectives of the Nantes Centre in Plant Science first of all concerns the biosynthesis of macromolecules and assemblies *in planta*, secondly their structural characteristics and related physico-chemical and functional properties, and thirdly with their behaviour in multiphasic systems in relation to end-uses in food and non-food applications. In addition, human nutrition is also considered.

One of the challenges for our scientific community over the next few years will be to improve how we respond to social demand, which has changed considerably over the last three years, and to hand down to our children a sustainable world. The situation today is indeed characterised by accelerating changes in many aspects concerning human lifestyles, social order, the environment and agricultural and industrial production systems. Increasing consumer suspicion of industrially-prepared foods, for example, means that more efforts will have to be made to guarantee the safety of these products. Consumer requirements are rather similar in most other industrial sectors like, for example, in the case of cosmetics and pharmaceutical products for which the use of animal-derived products is questionable. On the other hand, much of mankind is still suffering from de- or malnutrition, and questions concerning agricultural production and the availability of foodstuffs are still pregnant issues in many countries. To summarise, we have to take into account more accurately in our research programmes the demands made by citizens and consumers for sustainable agriculture, environmentally-friendly processes and new technology requirements and safe products. We also have to satisfy more adequately the great diversity of cultural, social, economical or age situations of consumers in developed countries, without ignoring demands from the so-called 'southern' countries. By considering the issue of safety, the increased development and use of plant-based products could partly satisfy this demand, even if we cannot ignore the controversial GMO debate.

To meet these consumer requirements for higher quality and less-transformed products more efficiently, I am convinced that we need to consider the production-transformation chain from a more integrated approach than in the past. In contrast with the logic of the conventional upstream approach 'do your best with the existing and sometimes overproduced crops' re-engineering the chain *via* a downstream analysis from the end-products to plant production should make it possible to improve how the industrial processes, and also crop production, are adapted to consumer requirements regarding the sustainability of agriculture, the development of environmentally-friendly processes as well as the safety and diversity of end-products. A good example may be the following:

– We can expect that increased knowledge of the biosynthetic pathways of plant biopolymers may lead through to their composition, structure and interactions in biological material being controlled more effectively so as to enhance technological functionality and nutritional quality.
– Improved control of component localisation and tissue organisation should also improve the efficiency of separation processes.

Scientifically, this downstream approach needs to monitor the end-products as multiphasic systems at the various organisational levels more closely: molecular, supramolecular and mesoscopic. It requires more basic information on the role of biopolymers in such systems and on how they induce, by interacting with each other or with smaller molecules, assemblies or phase separation processes, and on how they act as surface active components at interfaces. Some of this knowledge exists already but is not efficiently integrated into strategies dealing with improving the quality of starting plant material.

We have to find a link between end-product properties and the biosynthesis, tissue localisation and supramolecular organisation of macromolecules *in planta*. The genetic engineering and modelling of these biopolymers at the molecular and supramolecular levels are key tools for establishing this link. The recent advances made in material science on the one hand, especially dealing with self-assembly technology, and on genomic, proteomic and structural biology on the other, have drastically changed the scientific landscape in our research fields. Based on material science developments, it seems easier today than in the past to study the molecular and supramolecular organisation of biopolymers in solid or condensed states, at least in model systems. *In planta*, the exploding knowledge in genomics and proteomics should make it possible to control biosynthetic pathways and tissue organisation more effectively. Consequently, the integration of both these scientific fields applied to our area of research should make this re-engineered downstream approach successful.

The scientific programme of this workshop was established in the light of this new approach, considering that scientists from different areas were able to rub shoulders during the whole meeting. The different sessions were defined to integrate all the chains, starting with the *Biosynthesis of Macromolecules* and *Biopolymer Design*, then dealing with model systems such as *Biopolymer Assemblies, Interfaces and Interphases* and ending with more applied studies concerning *Multiphasic Systems*. In each area, the keynote speakers were asked to provide a fairly broad overview and, when possible, to pitch their talk at a level that could be understood by all the conference's attendees. These different sessions were defined with the help of the scientific committee which also selected the research contributions, lectures and posters.

Both the venue, Nantes Congress Centre, and the weather, proved to be first class, and the meals were excellent. Special thanks are due to P. Lefer as secretary of the congress, to INRA for financial support and the City of Nantes for its help in subsidising hospitality during this meeting, which was attended by over 120 delegates from all over the globe. Indeed, the meeting was judged to be so successful that plans for a follow-up meeting in autumn 2004 are already under way.

Jacques Guéguen
Honorary President of the Workshop on Plant Biopolymer Science 2001
Organising Committee
(INRA Nantes, France)

Contents

Biosynthesis

Biochemical Mechanisms of Synthesis of $(1\rightarrow3),(1\rightarrow4)\beta$-D-Glucans:
Cellulose Synthase with an added Twist?   3
B.R. Urbanowicz, C.J. Rayon, N.C. Carpita, M.A. Tiné and M.S. Buckeridge

The Future Prospects for Broadening Soybean Utilization by Altering
Glycinin   13
N.C. Nielsen and E. Herman

Heterogeneity of Genes Encoding the Low Molecular Weight Glutenin
Subunits of Bread Wheat   24
T. Chardot, T. Do, L. Perret and M. Laurière

Identification, Cloning and Expression of a Ferulic Acid Esterase from
Neurospora crassa   31
V.F. Crepin, C.B. Faulds and I.F. Connerton

Biopolymer Design

Influence of Polysaccharide Composition on the Structure and
Properties of Cellulose-based Composites   39
M.J. Gidley, E. Chanliaud and S. Whitney

Attempt to Produce Caffeoylated Arabinoxylans from Feryloylated
Arabinoxylans by Microbial Demethylation   48
V. Micard, T. Landazuri, A. Surget, S. Moukha, M. Labat and X. Rouau

Enzymatically and Chemically De-esterified Lime Pectins: Physico-
chemical Characterisation, Polyelectrolyte Behaviour and Calcium
Binding Properties   55
M.-C. Ralet, V. Dronnet and J.-F. Thibault

Incorporation of Unsaturated Isoleucine Analogues into Proteins
In Vivo 63
T. Michon, F. Barbot and D. Tirrell

Binding of Two Lipid Monomers by Plant Lipid Transfer Proteins, LTP1 73
J.-P. Douliez and D. Marion

Topographical Comparisons of Family 13 α-Amylases using Molecular
Modelling Techniques 79
G. Paës, G. André and V. Tran

Biopolymer Assemblies

Glutenin Macropolymer: A Gel Formed by Glutenin Particles 91
C. Don, W. Lichtendonk, J. Plijter and R. Hamer

Swelling and Hydration of the Pectin Network of the Tomato Cell
Wall 98
A.J. MacDougall and S.G. Ring

Self-assembly of Acacia Gum and β-Lactoglobulin in Aqueous Dispersion 111
C. Sanchez, C. Schmitt, G. Mekhloufi, J. Hardy, D. Renard and P. Robert

Creation of Biopolymeric Colloidal Carriers Dedicated to Controlled
Release Applications 119
*D. Renard, P. Robert, L. Lavenant, D. Melcion, Y. Popineau, J. Guéguen,
C. Duclairoir, E. Nakache, C. Sanchez and C. Schmitt*

Interfaces, Interphases

Polyelectrolyte–Surfactant Complexes at the Air–Water Interface:
Influence of the Polymer Backbone Rigidity 127
D. Langevin

Adsorption Layers of β-Casein at the Air/Water Interface: Effect of
Guanidine Hydrochloride 145
*A. Aschi, A. Gharbi, P. Calmettes, M. Daoud, V. Aguié-Béghin
and R. Douillard*

Adsorption and Rheological Behaviour of Biopolymers at Liquid
Interfaces 153
R. Miller, V. Fainerman, M. O'Neill, J. Krägel and A. Makievski

Dynamic Surface Tension and Surface Dilational Properties of an
Amphiphilic Polysaccharide 166
S. Guillot, D. Guibert and M.A.V. Axelos

Polymerization of Coniferyl Alcohol (Monomer of Lignins) at the
Air/Water Interface   173
B. Cathala, V. Aguié-Béghin, R. Douillard and B. Monties

Multiphasic Systems

Emulsion-stabilizing Properties of Depolymerised Pectin: Effects of pH,
Oil Type and Calcium Ions   181
M. Akhtar, E. Dickinson, J. Mazoyer and V. Langendorff

Mixed Biopolymer Gels of κ-Carrageenan and Soy Protein   190
R.I. Baeza, D.J. Carp, P. Martelli and A.M.R. Pilosof

The Properties of v/ι-Carrageenan: Implications for the Gelling
Mechanism of ι-Carrageenan   201
F. van de Velde, H.S. Rollema and R.H. Tromp

Films and Foams of Sparkling Wines   212
M. Vignes-Adler and B. Robillard

Structure–Texture Relationships of Starch in Bread   226
S. Hug-Iten, F. Escher and B. Conde-Petit

Behaviour of Amylose and Amylopectin Films   235
P. Forssell, R. Partanen, A. Buleon, I. Farhat and P. Myllärinen

Gel Formation by Soy Glycinin in Bulk and at Interfaces   241
T. van Vliet, A. Martin, M. Renkema and M. Bos

Interactions between Cellulose and Plasticized Wheat Starch – Properties
of Biodegradable Multiphase Systems   253
L. Avérous

Proteins Films: Microstructural Aspects and Interaction with Water   260
*C. Mangavel, N. Rossignol, A. Gerbanowski, J. Barbot, Y. Popineau and
J. Guéguen*

Pea: An Interesting Crop for Packaging Applications   267
J.J.G. van Soest, D. Lewin, H. Dumont and F.H.J. Kappen

In Situ Study of the Changes in Starch and Gluten during Heating of
Dough using Attenuated-total-reflectance Fourier-transform-infrared
(ATR-FTIR)   275
O. Sevenou, S.E. Hill, I.A. Farhat and J.R. Mitchell

Effect of D₂O on the Rheological Behaviour of Wheat Gluten 284
J. Lefebvre, Y. Popineau and G. Deshayes

The Influence of the Thickness on the Functional Properties of Cassava
Starch Edible Films 291
N.M. Vicentini, P.J.A. Sobral and M.P. Cereda

Subject Index 301

Biosynthesis

Biochemical Mechanisms of Synthesis of (1→3),(1→4)β-D-Glucans: Cellulose Synthase with an added Twist?

B. R. Urbanowicz,[1] C. J. Rayon,[1] N. C. Carpita,[1] M. A. Tiné[2] and M. S. Buckeridge[2]

[1]DEPARTMENT OF BOTANY AND PLANT PATHOLOGY, PURDUE UNIVERSITY, WEST LAFAYETTE, IN 47907 USA
[2]INSTITUTO DE BOTÂNICA, SEÇÃO DE FISIOLOGIA E BIOQUÍMICA DE PLANTAS, CP 4005, SÃO PAULO, SP, BRAZIL

1 Introduction

Synthases of all polymers containing (1→4)β-linked glucosyl, mannosyl and xylosyl residues have overcome a substrate-orientation problem in catalysis because this particular linkage requires that each of these sugar units be inverted nearly 180° with respect to its neighbors. We and others have proposed that this problem is solved by two modes of glycosyl transfer within a single catalytic subunit to generate disaccharide units, which maintain the proper orientation without rotation or re-orientation of the synthetic machinery in 3-dimensional space.[1–5] A variant of the strict (1→4)β-D-linkage structure is the mixed-linkage (1→3),(1→4)βD-glucan (β-glucan), a growth specific cell wall polysaccharide found in grasses and cereals and other members of the Poales.[6] β-Glucan is composed primarily of cellotriosyl and cellotetraosyl units linked by single (1→3)β-linkages. In reactions *in vitro* at high substrate concentration, a polymer composed of almost entirely cellotriosyl and cellopentosyl units is made.[4] These results support a model in which *three* modes of glycosyl transfer occur within the synthase complex, but the generation of odd numbered units demands that they are connected by (1→3)β-linkages and not (1→4)β. We propose that a central part of the β-glucan synthase complex is derived from an ancestral cellulose synthase, and that an additional glycosyl transferase associates with it to generate these odd numbered cellodextrin units. In contrast to xyloglucan and pectin synthases, which are completely enclosed within the lumen of the Golgi apparatus,[7,8] we provide evidence from limited proteolysis experiments that the catalytic domain of β-glucan synthase is oriented to the cytosolic side of the

3

Golgi membrane and extrudes β-glucan through a channel into the lumen. Thus, the β-glucan synthase is the topological equivalent of cellulose synthase.

1.1 The Unique Twisted Structure of β-Glucan

Mixed-linkage (1→3),(1→4)β-D-glucan (β-glucan) is a plant cell wall polysaccharide specific to grasses and cereals that appears during cell expansion. In all cereal endosperm walls, the ratio of cellotriosyl and cellotetraosyl units is between 2 and 3.[9] The polymer can be synthesized *in vitro* from enriched Golgi membrane fractions with UDP-Glc as a substrate.[10] However, whereas a polymer can be made that is similar to that synthesized natively at substrate concentrations between 100 and 250 μM, at high substrate concentrations nearing saturation, β-glucan is mostly composed of cellotriosyl units and the next higher odd numbered cellodextrin.[4]

1.2 Synthesis of β-Glucan at the Golgi Apparatus

As with all plant cell wall polysaccharides, the mechanism of synthesis of β-glucan is still not well understood. There is a steric problem associated with modeling (1→4)β-linked polymers, such as cellulose and β-glucan, because each sugar unit is inverted 180° with respect to its neighbor. One proposed mechanism of (1→4)β-D-glucan (cellulose) synthesis involves two glycosyl transferases operating opposite one another to add cellobiosyl units to the growing polymer.[3] The synthase properties of β-glucan resemble those of cellulose more closely than those of other Golgi-associated polysaccharide synthase complexes, leading us to believe that β-glucan synthase may have derived from an ancestral cellulose synthase gene.[4,11,12] The cellulose synthase genes (*Ces*A) encode polypeptides of 110 kDa that synthesize (1→4)β-D-glucosyl units. The proteins that are predicted to be encoded by these genes contain up to eight membrane-spanning domains and a large cytosolic active site with "U" motifs containing conserved aspartate residues and a QxxRW sequence.[13] They are conserved in genes encoding several processive glycosyltransferases in which repeating β-glucosyl structures are synthesized.[14]

A unique feature of *Ces*A genes is a plant-specific region, formerly called the Hypervariable Regions (HVRs).[15] However, these regions possess motifs of consecutive acidic and basic residues very well conserved in a sub-class-specific manner indicating a role in catalysis or regulation. Hence, we have now suggested that they be called Class-specific Regions (CSRs). A phylogenetic tree of the *Ces*A gene family based on their CSR showed that CesA sequences from cereals are more closely related to each other than to those from dicot species.[15] The diversity of polysaccharide structures encoded by the four U-motifs and CSR structures and the size of the *Ces*A gene family suggest that some of them encode synthases for polysaccharides other than cellulose. These analyses suggest that (1→3),(1→4)β-D-glucan synthase is one such candidate.

The mixed-linkage (1→3),(1→4)β-D-glucan synthase possesses more similarities to cellulose synthase than do synthases of other non-cellulosic polymers bearing $(1{\rightarrow}4)\beta$-linked backbones.[15] Buckeridge *et al.*[4] proposed a model of synthesis for β-glucan that builds off that for cellulose synthesis, which includes multiple sites of glycosyl transfer in the synthase complex and favoring a "three-site" model of substrate binding and glycosyl transfer that would add cellotriosyl units to the growing polymer. Proposed models for cellulose and β-glucan synthesis account for the synthesis of $(1{\rightarrow}3)\beta$-D-glucan (callose) if there is a loss of all but one site of glycosyl transfer.[4] Although this mode of synthesis has been indicated by our data, it is still not known how the β-glucan synthase complex is oriented topologically with respect to the Golgi membrane or how the components are organized. Muñoz *et al.*[7] and Neckelmann and Orellana[16] suggested that in pea Golgi a UDP-Glc transporter transports UDP-Glc into the lumen of the Golgi where Glc is then consumed in the synthesis of xyloglucans. They proposed further that all of the enzymes involved in xyloglucan synthesis reside entirely within the lumen of the Golgi. Taking a step further, if the β-glucan synthase borrowed a mechanism of synthesis directly from cellulose synthase, then its membrane topology would be expected to place the active site of UDP-Glc binding to the cytosolic face of the Golgi membrane and that, like cellulose synthase, the glucan product is extruded through a membrane channel into the lumen of the Golgi.

1.3 Metabolic Channeling of Glucose to Cellulose Synthase

Amor *et al.*[17] reported that sucrose synthase (SuSy) from cotton fibers is tightly associated with cellulose synthase where it channels UDP-Glc from sucrose to cellulose synthase in the plasma membrane.[18] In maize, SuSy is detected immunocytochemically in both plasma membrane-enriched and Golgi membrane-enriched fractions, whereas in soybean, the synthase is only detected in plasma membranes.[4] If β-glucan synthase is derived from an ancestral cellulose synthase and possesses a similar mechanism of synthesis, then part of this mechanism would include the association of SuSy to control the supply of UDP-Glc to the active site.

Limited proteolysis studies of vesicle fractions have been a useful method to probe topology of many membrane-associated protein complexes.[19] On the other hand, the use of detergents to solubilize β-glucan synthase, such as callose synthase, has been studied to understand the importance of the phospholipid environment in the activity of this membrane-associated synthase.[20–22] The susceptibility of the β-glucan synthase complex to protease digestion and detergent treatments, indicated by loss of synthase activity, will help us determine whether the synthase complex is oriented on the cytosolic face of the Golgi membrane or, like xyloglucan and polygalacturonic acid synthases, fully contained within the lumen.

2 Experimental

2.1 Plant Material and Membrane Preparation

Maize (*Zea mays* L.) seeds were soaked in the dark overnight at 30 °C in deionized water bubbled with air. They were then sown into moistened medium-grade vermiculite, and grown in darkness for 2 d. Fresh maize coleoptiles and etiolated shoots within were harvested into a chilled mortar and overlaid with an equal volume of ice cold homogenization buffer consisting of 100 mM HEPES[BTP], pH 7.4, 20 mM KCl, and 84% (w/v) sucrose. In addition, 1 g of activated charcoal per 10 g of plant material was sprinkled on coleoptiles before addition of homogenization buffer to absorb inhibitory phenolics released by the maize during mashing.[10] After mashing, the homogenate was squeezed through nylon mesh (47 μm^2 pores), and 20 ml of the homogenate was then pipetted into a 38.5 ml Ultraclear centrifuge tube (Beckman) and overlaid with 8 ml of 35% (w/v) sucrose in a gradient buffer containing 10 mM HEPES[BTP], pH 7.2, 7 ml of 30% (w/v) sucrose in gradient buffer, and 4 ml of 9.5% (w/v) sucrose in gradient buffer. The sucrose gradient was centrifuged at 140 000 g in a swinging bucket rotor (Model SW28, Beckman) for 60 min, and the interface enriched in Golgi membranes (35%/30% interface) was removed with a Pasteur pipette and used directly for β-glucan synthase reactions.

2.2 Glucan Synthase Reactions

Reactions were performed with 500 μl (342 μg) of freshly isolated membranes in 4 ml borosilicate glass vials containing 500 μl of reaction buffer (RB), at a final concentration of 250 μM UDP-Glc (Sigma) in 10 mM HEPES[BTP], pH 7.4, containing 1.08 M sucrose, 20 mM KCl, 15 mM MgCl$_2$ and 1 μCi of UDP-[U-^{14}C] -Glc (320 mCi/mmol; Amersham Pharmacia Biotech UK Limited), and placed at 30 °C for 2 h. Reactions were stopped by addition of 3 ml of ethanol, and the suspensions were heated at 105 °C for 5 min. The suspensions were then cooled to room temperature and centrifuged for 5 min at 10 000 g. The pellets were washed five times with 80% (v/v) ethanol and heated in a 105 °C oven for 5 min, and then were shaken, cooled and recentrifuged. The washed pellets were dried under a stream of nitrogen gas at 45 °C.

Reactions were performed the same as stated above with the addition of proteinase K (Sigma, EC 3.4.21.64) to the RB containing 250 μM UDP-Glc without radioactivity at concentrations indicated in a final reaction volume of 1 ml. At 30 min into the reaction, samples were placed on ice and 10 μl of 50 mM PMSF (in ethanol) and 1 μCi of UDP- [U-^{14}C]-Glc were added to the samples. They were then incubated for an additional 90 min at 30 °C.

For the experiments with detergents, 500 μl of RB containing Triton X-100 or CHAPS were added to an equal volume of Golgi membranes to reach final concentrations up to 0.1% (v/v) of Triton X-100 and up to 0.6% (w/v) of CHAPS. The incubation was carried out at 4 °C for 30 min. At 30 min, 1 μCi of UDP-[U-

[14C]-Glc was added to the samples, and they were incubated at 30°C for an additional 90 min.

2.3 Digestion with *Bacillus subtilis* $(1\rightarrow3),(1\rightarrow4)\beta$-D-Glucan endo-4-glucanohydrolase (β-Glucan endohydrolase) as a Specific Assay for β-Glucan

A *B. subtilis* β-glucan endohydrolase selectively cleaves a $(1\rightarrow4)\beta$-D-glucosyl linkage only when it is preceded by a $(1\rightarrow3)\beta$-linkage,[23] making it useful for quantifying the amount of polymer synthesized. The reaction products were resuspended in 100 μl of water and 50 μl of β-glucan endohydrolase in 20 mM sodium acetate and 20 mM NaCl, pH 5.5 were added.[10] The products released from the digestion of purified β-glucan yield mostly cellobiosyl- and cellotriosyl-$(1\rightarrow3)\beta$-D-glucose.[9,24] The samples were incubated for 3 h at 37°C and stopped by boiling for 2 min, cooled and microcentrifuged at 14 000 rpm for 5 min. The labelled oligosaccharides from β-glucan were separated by high performance anion-exchange chromatography pulsed amperometric detection (HPAEC-PAD), followed by liquid scintillation counting. The pellet was resuspended in 1 ml of water, and the radioactivity was determined by liquid scintillation spectroscopy.[10]

2.4 HPAEC-PAD Analysis

The oligomers from β-glucan endohydrolase digestion were separated with an anion-exchange column (Carbo-Pac PA1) and detected by PAD (Dionex). The column was first equilibrated in 0.2 M NaOH, and the samples were eluted in a linear gradient of 0.2 M NaOH/0.2 M NaOAc. Twenty-five μl of 0.1 mg/ml pre-digested barley β-glucan (Sigma) were added to each sample as an elution standard. The radioactivity in 0.5 ml fractions collected in 1 ml of 2 M acetic acid was then determined by liquid scintillation spectroscopy.

3 Results

3.1 Limited Proteolysis with Proteinase K

β-Glucan with a cellotriosyl to cellotetraosyl ratio similar to that observed *in vivo* is synthesized in standard reaction conditions with 250 μM UDP-Glc (Figure 1). The addition of proteinase K to intact, enriched Golgi membranes during *in vitro* synthesis of β-glucan lowered the amount of total β-glucan produced. A small increase in total β-glucan synthase activity was observed with 100 μg/ml proteinase K, but activities began to drop until reaching 20% of its original activity with higher concentrations of proteinase (Figure 2A). Callose synthase was strongly activated, by an average of 10-fold, by proteinase K (Figure 2B).

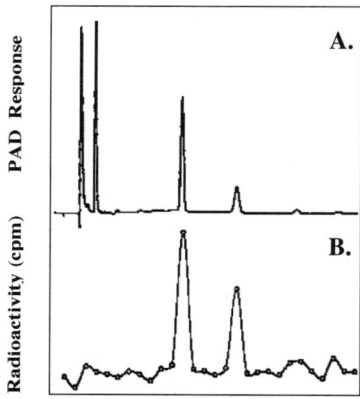

Figure 1 *Separation of β-glucan cellodextrin-(1→)β-D-glucose oligomers by HPAEC. (A) Detection by PAD of authentic oligomers from β-glucan added to the reaction product as an elution marker. (B) Determination of radioactivity in β-glucan oligomers synthesized* in vitro

3.2 Detergent Treatment

We investigated the ability of detergents to solubilize the mixed-linkage (1→3),(1→4) β-D-glucan synthase, yet preserve activity. The Golgi membranes exposed to Triton X–100, a non-ionic detergent, completely lose β-glucan and callose synthase activity (Figure 3). The synthase activities in Golgi membranes incubated with CHAPS, a zwitterionic detergent, behave quite differently. At low levels of CHAPS (0.03%–0.2%), the β-glucan synthase activity is unaffected (Figure 4A). Upon treatment of the Golgi membranes with 0.3% to 0.6% CHAPS, critical micellar concentrations are approached and β-glucan synthase activity drops to about 30% of original activity. The ratio of cellotriosyl units to cellotetraosyl units is unchanged regardless of the CHAPS concentration used (data not shown). The callose synthase activity is increased about 2-fold in low concentrations of CHAPS (Figure 4B). Above 0.2% CHAPS, the increase in callose synthesis parallels the loss of β-glucan synthase activity.

4 Discussion

The addition of proteinase K and non-ionic detergent each greatly lower the activities of mixed-linkage (1→3),(1→4)β-D-glucan. These results are consistent with those of Gibeaut and Carpita,[25] who showed that Triton X–100 and protonophores, both treatments that collapse membrane potentials and pH gradients, caused greatly lowered β-glucan synthase activity. The decrease in activity associated with the addition of proteinase K to the reaction indicates that something associated with the synthase mechanism is accessible to inactivation by protease digestion (Figure 2).

There are several ways that β-glucan synthase and its associated polypeptides may reside at the Golgi apparatus of maize. We have determined that intact

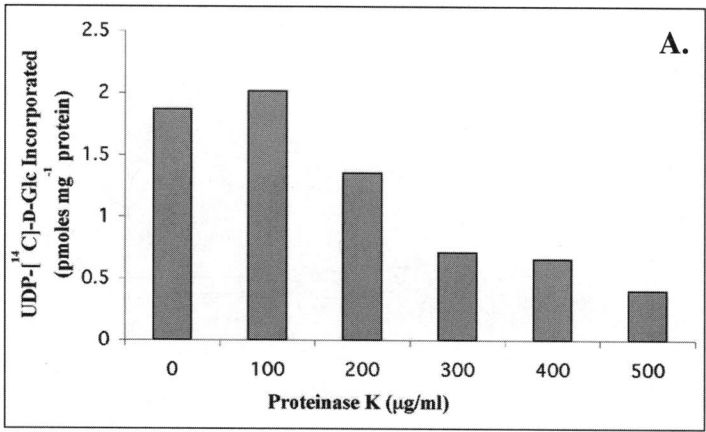

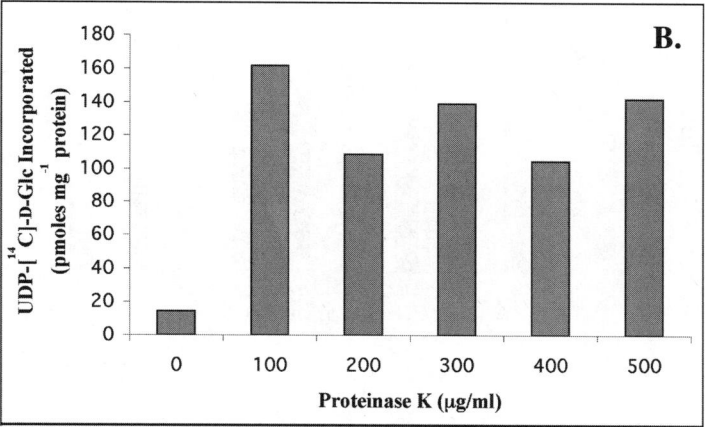

Figure 2 *The effect of limited proteolysis on β-glucan synthesis. (A) Total β-glucan activity (pmoles UDP-^{14}C-Glc mg^{-1} protein) after 30 min incubation with various concentrations of proteinase K. (B) Total callose synthesized in these same reactions*

Golgi vesicles treated with proteinase K greatly lowers the activity of β-glucan synthase during *in vitro* synthesis. These results are quite different from similar experiments done by Muñoz *et al.* on the xyloglucan synthase mechanism in the Golgi apparatus of pea. They found that protease treatment neither affects xyloglucan synthesis nor significantly alters the UDP-Glc transporter function.[7,16] Based on their results, their model for xyloglucan synthase predicts that it is encased completely within the Golgi lumen.

On the other hand, Buckeridge *et al.*[4] proposed that β-glucan synthase is an ancestral cellulose synthase that has an additional site of glycosyl transfer, giving it its unique ability to make cellotriose units. Our proteinase K results predict that, similar to cellulose synthase at the plasma membrane, β-glucan synthase would also share an active site on the cytoplasmic face of the Golgi membrane,

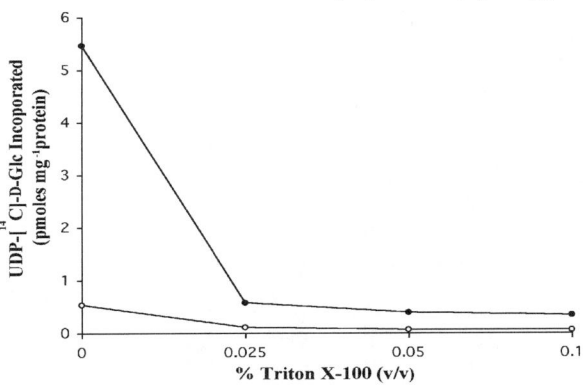

Figure 3 *The effect of Triton X-100 on total β-glucan and callose synthesis*

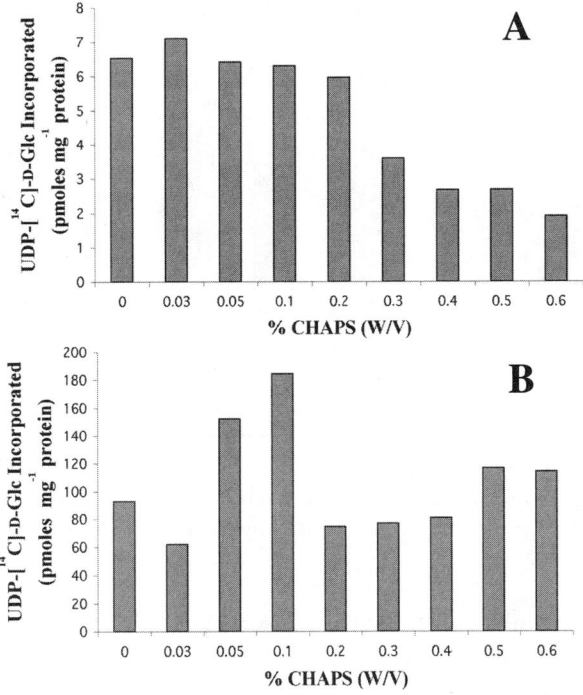

Figure 4 *Effect of CHAPS on β-glucan and callose synthase. (A) Total β-glucan activity (^{14}C-UDP-Glc mg^{-1} protein) after incubation in various concentrations of CHAPS. (B) Total callose activity (^{14}C-UDP-Glc mg^{-1} protein) after incubation in CHAPS*

allowing UDP-Glc to directly bind and be incorporated into the growing polymer. We have immunolocalized SuSy to the Golgi membrane of maize, which is unique and has not been found in other plants such as soybean.[4] SuSy is associated with the plasma membrane of cotton fibers engaged in cellulose synthesis,[17] where its proposed association with cellulose synthase constitutes part of a metabolic channeling of UDP-Glc from Suc directly into the active site of cellulose synthase.[26] If the core β-glucan synthase is indeed derived from an ancestral cellulose synthase, it would also be expected to have SuSy associated as part of the same mechanism at the Golgi apparatus. Alternatively, maize and other cereals may have incorporated SuSy into metabolic channeling of UDP-Glc to the UDP-Glc transporter, which is essential to provide substrate for synthesis of other non-cellulosic polysaccharides.[16,27]

Treatment of Golgi membranes with Triton X-100 inactivates both β-glucan and callose synthase activities. Triton X-100 removes phospholipids but does not solubilize callose synthase.[20] Sloan *et al.*[21] and Sloan and Wasserman[22] showed that the association of a callose synthase with phospholipids is essential for the active enzyme. Our data indicate that such an association may also be true for β-glucan synthase. The covalently bound phospholipids could serve as anchors for membrane-associated proteins and hence, serve a conformational role in the proper orientation of β-glucan synthase at the Golgi surface. The removal of phospholipids would result in loss of conformation and irreversible inactivation of the synthase. The incubation of the Golgi membranes with CHAPS also inhibits the biosynthesis of β-glucan and callose, but at higher concentrations of detergent that achieve critical micelle concentration. The decrease in activity of β-glucan synthase is correlated with a slight increase of callose synthesis when a concentration higher than 0.2% CHAPS is used. This correlation suggests that UDP-Glc provided by SuSy is still accessible to the complex, but at least one site of the β-glucan synthase is affected in such a way as to increase the production of callose. Thus, these data are consistent with a model that places β-glucan synthase at a site where part of the complex is exposed to the cytosolic face of the Golgi membrane.

References

1. N. Carpita, M. McCann and L. R. Griffing, *Plant Cell*, 1996, **8**, 1451.
2. M. Koyama, W. Helbert, T. Imai, J. Sugiyama and B. Henrissat, *Proc. Natl. Acad. Sci. USA*, 1997, **94**, 9091.
3. N. C. Carpita and C. E. Vergara, *Science*, 1998.
4. M. S. Buckeridge, C. E. Vergara and N. C. Carpita, *Plant Physiol.*, 1999, **120**, 1105.
5. M. S. Buckeridge, C. E. Vergara and N. C. Carpita, *Phytochemistry*, 2001, **57**, 1045.
6. B. G. Smith and P. J. Harris, *Biochem. System. Ecol.*, 1999, **27**, 33.
7. P. Muñoz, L. Norambuena and A. Orellana, *Plant Physiol.*, 1996, **112**, 1585.
8. F. Goubet and D. Mohnen, *Planta*, 1999, **209**, 112.
9. P. J. Wood, J. Weisz and B. A. Blackwell, *Cereal Chem.*, 1994, **68**, 530.
10. D. M. Gibeaut and N. C. Carpita, *Proc. Natl. Acad. Sci. USA*, 1993, **90**, 3850.
11. D. M. Gibeaut and N. C. Carpita, *FASEB J.*, **8**, 904.
12. N. C. Carpita, *Annu. Rev. Plant Physiol. Plant Mol. Biol.*, 1996, **47**, 445.

13. J. R. Pear, Y. Kawagoe, W. E. Schreckengost, D. P. Delmer and D. M. Stalker, *Proc. Natl. Acad. Sci. USA*, 1996, **93**, 12637.

14. I. M. Saxena, R. M. Brown, Jr., M. Fevre, R. A. Geremia and B. Henrissat, *J. Bacteriol.*, 1995, **177**, 1419.

15. C. E. Vergara and N. C. Carpita, *Plant Mol. Biol.*, 2001, **47**, 145.

16. G. Neckelmann and A. Orellana, *Plant Physiol.*, 1998, **117**, 1007.

17. Y. Amor, C. H. Haigler, S. Johnson, M. Wainscott and D. P. Delmer, *Proc. Natl. Acad. Sci. USA*, 1995, **92**, 9353.

18. D. P. Delmer and Y. Amor, *Plant Cell*, 1995, **7**, 987.

19. A. E. Wu and B. P. Wasserman, *Plant J.*, 1993, **4**, 683.

20. B. P. Wasserman and K. J. McCarthy, *Plant Physiol.*, **82**, 396.

21. M. E. Sloan, P. Rodis and B. P. Wasserman, *Plant Physiol.*, 1987, **85**, 516.

22. M. E. Sloan and B. P. Wasserman, *Plant Physiol.*, 1989, **89**, 1341.

23. M. A. Anderson and B. A. Stone, *FEBS Lett.*, 1975, **52**, 202.

24. R. G. Staudte, J. R. Woodward, G. B. Fincher and B. A. Stone, *Carbohydr. Polym.*, 1985, **71**, 301.

25. D. M. Gibeaut and N. C. Carpita, *Protoplasma*, 1994, **180**, 92.

26. D. P. Delmer, *Annu. Rev. Plant Physiol. Plant Mol. Biol.*, 1996, **47**, 445.

27. C. Wulff, L. Norambuena and A. Orellana, *Plant Physiol.*, 2000, **122**, 867.

The Future Prospects for Broadening Soybean Utilization by Altering Glycinin

Niels C. Nielsen[1] and Eliot Herman[2]

[1]USDA-ARS, DEPARTMENT OF AGRONOMY, PURDUE
UNIVERSITY, WEST LAFAYETTE, IN 47907, USA
[2]SOYBEAN GENOMICS AND IMPROVEMENT LABORATORY,
USDA/ARS, BELTSVILLE, MD 20705 USA

1 Abstract

Cultivated soybeans, that contain about 40% protein and 20% oil on a dry weight basis, are one of the richest sources for plant protein, albeit sulfur deficient, among crop species. Seven glycinin genes encode the 11S storage globulin gene family in soybean. Each gene has been cloned and sequenced. Six are functional genes expressed in varying amounts in seeds, while the seventh lacks the 3′-end of its coding region and is nonfunctional. The initial translation products from the glycinin genes in cotyledons are preproproteins. After co-translational removal of a signal peptide, and chaperon-mediated subunit folding in rough endoplasmic reticulum, the 3S proglycinin monomers assemble into 9S trimers. The trimers are transported *via* the endomembrane system through the Golgi apparatus to storage vacuoles where they are deposited after post-translational modification by a specific asparaginyl endopeptidase and transformed into the stored 11S hexamer oligomer. Soybeans and legumes in general are deficient in sulfur amino acids. Attempts to increase methionine content by expressing high-methionine engineered glycinin in tobacco seeds have resulted in post-translational instability indicating that there are hurdles to engineering and expressing improved variants of glycinin. One solution to this problem may be found in soybean knockouts of conglycinin, the 7S storage protein, resulting in up-regulation of one of the glycinin genes, and these seeds produce compensating levels of total seed protein in transgenic soybeans. The glycinin that is induced is sequestered in ER-derived protein bodies that may prove useful as a storage site for not only modified glycinin but also as a means to produce novel proteins in transgenic soybeans. The wide use of soybean protein as human and animal food presents a large potential to have broad impact on agriculture by engineering soybeans.

2 Seeds and Protein Storage

Plants store proteins in embryo and vegetative cells to provide metabolic resources for subsequent stages of growth and development (ref. 1, for review). These are critical stages in the life cycle of plants, and the proteins are used either to recover from a dormant phase or to support rapid growth. Seeds are perhaps the best studied example of this phenomena. Proteins synthesized and stored during seed maturation become mobilized during germination to support the rapid growth of the new seedling. Other parallel examples of this developmental pattern are the storage of proteins in bark, tubers and other organs that are required to support spring growth after over-wintering. Similarly many vegetative organs, for example soybeans, accumulate vegetative storage protein in leaves. These are mobilized following seed set to support the formation of constituents accumulated in seeds. In agriculture, it is the seeds that are the primary commodity of the large crops and the storage of protein in these seeds accounts for a large fraction of the available amino acids for both humans and animals throughout the world.

Mechanisms for protein storage and mobilization have been elaborated to serve many different developmental and physiological functions. For example, stored protein reserves provide building blocks for rapid growth following seed and pollen germination. Similarly, stored reserves in vegetative cells provide the building blocks for seed and fruit set during reproductive growth and for rapid expansion of vegetative structures following periods of dormancy. In agriculture, protein storage in seeds and vegetative tissues accounts for much of the protein consumed directly as food by humans and livestock. Consequently, the biochemistry of storage proteins and the cellular and physiological mechanisms regulating their synthesis are of practical as well as academic interest.

3 Soybean Seed Proteins

Eight major oilseed crops traded in international markets account for more than 95% of the world's vegetable oil. These include soybean, cottonseed, peanut, sunflower, rapeseed, flaxseed, copra and palm kernel. Soybean is unique among these oilseed crops in that it also contains about 40% dry weight protein. Because of the large amounts of both oil and protein, soybean has dominated world oilseed production. The US, Brazil and China produce most of the world's soybean crop and as a widely traded commodity it is available throughout the world as a cheap source of quality protein and oil.

The center of origin of soybean is considered to be China. They are described in the book Materia Medica 'She-non' written over 5000 years ago.[2] It is referred to as 'tchouan' in the Chinese dictionary of Sui Sham. The dictionary of 'Kouangia,' which dates to the beginning of the Christian Era, refers to it as ta-teou (grand pea) and also sou. The names soi, soy, soya and soja, which are used to describe the soybean today, all were probably derived from this ancient Chinese word sou. Soybeans first appeared in Europe in the Middle Ages as a result of trade with China. Although soybeans were first planted in the US before the American Revolution and used as a forage crop, it was not until the 1900s that soybeans began to emerge as an important oil crop. It was used as a replacement for cottonseed oil when Boll Weevil attacks on cotton drastically reduced that source of oil. Soybean production in the US greatly expanded during the Second World War and has continued to grow to the present. By the 1970s Brazil and Argentina joined the US as major producers and soybeans are now one of the great crops widely traded and utilized throughout the world.

During seed maturation a small set of seed storage proteins (Figure 1) are accumulated in parallel with the oil and non-structural carbohydrates. Mature soybeans by weight average about 40% protein, 20% oil and 12% non-structural carbohydrate, much of which is in the form of raffinose-series oligosaccharides that cannot be completely digested by humans. The overall composition and ratio of storage compounds in seeds is exquisitely sensitive to environmental conditions following seed set.[3] Environmental stress usually favors oil accumulation at the expense of protein. Although revenues derived from soy oil continue to be greater than those from the protein, the protein fraction continues to increase in importance. Some soybean cultivars and wild accessions contain

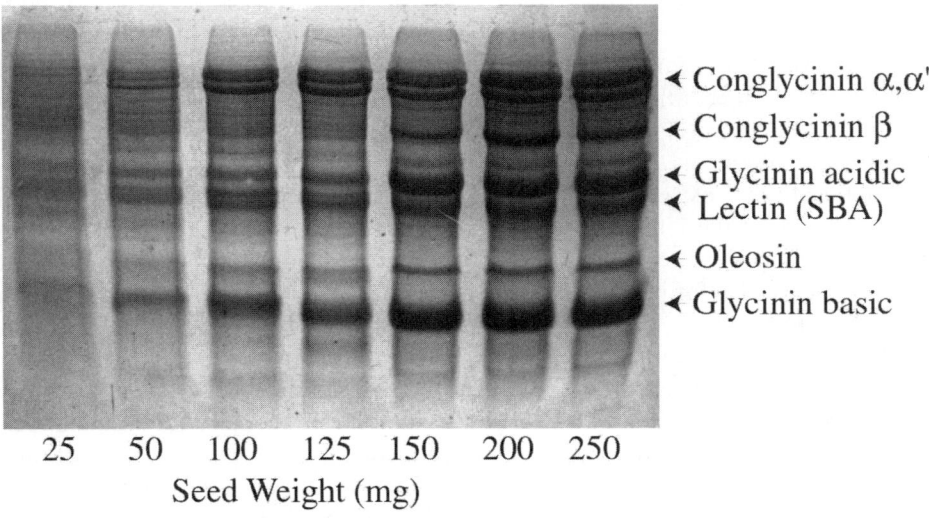

◄ Conglycinin α,α'
◄ Conglycinin β
◄ Glycinin acidic
◄ Lectin (SBA)
◄ Oleosin
◄ Glycinin basic

25   50   100   125   150   200   250
Seed Weight (mg)

Figure 1 *The accumulation of soybean polypeptides during seed development as visualized by SDS/PAGE. The samples shown from 25 mg seed fresh weight through 250 mg seed fresh weight encompass the period just before the onset of storage protein gene expression and accumulation through late maturation when storage protein accumulation is complete. The primary soybean storage proteins of soybean are the 7S conglycinin and 11S glycinin proteins that are accumulated at high levels from mid to late seed maturation. Soybean lectin (SBA) is prominent among the minor proteins accumulated in the protein storage vacuoles. Oleosin is the protein that encases the oil bodies that are accumulated in parallel with the storage protein vacuoles*

nearly 50% protein, although the physiological characteristics of these lines usually do not permit them to be produced commercially. Nonetheless, because soybeans with high protein content are important economically, many attempts have been made to develop high protein soybean lines (ref. 4, for example). Unfortunately, because of the low yields of these high protein lines compared to other readily available varieties, farmers have not readily accepted such varieties into production agriculture. A systematic evaluation of soybean breeding lines by plant breeders has revealed that protein content is inversely correlated with yield statistically. Negative statistical correlations are also observed between soybean oil or carbohydrate and protein content. These compositional changes are undoubtedly the consequence of differences in the partitioning of carbon among the substrates required for their synthesis, and this in turn is controlled at the genetic level by seed developmental processes. A detailed understanding of the developmental controls that modulate seed development is probably as important for the production of improved soybean varieties as efforts to manipulate quality *via* genetic engineering.

4 Soybean Seed are Widely used in Processed Foods and Animal Food

Soybean proteins are widely used as food additives in European derived societies, primarily in processed foods, and this trend continues to grow annually. This makes soybean proteins a pervasive component of the human diet in industrialized countries. Solvent extracted soybean meal is also widely used as an animal feed additive (ref. 5, for review), because it is an inexpensive source of high quality protein that contains more of essential amino acids lysine and tryptophan than most cereal crops. Combined with corn, the other primary feed grain used in the United States, a ration can be assembled that is adequate in both sulfur amino acid and lysine contents, and provides a high protein diet that is well balanced for poultry and pigs.

The widespread use of soybean is not without difficulty, because both humans and animals develop immune responses to soybean (ref. 6 for review). For example, the presence of raffinose-series oligosaccharides is responsible for the development of flatulence and the trypsin inhibitors cause physiological difficulties in the gut that affect digestion of foods. Although breeders have selected varieties with reduced protease inhibitors and the raffinose-series oligosaccharides, neither trait has been introduced into crops that are widely cultivated. Further improvements can be achieved either by generating mutants or by disrupting gene expression using biotechnological approaches.

What is not often appreciated is the extent to which immune responses to soybean proteins influence the growth and development in both humans and livestock. Three to eight percent of children develop immune responses to soybean proteins, sensitivities that usually result in gastrointestinal disturbances that are outgrown upon reaching puberty. Nonetheless, as much as 2% of adults present with a lasting sensitivity to soybean products (see refs. 7, 8 for review).

Livestock, particularly swine[9] and preruminant calves,[10] also exhibit sensitivity to soybean-based rations, and immune responses to soy have been described in rainbow trout[11] and chicken. In domesticated animals such as these, the sensitivity is usually manifested by poor rates of gain and gastrointestinal difficulties.

Soybeans are included among the group of common food allergies, along with wheat, milk and eggs. These foods make a major contribution to most human diets. Although other food allergies such as peanut, shellfish and nut allergies are widely considered to be major problems because of the potential lethal complications of anaphylaxis, the allergens responsible for these immune responses are usually minor components of the average diet, and avoidance of foods that contain them is the most effective means for prevention. In contrast, the major food allergies present problems due to the significance of these foods in the diet even though they cause less severe immune responses. The allergenic response to soybeans is due to sensitivity to the major storage proteins and a few minor ancillary seed proteins,[12] primarily protease inhibitors, seed lectin and a cysteine protease called P34.[13] The P34 is an immunodominant allergen and although a minor protein in abundance in the seed it induces a strong response among soy-sensitive people. The human allergenic response to the 11S storage protein glycinin and P34 has been characterized.[14–17] The epitopes on the P34 and glycinin that bind IgEs have been mapped to the primary sequence of both proteins. Moreover, the three-dimensional protein structures of glycinin and P34 have been modeled, and this revealed distinct antibody binding sites on the surface of each molecule. Allergic responses remain a significant and growing problem for people, and careful characterization of the epitopes involved will permit a focus on the structures involved, which in turn may permit a better understanding of the molecular processes that lead to IgE production. The experiments may lead to ways to combat allergy and ways to defuse concerns expressed about the potential of factors in transgenic plants to elicit immune responses.

5 Glycinin Synthesis and Assembly

The genes encoding glycinin, their expression and assembly, together with the accumulation of these proteins in seeds, have been described in great detail in the primary literature and in several reviews.[18–21] A small family of nuclear genes encodes glycinin subunits, a family that presently includes six functional and one pseudogene. Each of these genes has been cloned and sequenced[19] (Beilinson *et al.*, TAG, submitted). The six functional genes are expressed as seed-specific proteins, while the seventh is a pseudogene that lacks the 3'-end of its coding region and is nonfunctional. Glycinin genes *Gy1*, *Gy2* and *Gy6* are tandemly arranged on chromosome 3 in a region that spans more than 50 kb, and is homoeologous to a region on chromosome 19 that includes *Gy3* and *Gy7*. A summary of this data is shown in Figure 2. This result is consistent with the complex history of duplications that marked the evolution of the soybean genome. The other two genes, *Gy4* and *Gy5*, segregate independently both from

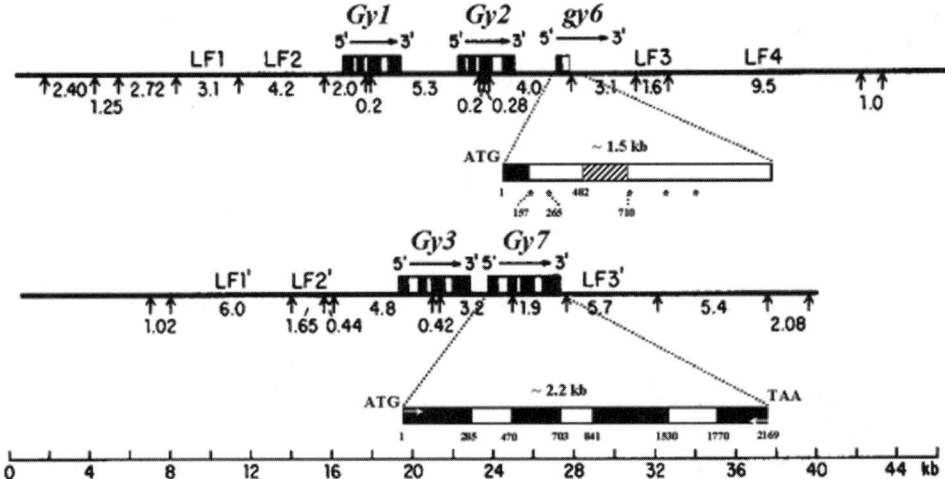

Figure 2 *Gene structure in the region containing the glycinin genes. Gy1, Gy2 and Gy6 genes of chromosome 3 are located on homoeologous regions of chromosome 19 that contains Gy3 and Gy7. The adjacent leaf protein genes LF1, LF2 and LF3 bracket Gy1 and Gy2 and paralleled by LF1', LF2' and LF3' that bracket Gy3 and Gy7. The similar gene structure of glycinin genes in chromosomes 3 and 19 indicates that there was a duplication in the soybean genome*

one another and from the homoeologous regions that include the other five glycinin genes. The chromosomal localization of *Gy4* is unknown, but it is mapped to Linkage Group O. *Gy5* has been mapped to soybean chromosome 13.

The expression of the glycinin genes is complex. The genes are subject to seed specific regulation during mid to late seed maturation, and this involves motifs found upstream from the gene coding regions in the genes. Individual glycinin genes appear to be capable of differential regulation. Although this has not yet been investigated in great detail, present evidence indicates that environmental and developmental events induce the differential regulation. The initial translation products from the glycinin mRNAs in cotyledons are preproteins.[20] After co-translational removal of a signal peptide the proglycinin is transiently sequestered in the ER lumen where it is folded in ATP dependent reactions that involve molecular chaperones.[22] Intramolecular disulfide bridges are formed while the proglycinin subunits are in the ER lumen, presumably with the aid of PDI.[23] The high level of glycinin production in the cotyledons and the requirement of BiP in posttranslational folding is associated with high levels of BiP expression and protein content in the maturing cotyledons,[24] but the precise details of the molecular interactions between proglycinins and chaperones remains to be determined. Simultaneously with folding, the 3S proglycinin monomers assemble into 9S trimers, the transport competent conformation of the protein.[22] The trimers are transported *via* the endomembrane system through the Golgi apparatus to protein storage vacuoles where they are deposited and accumulated. Each proglycinin monomer is considered to have a vacuolar targeting sequence(s), although these sequence(s) are have not been elucidated. The proglycinin is cleaved at a single site on the carboxyterminal side of a conserved aspargine residue found in all 11S proteins from species from both gymno- and angiosperms (ref. 20, for review, refs. 25, 26). The cleavage of the glycinin monomers is required to form 11S hexamer oligomers, a form of glycinin that is purified from seeds.

6 Improving Soybeans and Glycinin by Biotechnology

Although genetically engineered soybeans currently account for over half the US crop, these alterations have largely been focused on improving the agronomic performance of the plants by adding genes that confer herbicide and insect resistance. The first steps of exploiting biotechnology to improve soybean seed products have already occurred. E. I. DuPont de Nemours has produced a soybean line that cosuppresses a fatty acid desaturase.[27,28] The lack of this enzyme causes soybean oil to have a saturation similar to that achieved after chemical processing of extracted soybean oil. Reducing fatty acid saturation dramatically reduces the oxidation and rancidity of oil pressed from the recombinant seeds. By carrying out this process *in planta*, large savings in transportation, processing and energy will be reduced.

7 Engineering Essential Amino Content Improvement in Soybeans

Efforts to alter protein content have proceeded slowly and cautiously due in part to the results of early attempts to increase sulfur amino acid content using recombinant gene technology that identified several problems. It has been widely recognized that foreign gene expression could add proteins rich in essential sulfur amino acids to soybeans yielding a completely balanced protein product. Because soybean transformation has lagged behind the development of other plants, model plants such as tobacco have been used to test the consequence of introducing changes into glycinin to improve the sulfur amino acid content. These initial experiments identified a major problem of post-translational instability of foreign gene products. For example, an early attempt was to increase the methionine content of the 7S storage protein of *Phasolus vulgaris* by inserting an oligonucleotide in the cDNA encoding an alpha-helix rich in methionine.[29] Although this transgene was highly expressed, the protein synthesized was degraded rapidly and did not accumulate in the seed. Subsequent elucidation of the three dimensional structure suggested that the protein's structure was opened by the insertion, thus making it susceptible to protease attack. Further investigations indicated that the site of the mutant protein's degradation was the seed protein storage vacuole.[30] Such a result suggested there would be a major problem for expressing foreign and modified proteins in plants because all of these ideally should be targeted to accumulate in the storage compartment. Vitale and colleagues investigated the effects of storage protein oligomer formation on protein stability.[31] They produced a mutant 7S protein in which the carboxyterminal was deleted. The mutant 7S storage protein was unable to form trimers, and was unstable post-translationally. The site of its degradation remains to be determined, although it appears that it is not transported to the vacuole *via* normal progression through the endomembrane system.[32] Instead, this protein appears to be degraded close to its site of synthesis by the ER, an observation that indicates there are likely multiple sites of post-translational instability, and each of these presents risks of synthesis or accumulation failure of seed products produced by recombinant genetic technology. During the next few years, further advances in protein modeling will likely improve the ability to predict stability of engineered proteins.

Other attempts to modify seed proteins by changing individual amino acid residues to those that encode methionine has met with mixed results. Modification of glycinin to increase sulfur amino acid content been accomplished and these gene products are capable of normal assembly into trimers *in vitro*.[26] However, the modified glycinin subunits do not accumulate in seeds of transgenic tobacco, and are rapidly degraded *in vitro* by the asparaginyl endopeptidase responsible for post-translational modification of proglycinin. Apparently, the modifications used to improve methionine content elicit small changes in protein conformation, and these render the mutant proteins sensitive to seed peptidases co-located in the storage vesicles. Despite these disappointing results, there is evidence that some proteins might tolerate changes in amino acids to methionine better than others. For example, mutation of the *Phaseolus vulgaris* phytohemagglutinin protein to include added methionine was more successfully accumulated than glycinin.[33] Permissive sites were identified that greatly increased methionine content and the mutant protein expressed in tobacco seeds was accumulated. This result indicates that while post-translational instability is a problem, it can be solved by either tests of many constructs or by better understanding of which amino acids introduce adverse conformational changes in storage proteins. Improvements in technology that will permit the rapid production and analysis of many variations will make culling unstable proteins easier. Knowledge of how proteins fold and the critical amino acids important in this process will also make modeling stable mutants of glycinin easier.

An alternative approach of introducing a gene encoding a seed storage protein that is rich in methionine was one of the first approaches attempted and the results of this created much of the public concern about GMOs. The major storage protein of Brazil nuts is highly enriched in sulfur amino acids and if expressed in sulfur deficient legume seeds, even at moderate levels, could completely correct the nutritional deficiency. The isolation and transfer of the gene to model tobacco seeds and later soybeans was successful;[34,35] however, the research was terminated when it was recognized that this protein was a major allergen to people with a sensitivity to nut proteins. If such a protein was widely distributed in processed foods, there would be a high degree of risk for anaphylaxis by sensitive to people. Interestingly, even though this product never entered the food supply because scientists responsibly evaluated the protein for potential immunological responses, the possible risk presented by this research is often cited by critics of genetic modification of crops as an example of its risk.

There are emerging concerns that foreign and engineered proteins may result in newly introduced food allergies. The widespread use of soybean in processed foods makes this an especially sensitive issue. However, current techniques can easily identify and characterize transgene products that elicit allergenic reponses. Soybeans are intrinsically allergenic for two of the known seed protein allergens, P34, a cysteine protease, and glycinin. Linear epitopes in these proteins have been identified using IgE from individuals sensitive to soybean products.[14-17] This approach can be extended to evaluate engineered and foreign proteins to determine whether the new gene products present any increased potential for allergenic responses. Many studies have shown that food allergies tend to be directed at a broad spectrum of foods and solid phase IgE assays of plant materials or foods can assay whether of not there is any sensitivity. These studies can be performed by heterologous synthesis of the proteins and direct testing with sera. Much of the analysis of food allergenic responses has been conducted with the sera from sensitive people. Animal models based on mice and pigs are under development that would permit a more rigorous evaluation of the allergenic potential of engineered and foreign proteins. The synthesis and assembly assays processing mutants and high methionine versions of 11S glycinin provide a model for how these studies can be conducted. Using this approach it should prove feasible to design improved and modified versions of glycinin that can be accumulated in the seed, provide enhanced nutrition for both people and animals while avoiding parallel complications due to immune responses.

A recent paper published by Kinney *et al.*[36] suggests one possible approach to improve the composition of soybeans is by not directing proteins to the vacuole where they may be susceptible to proteolysis. Transgenic soybeans were developed that suppress the 7S storage protein conglycinin. The recombinant soybeans still possess the same protein and oil content as the wild type due to the accumulation of additional glycinin. In this case, G2 subunits produced from *Gy2* replaced the conglycinin that was absent from the mutant seeds (Figure 3). The increased G2 isoform did not accumulate in the protein storage vacuole, but instead accumulated in ER-derived protein bodies as the trimeric precursor. The ER-derived protein bodies were stable, accumulated in maturing seeds, and persisted in the seed into germination. These protein bodies are metabolically inert in that they possess an ER-derived limiting membrane that is not osmotically active. As inert compartments, the ER-derived protein bodies do not appear to possess any active hydrolases, a result that suggests any protein accumulated in this compartment would likely be stable. The mechanism that causes the G2 isoforms in the mutant to be directed to aggregate in the ER-derived protein bodies, rather than progressing through the endomembrane system to the protein storage vacuole with the other isoforms, is not known. The G2, when produced at normal amounts in wild-type seeds, is not sequestered into ER-derived vacuoles, and presumably traffics through the endomembrane system correctly. If engineered and/or foreign proteins can routinely be targeted to ER-derived protein bodies, that may solve, at least in part, the post-translationally instability problems. The establishment of ER-derived vacuoles could

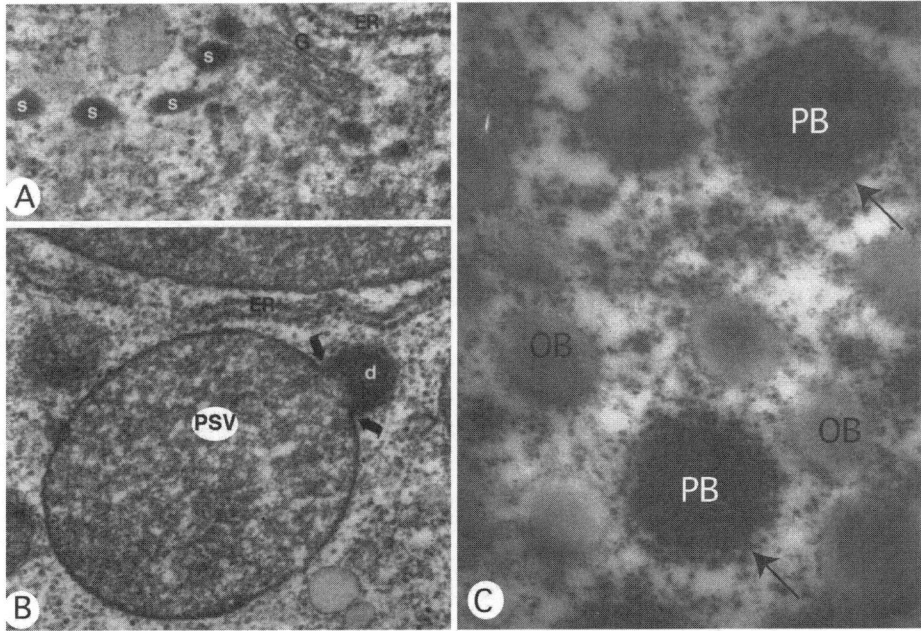

Figure 3 *Electron micrographs that show the cellular mechanism of storage protein in wild-type soybean cotyledons and in transgenic soybeans that cosuppress conglycinin. Micrographs A and B show the pathway of storage protein accumulation in wild-type soybeans by the formation of dense protein-containing secretion vesicles by the Golgi (A) that then fuse with the protein storage vacuoles (PSV) releasing the storage proteins into the organelle's matrix. In transgenic soybeans that cosuppress conglycinin the glycinin Gy2 isoform accumulates to compensate for the lost conglycinin. The Gy2 gene product is accumulated in ER-derived protein bodies (PB). Other organelles shown include ER (endoplasmic reticulum) and oil bodies (OB)*

permit the efficient use of plants as protein factories for the production of largely improved and novel products.

References

1. E. M. Herman and B. A. Larkins, *Plant Cell*, 1999, **11**, 601.
2. Y.-Y. Li and L. Grandvoinnet (1911–12) *Le Soja: L'Agriculture Pratique*, v11, Nos. **102**, 176–196.
3. M. E. Westgate, E. Piper, W. D. Batchelor and C. Hurburgh, Jr. in J. K. Drackley, *Soy in Animal Nutrition*, 2000, Federation of Animal Science Societies, Savoy, IL, USA, pp. 75–89.
4. R. W. Yaklich, *J. Ag. Food Chem.*, 2001, **49**, 729.
5. K. C. Ree in J. K. Drackley, *Soy in Animal Nutrition*, 2000, Federation of Animal Science Societies, Savoy, IL, USA, pp. 46–55.
6. I. E. Liener in J. K. Drackley, *Soy in Animal Nutrition*, 2000, Federation of Animal Science Societies, Savoy, IL, USA, pp. 13–45.

7. H. A. Sampson. *JAMA*, 1997, **278**, 1888.
8. E. Young, M. D. Stoneham, A. Petruckevitch, J. Barton and R. Rona, *Lancet*, 1994, **343**, 1127.
9. J. P. Lalles, D. Dreau, H. Salmon and R. Toullec, *Res. Vet. Sci.*, 1996, 111.
10. D. F. Li, J. L. Nelssen, P. G. Reddy, F. Blecha, J. D. Hancock, G. L. Hancock, G. L. Allee and R. D. Goodband, *J. Anim. Sci.*, **68**, 1790.
11. T. Storebakkken, S. Refstie and B. Ruyter in J. K. Drackley, *Soy in Animal Nutrition*, 2000, Federation of Animal Science Societies, Savoy, IL, USA, pp. 127–170.
12. A. M. Herian, S. L. Taylor and R. K. Bush, *Int. Arch. Allergy Appl. Immunol.*, 1990, **92**, 193.
13. A. J. Kalinski, D. L. Melroy, R. S. Dwivedi and E. M. Herman, *J. Biol. Chem.*, 1992, **267**, 12068.
14. R. M. Helm, G. Cockrell, E. Herman, A. W. Burks, H. A. Sampson and G. A. Bannon, *Int. Arch. Allergy Immunol.*, 1998, **117**, 29.
15. R. M. Helm, G. Cockrell, C. M. West, E. M. Herman, H. A. Sampson, G. A. Bannon and A. W. Burks, *J. Allergy Clin. Immunol.*, 2000, **105**, 378.
16. R. M. Helm, G. Cockrell, C. Connaughton, H. A. Sampson, G. A. Bannon, V. Beilinson, D. Livingstone, N. C. Nielsen and A. W. Burks, *Int. Arch. Allergy and Immunol.*, 2000, **123**, 205.
17. R. M. Helm, G. Cockrell, C. Connaughton, H. A. Sampson, G. A. Bannon, N. C. Nielsen and A. W. Burks, *Int. Arch. Allergy and Immunol.*, 2000, **123**, 213.
18. N. C. Nielsen and Y.-W. Nam in P. R. Shewry, R. Casey, eds. *Seed Proteins*, Kluwer Academic Publishers, Dordrecht, The Netherlands, 1999, pp. 285–313.
19. N. C. Nielsen, C. D. Dickinson, T. J. Cho, V. H. Thanh, B. J. Scallon, R. L. Fischer, T. L. Sims, G. N. Drews and R. B. Goldberg, *Plant Cell*, 1989, **1**, 313.
20. N. C. Nielsen, R. Jung, Y. Nam, T. W. Beaman, L. O. Oliveira and R. B. Bassuner, *J. Plant Physiol.*, 1995, **145**, 641.
21. D. W. Meinke, J. Chen and R. N. Beachy, *Planta*, 1981, **153**, 130.
22. Y.-W. Nam, R. Jung and N. C. Nielsen, *Plant Physiol.*, 1997, **115**, 1629.
23. R. Jung, Y.-W. Nam, I. Saalbach, K. Muntz and N. C. Nielsen, *Plant Cell*, 1997, **9**, 2037.
24. A. J. Kalinski, D. L. Rowley, D. S. Loer, C. Foley, G. Butta and E. M. Herman, *Planta*, 1995, **195**, 611.
25. C. D. Dickinson, E. H. Hussein and N. C. Nielsen, *Plant Cell*, 1989, **1**, 459.
26. R. Jung, M. P. Scott, Y.-W. Nam, T. W. Beaman, R. Basssuner, I. Saalbach, K. Muntz and N. C. Nielsen, *Plant Cell*, 1998, **10**, 343.
27. A. J. Kinney, *J. Food Lipids*, 1996, **3**, 273.
28. A. J. Kinney and S. Knowlton in S. Roller and S. Harlander, eds., *Genetic Modification in the Food Industry*, Blackie, London, 1998, pp. 193–213.
29. L. M. Hoffman, D. D. Donaldson and E. M. Herman, *Plant Molec. Biol.*, 1998, **11**, 717.
30. J. J. Pueyo, M. J. Chrispeels and E. M. Herman, *Planta*, 1995, **196**, 586.
31. A. Vitale, A. Bielli and A. Ceriotti, *Plant Physiol.*, 1995, **107**, 1411.
32. L. Frigerio, A. Pastres, A. Prada and A. Vitale, *Plant Cell*, 2001, **13**, 1109.
33. S. Kjemtrup, E. M. Herman and M. J. Chrispeels, *Eur. J. Biochem.*, 1994, **226**, 385.
34. S. B. Altenbach, K. W. Pearson, G. Meeker, L. C. Staraci and S. M. Sun, *Plant Mol. Biol.*, 1989, **13**, 513.
35. J. A. Nordlee, S. L. Taylor, J. A. Townsend, L. A. Thomas and R. K. Bush. *N. Eng. J. Med.*, 1996, **334**, 688.
36. A. J. Kinney, R. Jung and E. M. Herman, *Plant Cell*, 2001, **13**, 1165.

Heterogeneity of Genes Encoding the Low Molecular Weight Glutenin Subunits of Bread Wheat

T. Chardot, T. Do, L. Perret and M. Laurière

UMR DE CHIMIE BIOLOGIQUE, INRA INA-PG CENTRE DE
BIOTECHNOLOGIE AGRO-INDUSTRIELLE, F-78850
THIVERVAL-GRIGNON

1 Abstract

Among the wheat storage proteins, the low molecular weight glutenin subunit (LMW-GS) group is probably the more complex. It encompasses several subgroups which differ by their N-terminal amino acid sequences. To get more information on the diversity of the genes encoding the LMW-GS, we attempted to clone some of them. Two different strategies were used to amplify their coding sequences from genomic DNA using PCR techniques.

The first strategy used primers corresponding to the N- and C-terminal sequences of already published mature peptides. Several DNA fragments with various sizes were amplified and more than 2500 clones were obtained. Among them 95 clones were further analyzed. The size of their insert ranged from 1200 to 800 and less bp. Restriction analysis of these inserts showed that they can be gathered in ten families. 13 Clones have been sequenced and all exhibited the features of LMW-GS coding sequences. For eight of the sequences, the coding frame was interrupted by 1 to 10 stops codons, for three sequences only parts of a LMW-GS coding frame could be recognized. The two last sequences were encoding for LMW-GS which were 270 and 327 residues long.

The second strategy used primers anchored in the promoters and in the C-terminal part. We obtained four different clones with complete coding sequence and part of the promoting sequence. They were LMW-GS coding sequences (852 bp, 284 amino acid residues) and did not contain any interruption of the coding frame.

The use of PCR techniques is a convenient way for investigating the complexity of the wheat LMW-GS multigenic family. The existence of numerous pseudo genes gives interesting insights into the evolution of this multigenic family. Extinguished LMW-GS may be a reservoir for the future?

2 Introduction

The sequence of the wheat genome is still unknown. Its size, much larger than that of the human genome, is a roadblock to the study of the organization of the genes. The knowledge of that organization would be useful from agronomic, nutritional and health points of view. Concerning the genes encoding the storage proteins expressed constitutively in the grain, little is known about their precise number, their heterogeneity and their organization at the molecular level. They are encoded by clusters of homologous genes on the homologous chromosomes 1 and 6. Among them, those encoding the low molecular weight glutenin subunits (LMW-GS) are probably the more heterogeneous and the less known. As other prolamins they are secretory proteins with a signal peptide of 20 amino acids. Depending on the N-terminal sequence of their mature peptide, several subgroups were identified. They are the s- and m-LMW-GS, whether they have serine or methionine, as N-terminal or the LMW-GS with sequences homologous to alpha-, beta- and gamma-gliadins. The s-LMW-GS appear heterogeneous and highly expressed in the endosperm. Only two complete sequences were described.[1,2] Using different primer sets, we tried to clone these genes using PCR. DNA fragments were cloned, sequenced and analyzed. We discuss here the heterogeneity of the sequences we obtained.

3 Material and Methods

3.1 Genomic DNA Preparation

DNA was prepared according to Benito *et al.*[3] from 250 mg of ethiolated seedlings, 3 days old, of var. Nepawa. It was stored at $-20\,^{\circ}$C until used.

3.2 DNA Amplification and Cloning

LMW-GS primers for PCR were synthesized by MWG Biotech (Germany). They are described in Table 1.

PCR reaction (25 μL final volume) contained Taq polymerase, 100 pmoles of the oligonucleotides primers, 2500 pmoles dNTP, *T. aestivum* genomic DNA (20 to 100 ng). 25 to 30 cycles were performed. A 1 min denaturation step at 94 $^{\circ}$C was followed by a hybridization step at 60 or 62 $^{\circ}$C for 0.5 to 2 min, followed by a 1.5 min elongation step at 72 $^{\circ}$C. The experiment was terminated by a 4 min step at 72 $^{\circ}$C. The PCR fragments were purified using Wizard PCR prep kit, either directly or after electrophoresis separation, and were cloned in the PGEM-T vector. The plasmidic DNA were first restriction analyzed, and then sequenced by MWG Biotech (Germany).

Table 1 *Oligonucleotides used in this study. Primers LMW0004-1; LMW0006-1 LMW0002-2; LMW0006-2 and LMW0007-2 were aimed at amplifying the sequences coding for LMW-GS using genomic DNA as a template*

```
LMW004-1 : (Adapted from ref 14)
Nucleic sequence :        5  AGC CAC ATC CCT GGT TTG GA 3
Proteic translation:   Nterm    S   H   I   P   G   L         Cterm

LMW006-1 : (from ref 15)
Nucleic sequence :        5  CGA GCA TAT CCT AAC AGC CCA 3

LMW002-2 : (from ref 15)
Nucleic sequence :        3  G TGG CCT CAA CCA CGG ATG 5
Proteic translation :  Nterm     T   G   V   G   A   Y        Cterm

LMW006-2 : (from ref 15)
Nucleic sequence :        3  CG TGG CCT CAA CCA CGG ATG 5
Proteic translation:   N term  G   T   G   V   G   A   Y      Cterm
```

3.3 Sequence Alignments

Sequences were analyzed using the BioEdit software V 4.7.1 by Tom Hall. The Blosum62 matrix was used for sequences alignments.

4 Results

Figure 1 summarizes the strategy used to amplify LMW-GS coding and non-coding sequences and the length of the clones studied.

4.1 Amplification of DNA Sequences using Primers Anchored at the N- and C-Terminal Sequences of Mature s-LMWG-GS

Numerous DNA fragments with various sizes were amplified (not shown) using oligonucleotide primers LMW004-1/LMW002 and LMW006-1/LMW006-2 designed from Masci *et al.*[2] They have been cloned into the pGEM-T vector. More than 2500 clones were obtained, which were white/blue screened. Among the positive clones, 95 were further restriction analyzed using either Bstx I + Sph I, to excise the full DNA fragment inserted into the pGEM-T plasmid, or Nde I + Xho I, to excise clones containing XhoI site. This allowed us to distinguish ten different families, according to their restriction maps. The DNA of 13 clones, representing the different families, were sequenced. All the sequences obtained exhibited the features of LMW-GS coding sequences as deduced from Blast searches or from translations. Height sequences had a coding frame interrupted by 1 to 10 stop codons. For three sequences, only parts of a LMW-GS coding frame could be recognized from the sequence. Figure 2 shows the alignment of the amino acid sequences of the mature peptide derived from the longest DNA

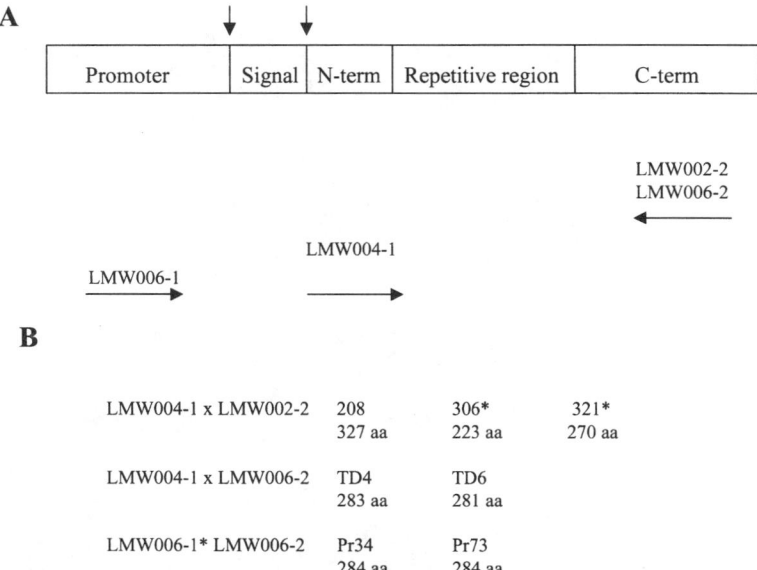

Figure 1 (*A*) *Schematic representation of a LMWG gene, with the nucleotidic primers used in this study. Vertical arrows indicate the beginning of the coding frame, and of the mature peptide. Central sub-domains were not represented.* (*B*) *Summary of clones of interest obtained in this study.* * indicates pseudo genes

sequences. Clone 208 exhibited the longest size with 327 residues. Clones TD4 and TD6 were respectively 284 and 281 residues long. Clone 321 (270 residues) is shown because it contained only one stop codon. Their N-terminal acid sequence (SHIP), imposed by the primers used, make them putative s-type LMW-GS. They are highly homologous with highly conserved N-terminal (over 13 residues) and C-terminal (over 54 residues). The central domain which exhibited repetitive stretches rich in glutamine and proline, ranged from 196 (clone 321) to 250 residues (clone 208). It can be divided according to Cassidi *et al.*[4] into three subdomains II to IV. Domains II and IV are mainly responsible for the differences of the size of the clones. Domain III limited stretches of glutamine and is strongly conserved. All clones contained the six conserved cysteines (white stars in domains III and V), which are assumed to be engaged in disulfide bridges. Only one additional cysteine was found at a position that can slightly differ in domain IV (black stars) and which may have a role in polymer formation.[5]

4.2 Amplification of DNA Sequences using Primers Anchored at the Promoter and C-Terminal Sequences of LMWG-GS

Using primers anchored in the promoters and in the C-terminal part of LMW-GS, we amplified, using PCR techniques, DNA molecules with size ranging from 600 to 1200 bp, as deduced from agarose gel electrophoresis. The major 1200 bp fragment was cloned into the pGEM-T plasmid. 240 colonies were obtained, and

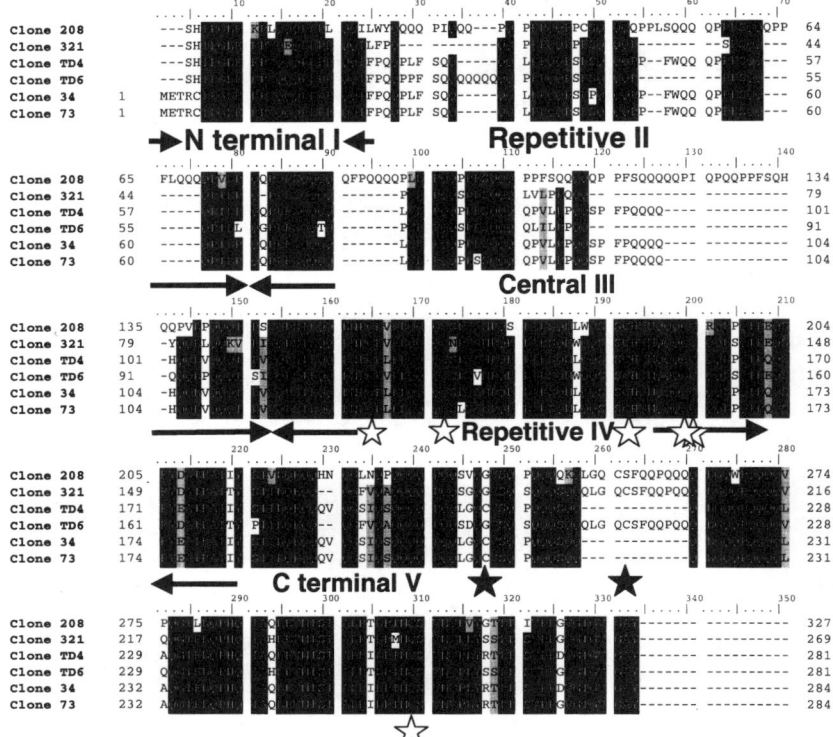

Figure 2 *DNA fragments obtained upon amplification of genomic DNA from wheat (var. Neepawa) with oligonucleotide primers LMW0004-1 x LMW0002-2; LMW0004-1 x LMW0007-2 and LMW0006-1 x LMW0006-2 have been inserted into pGEMt plasmid. DNA from clones TD6r, 321, TD4, 34, 73 and 208 have been sequenced and the nucleotide sequence translated. Proteic sequences have been aligned. Amino acid residues on black background are conserved in all sequences. Amino acid residues on gray background are homologous. White stars represent cysteines conserved in all glutenin coding sequences. Black stars represent cystein with variable location*

white/blue screened. 20 colonies were PCR screened using proper oligonucleotide. All cloned were positives. They exhibited similar sizes, as estimated from 1.5% agarose electrophoresis. Four clones were selected among the positive colonies. They all contained a 1350 bp insert. These clones have been sequenced.

4.2.1 Analysis of the Non-coding DNA Sequences. The partial promoting sequences of the four clones showed strong homologies and typical boxes (results not shown). The endosperm box[6–8] was located at position −311. It was almost immediately followed by a GCN4 motif (−292), controlling endosperm specific expression.[9] The CAT box was found at position −169. AGGA box was found at −100. The AACA motif was found at position −165. TATA boxes were found at position -62 and -79. Initiator codon was found in the CCACCATGA consensus sequence.[8,10]

4.2.2 Analysis of the Coding Sequences. All sequences of the previous clones were coding for a precursor protein, which all belonged to the METRCIP LMW-GS subfamily. They were 284 amino acid residues long, differing only by three residues. The four clones sequences exhibited a 20 amino acid residue signal peptide, followed by the N-terminal domain with the conserved 13 residues, a 210 residues repetitive domain. Compared to the above described clones, they differ by repetitive stretches in domain II and by the domain IV which was 12 residues shorter. It shifted the non-conserved cysteine found in this last domain. The 54 residues C terminal domain was conserved (Figure 2).

5 Discussion

Using N and C terminal directed oligonucleotide sequences in PCR technique, allowed us to amplify and to clone various types of DNA sequences. Some of them exhibited typical LMW-GS coding sequences potentially expressed. Others showed various numbers of stop codons, up to 10 in the typical LMW-GS open reading frame or only partially LMW-GS coding sequences features. That makes them putative pseudo genes. Such DNA sequences are already described.[11-13] Their existence is likely to be common in different wheat species. They are certainly not expressed *in vivo*, due essentially to the number of stop codons found in the DNA sequences. They may represent reservoirs of variability that could be revived by mutagenesis. Our goal was to clone s-type LMW-GS which are largely expressed in bread and durum wheat.[1,2] We used primers corresponding to the SHIP sequences at the N-terminal. The amplified DNA sequences with SHIP N-terminal coding sequences corresponded to potentially expressed LMW-GS. These DNA sequences were only putative s-LMW-GS, due to the PCR technique used, which dictates the 5' and 3' terminal sequences. It is interesting to note their high homology to the clone obtained with a primer anchored in the promoter. Another interesting feature is that, with this last strategy we elicited only m-LMW-GS, unlike Masci *et al.*[1,2] Presently 31 LMW-GS sequences were described in EMBL. Our work underlines again the difficulty to isolate s-type LMW-GS sequences despite the fact that we obtained more different DNA sequences than Masci *et al.*[1,2] using the same strategy. However, with this strategy using primers anchored in the promoter, we amplified DNA sequences with no stop codons in their coding sequences unlike with use of primers anchored at the N-terminal. Nevertheless PCR technique using both strategies is an attractive method to investigate the diversity and complexity of wheat storage protein genes in the absence of the complete sequencing of the wheat genome.

This work was supported by a grant from INRA (Action Nouvelle Soutenue to TC and ML).

References

1. S. Masci, R. D'Ovidio, D. Lafiandra and D. D. Kasarda. *Theor. Appl. Genet.*, 2000, **100**, 396.
2. S. Masci, R. D'Ovidio, D. Lafiandra and D. D. Kasarda, *Plant Physiol.*, 1998, **118**, 1147.
3. C. Benito, M. Figueiras, C. Zaragoza, F. J. Gallego and A. de la Peña, *Plant. Mol. Biol.*, 1993, **21**, 181.
4. B. G. Cassidy, J. Dvorak and O. D. Anderson. *Theor. Appl. Genet.*, 1998, **95**, 743.
5. P. R. Shewry and A. S. Tatham, *J. Cereal Sci.*, 1997, **25**, 207.
6. J. Forde, J. M. Malpica, N. G. Halford, P. R. Shewry, O. D. Anderson, F. C. Greene and B. J. Miflin, *Nucleic Acids Res.*, 1985, **19**, 6817.
7. M. Kreis, B. G. Forde, S. Rahman, B. J. Miflin and P. R. Shewry, *J. Mol. Biol.*, 1985, **183**, 499.
8. V. Colot, D. Bartels, R. Thompson and R. Flavell, *Mol. Gen. Genet.*, 1989, **216**, 81.
9. Y. Onodera, A. Suzuki, C. Y. Wu, H. Washida and F. Takaiwa. *J. Biol. Chem.*, 2001, **276**, 14139.
10. M. Kozak, *Cell*, 1986, **44**, 283.
11. M. Tanurdzic, D. Obreht and L. J. Vapa, unpublished.
12. M. Benmoussa, L. P. Vézina, M. Pagé, S. Yelle and S. Laberge. *Theor. Appl. Genet.*, 2000, **100**, 789.
13. R. D'Ovidio, S. Masci, C. Marchitelli, P. Tosi, M. Simeone and E. Porceddu, unpublished.
14. R. D'Ovidio, C. Marchitelli, L. Ercoli Cardeli and E. Porceddu, *Theor. Appl. Genet.*, 1999, **98**, 455.
15. R. D'Ovidio, M. Simeone, S. Masci and E. Porceddu, *Theor. Appl. Genet.*, 1997, **95**, 1119.

Identification, Cloning and Expression of a Ferulic Acid Esterase from *Neurospora crassa*

V. F. Crepin[1] C. B. Faulds[2] and I. F. Connerton[1]

[1]DIVISION OF FOOD SCIENCES, SCHOOL OF BIOSCIENCES, UNIVERSITY OF NOTTINGHAM, SUTTON BONINGTON CAMPUS, LOUGHBOROUGH LE12 5RD, UK
[2]NUTRITION, HEALTH AND CONSUMER SCIENCES DIVISION, INSTITUTE OF FOOD RESEARCH, COLNEY, NORWICH NR4 7UA, UK

1 Introduction

The plant cell is a complex architecture of polysaccharides. For the complete hydrolysis of these polysaccharides, microorganisms require a battery of enzymes. Cross-linking of ferulic acids to cell wall components influences the cell wall properties such as extensibility, plasticity and digestibility. Cinnamoyl esterases, a subclass of carboxylic acid esterases (E. C. 3.1.1.1), are able to hydrolyse the ester bond between the ferulic acid and sugars present in plant cell walls. Ferulic acid esterases are classed as type A or B dependant on their specificity for the substrate aromatic moiety, specificity for the linkage to the primary sugar and ability to release dehydrodiferulic acids from esterified substrates. Esterases are novel enzymes with considerable potential for agri–food processing applications. This study reports the identification, the cloning and the expression of a type B ferulic acid esterase from *N. crassa* (*fae-1* gene, data base accession number: AJ293029).

2 Materials and Methods

The *Neurospora crassa* wild type ST *A* (74*A*) was used throughout. Genomic DNA was extracted according to Stevens and Metzenberg[1] and RNA was extracted according to Sokolovsky *et al.*[2]

Northern blot: RNA was electrophoresed in formaldehyde denaturing agarose gel and blotted by capillary transfer to Hybond-N$^+$ membrane (Amersham

Pharmacia Biotech). The homologous probe (*fae-1* gene) was labelled using the non-radioactive DIG-labelling system (Boehringer Mannheim).

Expression in *E. coli*: the cDNA of the *N. crassa fae-1* gene was cloned into the expression vector pET3a. *E. coli* BL21 (DE3) pLysS cells were transformed with the pET3a/*fae-1* constructed vector. Induction was performed in presence of IPTG.

Expression in *Pichia pastoris* GS115: the cDNA of the *N. crassa fae-1* gene containing the native signal sequence was cloned into the expression vector pPIC3.5K under the control of the alcohol oxidase promoter (*AOX*1). *P. pastoris* was transformed with pPIC3.5K/*fae-1* vector linearised by *Dra*I digestion. The transformation was performed following the manufacturer's procedure using the spheroplast method (Invitrogen). Induction of expression was performed in buffered complex methanol medium (BMMY).

3 Results

3.1 Identification of a Type B Ferulic Acid Esterase from *N. crassa*

Using a differential hybridisation screen, a previously unrecognised gene was identified following growth of *N. crassa* with sugar beet pulp as sole carbon source.[3] Translation of the genomic sequence revealed an open reading frame with a single putative intron that encodes a protein with 45% identity with an acetyl-xylanesterase reported from *Aspergillus awamori* and recently identity with a type B ferulic acid esterase from *P. funiculosum* (Figure 1).[4,5]

Consistent with the hybridisation protocol, Northern blots confirmed that the *N. crassa* gene was expressed on sugar beet pulp and absent on sucrose and wheat bran. This expression profile is similar to the characteristics of other type B ferulic acid esterases.[5,6]

A Southern blot was performed with genomic DNA extracted from several fungi in order to determine the presence of *fae-1* homologous genes in other fungal species. The sequence homologies detected between *N. crassa fae-1* gene and the various fungal genomes reflect their phylogenetic relationships (Figure 2).

3.2 Over-expression of a Type B Ferulic Acid Esterase from *N. crassa*

3.2.1 Over-expression in E. coli. The reading frame of *fae-1* was expressed in *E. coli* using the pET expression system. Various temperatures and IPTG concentrations were tested to find the best conditions in order to produce the largest quantity of soluble protein. However, the recombinant feruloyl esterase proved to be insoluble (Figure 3) and non-active irrespective of the conditions employed.

3.2.2 Over-expression in P. pastoris *GS115.* *P. pastoris* was transformed with pPIC3.5K/*fae-1* vector. Transformant were grown as small-scale cultures in

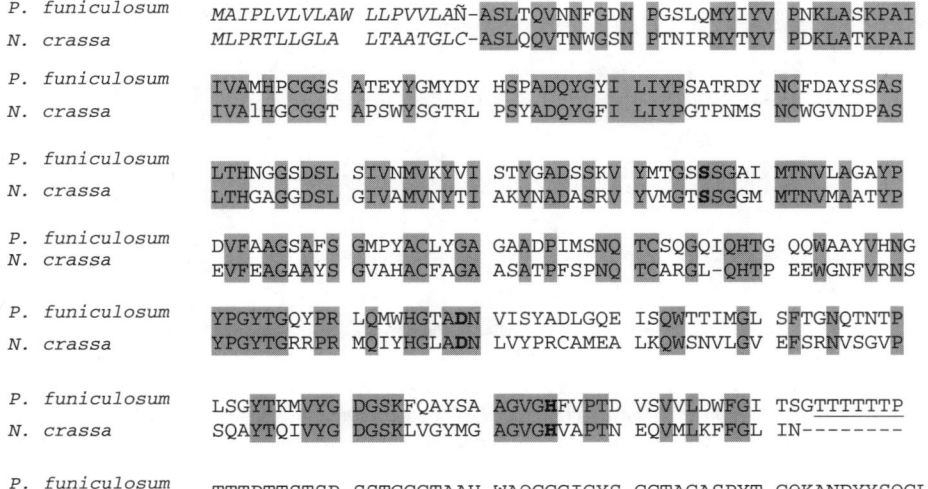

```
P. funiculosum   MAIPLVLVLAW LLPVVLAÑ-ASLTQVNNFGDN PGSLQMYIYV PNKLASKPAI
N. crassa        MLPRTLLGLA  LTAATGLC-ASLQQVTNWGSN PTNIRMYTYV PDKLATKPAI

P. funiculosum   IVAMHPCGGS ATEYYGMYDY HSPADQYGYI LIYPSATRDY NCFDAYSSAS
N. crassa        IVAlHGCGGT APSWYSGTRL PSYADQYGFI LIYPGTPNMS NCWGVNDPAS

P. funiculosum   LTHNGGSDSL SIVNMVKYVI STYGADSSKV YMTGSSSGAI MTNVLAGAYP
N. crassa        LTHGAGGDSL GIVAMVNYTI AKYNADASRV YVMGTSSGGM MTNVMAATYP

P. funiculosum   DVFAAGSAFS GMPYACLYGA GAADPIMSNQ TCSQGQIQHTG QQWAAYVHNG
N. crassa        EVFEAGAAYS GVAHACFAGA ASATPFSPNQ TCARGL-QHTP EEWGNFVRNS

P. funiculosum   YPGYTGQYPR LQMWHGTADN VISYADLGQE ISQWTTIMGL SFTGNQTNTP
N. crassa        YPGYTGRRPR MQIYHGLADN LVYPRCAMEA LKQWSNVLGV EFSRNVSGVP

P. funiculosum   LSGYTKMVYG DGSKFQAYSA AGVGHFVPTD VSVVLDWFGI TSGTTTTTTP
N. crassa        SQAYTQIVYG DGSKLVGYMG AGVGHVAPTN EQVMLKFFGL IN--------

P. funiculosum   TTTPTTSTSP SSTGGCTAAH WAQCGGIGYS GCTACASPYT CQKANDYYSQCL
```

Figure 1 *Peptide sequence homology between a type B ferulic acid esterase from* P. funiculosum *and* N. crassa. *The signal sequence is in italic and the catalytic triad is shown in bold. Grey boxes show identical amino acids. The extended* P. funiculosum *sequence features a C-terminal cellulose-binding domain preceded by a linker region (underlined)*

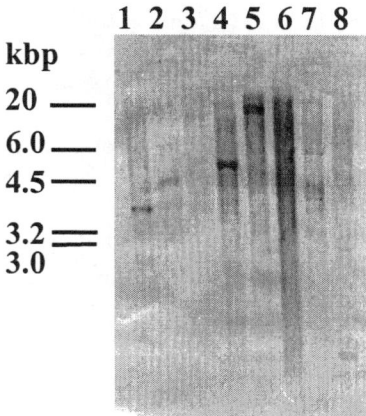

Figure 2 *Southern blot. Fungal genomic DNA were digested by* EcoRI *and hybridised with a probe homologous to* fae-1. *1:* T. thermophilus, *2:* T. stipitatus, *3:* P. sajorcaju, *4:* P. funiculosum, *5:* N. crassa, *6:* H. grisea, *7:* A. oryzae, *8:* A. niger

methanol induction medium (2 ml BMMY medium) for three days. The culture supernatants (10 µl) were then analysed by SDS-PAGE for feruloyl esterase expression. A colony transformed with the parental vector was used as a control of the background secretion levels (Figure 4).

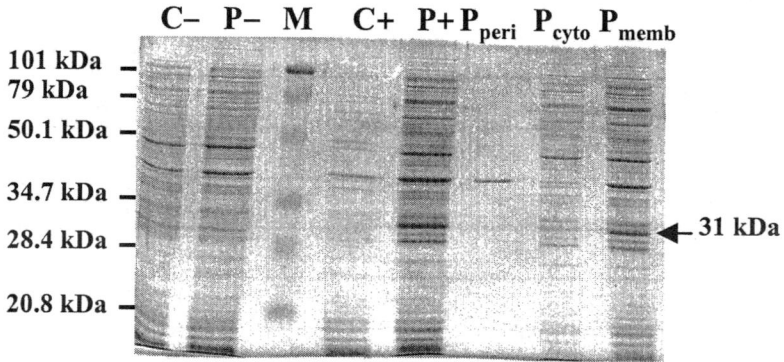

Figure 3 *Expression of a ferulic acid esterase from* N. crassa *in* E. coli. *Cells were grown at 38°C and induced at* $OD_{600} = 0.4$ *with 0.4 mM IPTG. C: control strain transformed with pET3a. P: protein expression strain transformed with pET3a/fae-1. Pperi: periplasmic fraction. Pcyto: cytoplasmic fraction. Pmemb: membrane and inclusion bodies fraction. M: standard protein molecular weight.* —: *pre-induction.* +: *post-induction*

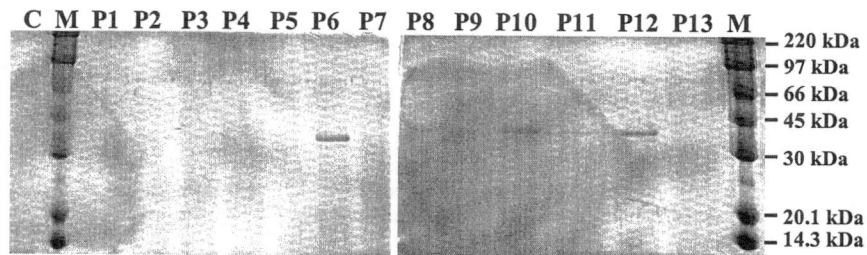

Figure 4 *Expression of* fae-1 *from* N. crassa *in* P. pastoris. *M: Standard protein molecular weight. C: Control strain. P1–P13: pPIC3.5K/fae-1 transformants*

Five transformants produced a major secreted protein band, with an estimated molecular weight of 40 kDa and no protein was detected in the control. The estimated molecular weight is greater than that predicted or observed from *E. coli*, which is likely due to post-translational modification. P6 and P10 transformants were retained to perform enzymatic assays. The culture supernatants were assays for activity against methyl caffeate (MCA) and methyl ferulate (MFA). The recombinant proteins were shown to be active as a feruloyl esterase and show the characteristics of a type B ferulic acid esterase.[6] Feruloyl esterase activity is reported in Table 1.

4 Conclusions

The identification of a type B feruloyl esterase from *N. crassa* was based on: (1) the homology found between the peptide sequence of a *P. funiculosum* type B feruloyl esterase and the translation of *N. crassa fae-1* gene; (2) the gene was

Table 1 *Feruloyl esterase activity of* P. pastoris *culture supernatants*

Clone	MFA (Activity expressed as U/ml of supernatant)	MCA (Activity expressed as U/ml of supernatant)
P-6	1.42	2.08
P-10	5.07	4.33
C (control)	0	0

specifically induced in presence of sugar beet pulp, a characteristic of type B ferulic acid esterase from *A. niger* and *P. funiculosum*; (3) the substrate activity profile of the recombinant enzyme with respect to methyl caffeate and methyl ferulate, specific substrates of type B ferulic acid esterase.

The *P. pastoris* expression system produced high yields of secreted protein. The recombinant protein represents the majority of the protein supernatant without further purification. The *E. coli* expression system was found to be unsuitable for expression of soluble and active ferulic acid esterase from a fungal source.

Acknowledgements

We thank the BBSRC, DTI and the LINK programme for funding this work and all the people who have been related to this project. Thank you to Nicola Cummings for her continued advice.

References

1. J. N. Stevens and R. L. Metzenberg, *Neurospora Newslett.*, 1982, **29**, 27.
2 V. Sokolovsky, R. Kaldenhoff, M. Ricci and V. E. A. Russo, *Neurospora Newslett.*, 1990, **37**, 41.
3. G. H. Thomas, I. F. Connerton and J. R. S. Fincham, *Mol. Microbiol.*, 1988, **2**, 599.
4. T. Koseki, S. Furuse, K. Iwano, H. Sakai and H. Matsuzawa, *Biochem. J.*, 1997, **326**, 485.
5. P. A. Kroon, G. Williamson, N. M. Fish, D. B. Archer and N. J. Belshaw, *European J. Biochem.*, 2000, **267**, 6740.
6. P. A. Kroon, C. B. Faulds and G. Williamson, *Biotechnol. Appl. Biochem.*, 1996, **23**, 255.

Biopolymer Design

Influence of Polysaccharide Composition on the Structure and Properties of Cellulose-based Composites

M. J. Gidley, E. Chanliaud and S. Whitney

UNILEVER RESEARCH COLWORTH, COLWORTH HOUSE,
BEDFORD MK44 1LQ, UK

Plant cell walls are Nature's most abundant source of organic polymers, making them a prime candidate for technologies based on renewable raw material resources. The bulk of cell wall material is in the form of so-called secondary cell walls, characteristic of woody tissues and used in, *e.g.* construction and paper industries. Primary cell walls are characteristic of growing and fleshy plant tissues, and contain several different polymer types that find specialised uses in food and other applications. As the mysteries of biological control of cell wall composition start to be unravelled by functional genomics and related approaches, it is timely to consider how best to make use of the tremendous natural resource presented by plant cell walls. This contribution will address (i) compositional diversity, (ii) principles of wall assembly and (iii) effects of individual polymers on primary cell wall material properties, taking advantage of a model system based on bacterial cellulose.

1 Polymer Composition in Primary Cell Walls

The major component of primary plant cell walls is water, which typically accounts for 70–90% by mass or volume. Most of the remaining components are polysaccharides belonging to one of three families, traditionally termed cellulose, hemicellulose and pectin. Of these, only cellulose is well-defined chemically, being a pure linear polymer of glucose in the β-configuration, linked *via* glycosidic (hemiacetal) bonds between positions 1 and 4 on adjacent residues [β-(1→4)-glucan]. Hemicelluloses are also sometimes referred to as 'glucan-binding' polymers[1] indicating their propensity to interact at the molecular level with cellulose. The common feature, and the reason why binding to cellulose is efficient, is that all hemicelluloses have backbones with conformational similarity to cellulose. All backbones have β-1→4 glycosidic links based on (i) glucose with a high level of further carbohydrate substitution (*xyloglucan*), (ii) mannose (which differs from glucose only by the orientation of one hydroxyl group) with variable substituents (*mannan*), or (iii) xylose (which differs from glucose by the absence of an exocyclic hydroxymethyl group) with variable substituents (*xylan*). It is now becoming clear that the nature of the backbone together with the type and extent of substitution of the backbone are both determinants of the strength of interaction with cellulose (see below).

Pectins are a complex group of polymers, typified by the presence of 1→4 linked α-galacturonic acid residues. There are three major classes of pectin components with characteristic structural features.[1] *Homogalacturonan* has contiguous galacturonic acid residues that are partially in the methyl ester form thereby affecting charge density. *Rhamnogalacturonan I* has a backbone of alternating galacturonic acid and rhamnose with frequent substitution of rhamnose residues with sidechains of polymeric arabinose (arabinan) and/or galactose (galactan). *Rhamnogalacturonan II* is a very complex polymer containing many unusual sugars and linkages, and has the potential to be cross-linked with borate ions. It is not yet clear whether the three types of pectin polysaccharide exist as regions within single polymers or as individual entities. There is much circumstantial evidence in support of covalent linkages between pectin and hemicellulose polymers, but unequivocal proof of their existence within the native cell wall remains elusive. In some tissues there are additional features, such as phenolic substituents on either pectin or hemicellulose, that are capable of being cross-linked with catalysis by peroxidases. Further to the three general polysaccharide classes, some tissues contain other polysaccharides, glycoproteins and proteins with putative roles in both structure and signalling.

The enormous chemical complexity of primary plant cell walls appears daunting from the point of view of linking polymer composition to functionality, but there are two counterbalancing factors. One is that, for many cell walls, the bulk of the polymers are present as a subset of the total possible. For instance, onion bulb cell walls primarily contain cellulose, homogalacturonan, rhamnogalacturonan with linear galactan sidechains, and a small amount of xyloglucan. This is sufficient simplification to at least construct testable hypotheses for the relationships between properties at the polymer level (nm) and materials level (mm). The second factor is that, for many polymer types, there are seed sources of relatively pure polymers. In these examples, the role of starch as an energy reserve has been replaced or augmented by relatively massive deposition of cell wall polysaccharides. Examples include (galacto)mannans from carob, date or guar seeds, xyloglucans from tamarind or nasturtium seeds and pectic galactan from lupin seeds. These seeds provide both a ready source of large quantities of polymer, and a biological system that is focussed on synthesis and deposition of a single polymer, hence providing a good starting point for purification of biosynthesis/turnover enzymes and genes.[2]

2 Principles of Wall Assembly

As described above, Nature has a large toolbox of polymers with which to construct cell walls for the particular requirements of a specific tissue location and developmental stage. Although the details are not fully elucidated, several of the principles underlying wall assembly are understood. A major influence is the fact that wall polymers are produced at two different locations within plant cells. Cellulose is 'extruded' directly through the outer (plasma) lipid membrane of the cell into the extracellular environment at sites known as 'terminal complexes'. Each of these sites produces a small number of chains in a parallel synthetic process. These chains self-associate close to the point of synthesis producing an aggregate held together by numerous specific hydrogen bonds and excluding water. Coalescence of many of these aggregates results in long persistent ribbons with, typically, *ca.* 10 nm cross-sectional dimensions.

All other polymers are synthesised inside the cell and transported in lipid vesicles that fuse with the plasma membrane and release their contents into the extracellular space. It is likely that this process occurs alongside cellulose deposition, and hence wall assembly is likely to be influenced by the interaction between cellulose and non-cellulosic polymers. After initial deposition, cell walls can be further modified both by integration of further secreted polymers and by the action of modifying enzymes such as pectin methyl esterase, xyloglucan endo-transglycosylase and peroxidase.

A current model for the primary plant cell wall is based on microscopic observations[3] and chemical extraction studies[4] and is represented in the cartoon below (Figure 1) based on McCann and Roberts.[5]

The essential elements of this model are that cellulosic 'rods' are cross-linked with other glycans to form a connected network that is inter-penetrated by a network of pectin. In the middle lamella region (the boundary between two adjacent cells), there is a lack of cellulose and an abundance of pectin. Although xyloglucan is identified as cross-linking cellulose in Figure 1, in reality a range of 1,4 linked β-glycans (*e.g.* glucans, mannans, xylans) are thought to be capable of performing this role.[1] Pectin structure is also complex as described earlier. There is little knowledge on the disposition of different pectic sub-types within walls. In addition, there is considerable heterogeneity both within single cell walls and between walls from different cells and tissues. This makes it difficult to disentangle the functional role played by individual polymers in the architecture and properties of walls. To study the roles of individual polysaccharides, we have made use of a bacterial system (*Acetobacter xylinus* 53524) that produces pure extracellular cellulose in an analogous manner to plant cells (Figure 2).

As *Acetobacter xylinus* can be fermented in liquid culture, other polymers can be added to the fermentation system to simulate in a controlled way the effects of individual polymers on plant wall assembly. Some of the principles that have emerged for hemicelluloses are:

1. Xyloglucan polymers with a range of different sidechain substitution patterns show significant molecular level interactions (as judged by 13C CPMAS

Plant Primary Cell Wall Model

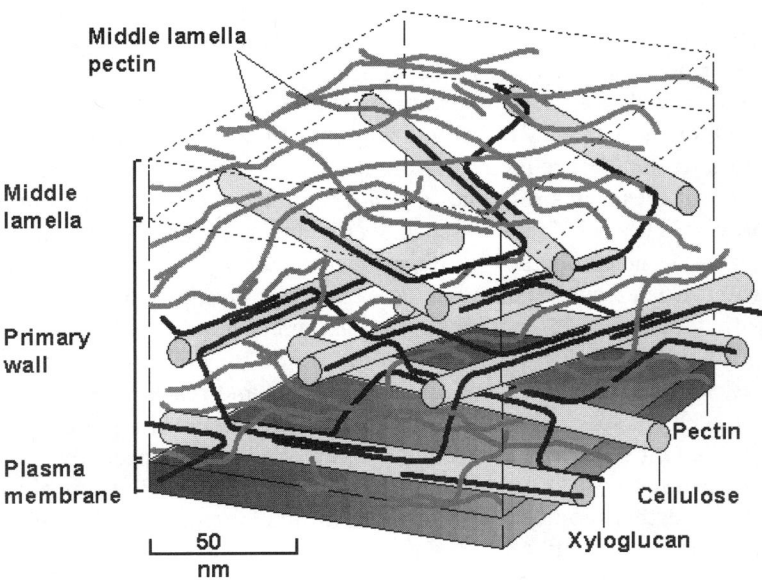

Figure 1 *Schematic and hypothetical model for primary plant cell wall structure (adapted from ref. 5)*

NMR) with cellulose and are probably partially trapped within cellulose ribbons (as judged by the requirement for strong alkali to release them). High molecular weight xyloglucans are able to extensively cross-link cellulose ribbons.[6]

2. Mannan-based polymers have a range of molecular interaction properties depending on the degree of substitution of the mannan backbone with sidechain galactose residues. Low levels of substitution lead to extensive molecular interaction with cellulose and the two polymers coalesce to form large aggregates. High levels of substitution lead to abolition of any detectable interactions. Intermediate levels result in a range of cross-linked structures.[7] Molecular composition in this case has a clear influence on cellulose composite architecture.

3. Xylan-based polymers show no convincing evidence for molecular interaction with, or cross-linking of, cellulose under the high dilution fermentation conditions. Whether this is also true in the less dilute cell wall environment remains to be determined. Description of hemicellulose polymers as 'glucan-binding' may therefore not always be accurate.

For pectins, there is no evidence for any direct molecular interaction with

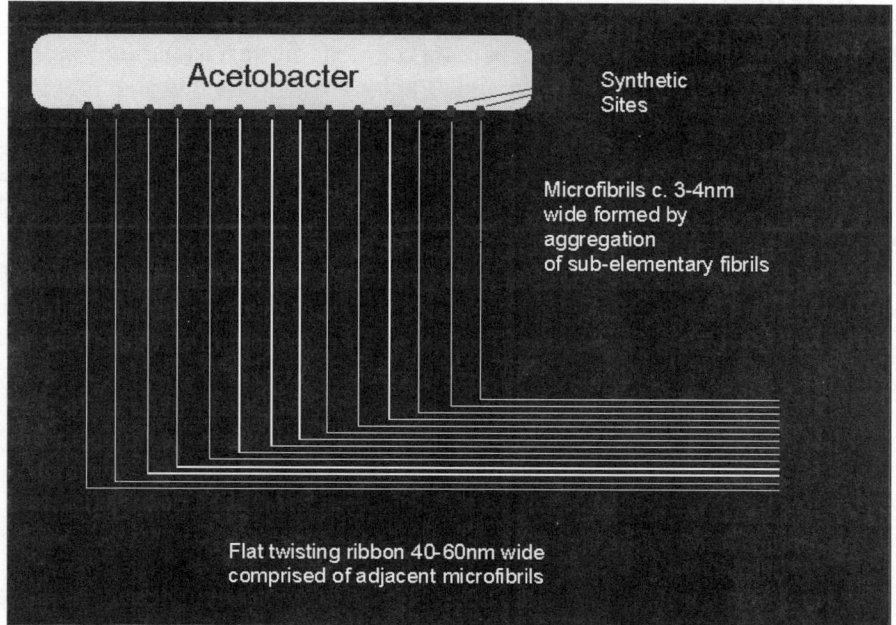

Figure 2 *Schematic representation of cellulose synthesis from* Acetobacter xylinus (*not to scale*). *Microfibrils of cellulose are secreted into the fermentation medium via 'terminal complex' transmembrane synthetic sites. In the extracellular medium, a number of elementary microfibrils coalesce to form a flat, twisting and highly persistent ribbon of cellulose. The presence of polysaccharides in the fermentation medium allows interactions to occur both before and after the assembly of microfibrils into ribbons. The right angle bend at the point of ribbon assembly is purely schematic*

cellulose, consistent with both their charged nature and the lack of shape complementarity. However, pectins are capable of self-association and network formation with divalent cations such as calcium. For fermentations in the presence of strong calcium/pectin networks, cellulose is unable to penetrate the network. For non-gelled pectin, there is no mechanism for retaining pectin in the vicinity of cellulose, and pectin is readily 'washed out' after fermentation. However, for relatively weak networks, a cellulose/pectin composite is formed that survives subsequent extensive washing provided calcium cross-links are retained.[8] A similar concept applies for other weakly networked polysaccharides such as certain xylan- and mannan-based polymers.[7]

What evidence there is for native plant cell walls suggests that these principles are reasonable, and can therefore stand as hypotheses for attempted disproving. Some of these principles also find technological use, for example in paper or textile sizing with xyloglucans or mannans.

3 Polymer Effects on Model Wall Material Properties

Characteristic material properties of plant cell walls all start from cellulose, but are capable of a large degree of modulation by the presence and state of other polysaccharides. To a first approximation, cell wall cellulose can be considered as an entangled assembly of stiff rods. The linear conformation of cellulose chains together with their aggregation into fibres/ribbons results in an exceptionally high axial ratio and consequently very weight-efficient structuring *via* hydrodynamic entanglements.

Under small deformation conditions, it appears that entanglement of stiff ribbons is the primary mechanism of structuring for a range of cell walls and model composites.[9] However, under larger deformation conditions, *e.g.* uniaxial or biaxial tension, non-cellulosic polymers have a marked influence. Figure 3 shows the very different results obtained on uniaxial extension of cellulose, cellulose/pectin and cellulose/xyloglucan composites. In both cases, incorporation of a second polymer makes cellulose more extensible and weaker. For pectin-containing composites, it can be shown that this altered mechanical behaviour is due entirely to the cellulosic component in the composite, as removal of the pectin (by chelation of calcium with CDTA) does not change the uniaxial mechanical properties (Figure 4).

In summary, based on studies of *Acetobacter* cellulose-based composites, the following effects of two polymer types have been found:

1. *Xyloglucan.* When xyloglucan molecular weights are high enough to extensively cross-link cellulose, uniaxial tensile stiffness is reduced and extensibility is increased.[9] Under biaxial tensile conditions, non-linear-elastic behaviour is

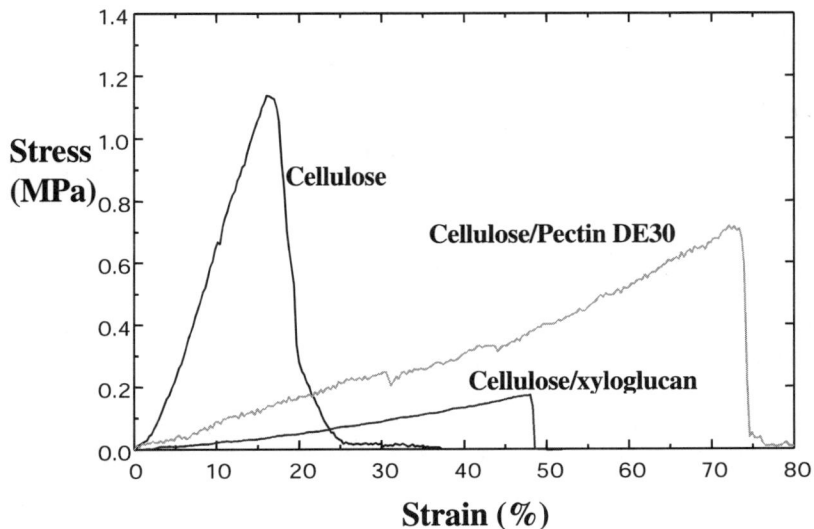

Figure 3 *Strain/stress curves for composites in a uniaxial tensile test*[8,9]

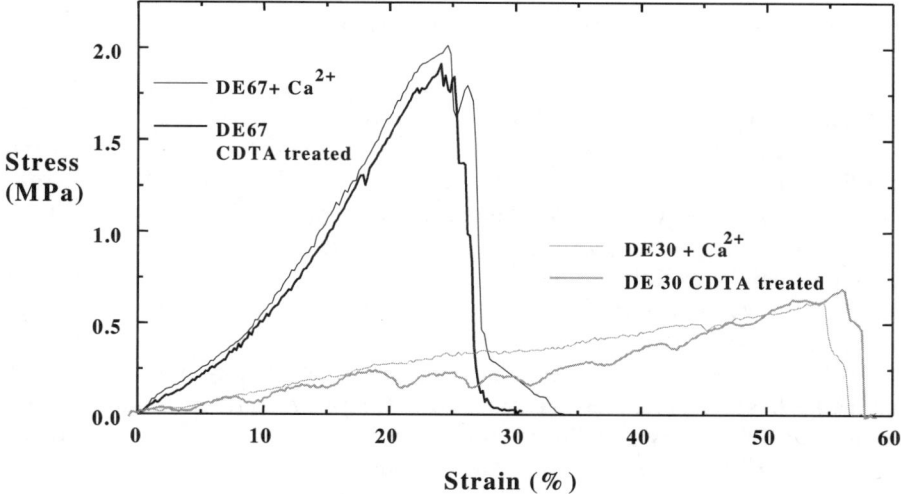

Figure 4 *Uniaxial tensile testing of two cellulose/pectin composites, before (thick lines) and after (thin lines) removal of pectin mediated by calcium chelation with CDTA*

found with very low apparent stiffness and pronounced creep behaviour.[10] This is interpreted as being due to cross-linking (thin) xyloglucan strands becoming partially responsible for load bearing, resulting in 'weaker' and more time-dependent mechanical behaviour.

2. *Pectin.* For pectin networks with cellulose deposited into them, uniaxial extensibility is enhanced and stiffness is decreased. This result is maintained after removal of the pectin component with calcium-chelating agents such as CDTA. It is suggested that the deposition of cellulose into a pre-formed pectin network results in more local alignment of cellulose ribbons, reducing the density of effective entanglements, and making them easier to pull apart on stretching.[8] In contrast, biaxial tensile results show only a modest weakening and no increase in extensibility. This is consistent with the proposed uniaxial mechanism as the geometrical constraints of the biaxial test geometry prevent the same mechanism from being operative.[10] Under tensile conditions, there is no evidence of a load-bearing role for the pectin component (*cf.* xyloglucan in [1] above).

From these results with just two of the non-cellulosic polymers present in cell walls, it is apparent that there is a richness of materials behaviour that Nature can call upon to suit the mechanical requirement of particular tissues at a given developmental stage. It is an intriguing possibility that, as the genetic machinery that controls production of cell wall polymers becomes clearer, it may be possible to select or design plant tissues to produce cell wall materials with specific and desirable mechanical performances.

4 Prospects for Future Exploitation

As a major source of renewable biomass, plant cell walls are a potentially attractive source of polymers from both economic and environmental standpoints. The challenges for exploitation are twofold. On the one hand, there is the challenge to make more (efficient) use of currently available materials. Alternatively, if desired polymers/composites can be specified at the molecular level, then biological approaches to the selection, modification or novel sourcing of materials can be exploited. One current limitation to the use of cell walls as a source of polymers is the relative difficulty in extracting specific molecules. However, there may be opportunities for better use of intact cell wall materials provided that desired properties can be obtained reproducibly and reliably. This challenge to specify complex materials to a similar level as isolated polymers is a major one, but success could lead to better utilisation of agricultural materials that are currently regarded as being of low value.

In the current era of functional identification of genes and their encoded proteins, it is likely that the factors controlling cell wall polymer synthesis will become apparent in a relatively short period of time. This is expected to result in the opportunity to select breeding lines and mutants with desired genes/mutations, as well as allowing the consequence of modulating the expression of specific genes to be investigated by transgenic approaches. At the time of writing, only three cell wall synthesis activities have had genes positively associated with them,[11-13] but it is likely that many more will be identified in the coming months and years. This will provide the plant biopolymer community with a new approach to designing and obtaining plant cell wall polysaccharides and their assemblies.

References

1. N. C. Carpita and D. M. Gibeaut, *Plant Journal*, 1993, **3**, 1.
2. J. S. G. Reid, *Curr. Opin. Plant Biol.*, 2000, **3**, 512.
3. M. C. McCann, B. Wells and K. Roberts, *J. Cell Sci.*, 1990, **96**, 323.
4. T. Hayashi, *Ann. Rev. Plant Physiol.*, 1989, **40**, 136.
5. M. C. McCann and K. Roberts, *The Cytoskeletal Basis of Plant Growth and Form*, C. W. Lloyd, ed., Academic Press, London, 1991, p. 109.
6. S. E. C. Whitney, J. E. Brigham, A. H. Darke, J. S. G. Reid and M. J. Gidley, *Plant Journal*, 1995, **8**, 491.
7. S. E. C. Whitney, J. E. Brigham, A. H. Darke, J. S. G. Reid and M. J. Gidley, *Carbohydrate Research*, 1998, **307**, 299.
8. E. Chanliaud and M. J. Gidley, *Plant Journal*, 1999, **20**, 25.
9. S. E. C. Whitney, M. G. E. Gothard, J. T. Mitchell and M. J. Gidley, *Plant Physiology*, 1999, **121**, 657.
10. E. M. Chanliaud, K. M. Burrows, G. Jeronimidis and M. J. Gidley, submitted for publication, 2001.
11. T. Arioli, L. C. Peng, A. S. Betzner, J. Burn, W. Wittke, W. Herth, C. Camilleri, H. Hofte, J. Plazinski, R. Birch, A. Cork, J. Glover, J. Redmond and R. E. Williamson, *Science*, 1998, **279**, 717.

12. M. E. Edwards, C. A. Dickson, S. Chengappa, C. Sidebottom, M. J. Gidley and J. S. G. Reid, 1999, *Plant Journal*, **19**, 691.
13. R. M. Perrin, A. E. DeRocher, M. Bar-Peled, W. Q. Zeng, L. Norambuena, A. Orellana, N. V. Raikhel and K. Keegstra, 1999, *Science*, **284**, 1976.

Attempt to Produce Caffeoylated Arabinoxylans from Feruloylated Arabinoxylans by Microbial Demethylation

V. Micard,[1] T. Landazuri,[1] A. Surget,[1] S. Moukha,[2] M. Labat[3] and X. Rouau[1]

[1]ENSAM/INRA ECOLE NATIONALE SUPÉRIEURE AGRONOMIQUE DE MONTPELLIER/INSTITUT NATIONAL DE LA RECHERCHE AGRONOMIQUE, UNITÉ DE FORMATION ET DE RECHERCHE TECHNOLOGIE DES CÉRÉALES ET DES AGROPOLYMÈRES (UFR-TCA), 2 PLACE P. VIALA, 34060 MONTPELLIER CEDEX 01, FRANCE
[2]INRA INSTITUT NATIONAL DE LA RECHERCHE AGRONOMIQUE, LABORATOIRE DE BIOCHIMIE DES CHAMPIGNONS FILAMENTEUX (LBCF), INSTITUT FÉDÉRATIF DE RECHERCHE EN BIOTECHNOLOGIE AGRO-INDUSTRIELLE DE MARSEILLE (IFR-BAIM), UNIVERSITÉS DE PROVENCE ET DE LA MÉDITERRANÉE, ESIL CP 925, 163 AV. DE LUMINY, 13288 MARSEILLE CEDEX 09, FRANCE
[3]IRD INSTITUT DE RECHERCHE POUR LE DÉVELOPPEMENT, UNITÉ DE RECHERCHE EN BIOTECHNOLOGIE MICROBIENNE POST-RÉCOLTE (UR 119)-INSTITUT FÉDÉRATIF DE RECHERCHE EN BIOTECHNOLOGIE AGRO-INDUSTRIELLE DE MARSEILLE (IFR-BAIM), UNIVERSITÉS DE PROVENCE ET DE LA MÉDITERRANÉE, ESIL CP 925, 163 AV. DE LUMINY, 13288 MARSEILLE CEDEX 09, FRANCE

1 Abstract

The ability of anaerobic and facultative aerobic bacteria to O-demethylate free ferulic acid and arabinoxylan ester-linked ferulic acid into caffeic acid was investigated. *Clostridium methoxybenzovorans* (strain SR3) was selected for free ferulic acid consumption and caffeic acid production criteria. To perform the reaction on ferulic acid esterified onto arabinoxylans, sonicated cellular extract of SR3 has been prepared and tested on free and arabinoxylan ester-linked ferulic acid. The sonicated cellular extract of SR3 was able to produce caffeic acid from free ferulic acid with an 84% yield. However, it was not able to *O*-demethylate ferulic acid linked to arabinoxylans.

2 Introduction

Water-extractable arabinoxylans (WEAX) from wheat flour carry ferulic acid ester-linked to some of the arabinose.[1,2] WEAX gels in the presence of free radical generating oxidants, by dimerisation of the ferulic acid.[3-5] The resulting covalent network exhibits high water-holding capacity.[1] Recently, special attention was paid to the cross-linking of polysaccharides with protein. According to Kato *et al.*,[6] proteins linked to polysaccharides may gain resistance to heat, proteolytic attack and organic solvents. In addition, new material with unexpected functional properties could be obtained. However, a recent study has demonstrated that no covalent linkage between feruloylated WEAX and proteins by the way of ferulic acid was obtained with the laccase and H_2O_2/horseradish peroxidase (POD) enzymatic oxidising system.[7] A covalent linkage between free ferulic acid (FA) and the tyrosine of a tripeptide has been recently obtained with the H_2O_2/POD system when a special FA/tyrosine ratio and kinetic control of the reaction were used.[8] According to Pierpoint,[9] a monophenol, like ferulic acid, is oxidised into semi-quinones which polymerise through free radical reactions but do not form adducts with compounds containing amino or thiol groups as do quinones issued from the oxidation of diphenols. A recent study has shown that cysteine could bind to the quinones of caffeic acid generated from the demethylation and oxidation of some of the WEAX ferulic acid by the manganese peroxidase oxidative system.[4] The conversion of ferulic acid into the diphenol caffeic acid could therefore be a way to covalently link arabinoxylans and proteins.

The objective of this work was to study the demethylation of WEAX ester-linked ferulic acid in order to produce caffeic acid ester-linked onto WEAX, possibly oxidable into quinone and able to link an amino acid of a protein in a second step. To our knowledge, if free ferulic acid can be *O*-demethylated intact by some microorganisms,[10-16] no microbial *O*-demethylation of linked methoxylated compounds has been reported in the literature. Three anaerobic bacteria (*Acetobacterium woodi*, *Clostridium methoxybenzovorans* and *Eubacterium callanderi*) and one facultative aerobic bacterium (*Enterobacter cloacae*) have been chosen in different phylogenetic groups for their ability to *O*-demethylate free ferulic acid.[10-13]

3 Materials and Methods

3.1 Materials

Acetobacterium woodii (strain WB1), *Clostridium methoxybenzovorans* (strain SR3) and *Eubacterium callanderi* (strain FD) were from the laboratory collection of LOMA-IRD (Marseille, France). *Enterobacter cloacae* (strain DG6) (ATCC

35929) is from the DSM collection (Deutsche Sammlung von Mikroorganismen und Zellkulturen GmbH, Braunschweig, Germany). WEAX, containing 10 μmoles/g of ester-linked ferulic acid, was prepared by the Unité de Formation et de Recherche Technologie des Céréales et des Agropolymères (EN-SAM/INRA, Montpellier, France) as described by Fincher and Stone[17] and modified by Rouau and Moreau.[18]

3.2 Methods

3.2.1 Growth of Bacteria Cells on Free Ferulic Acid Enriched Medium. The anaerobic techniques of Hungate[19] were used throughout this work for anaerobic strains. Inocula of bacteria strains were obtained from growing pre-culture grown on the basal medium described below. WB1, FD and SR3 were grown at 30, 31 and 37 °C, respectively, in a basal medium described by Mechichi *et al.*,[12] supplemented with 5 mM free ferulic acid. All experiments were performed in duplicate unless otherwise stated. Measuring optical density at 580 nm monitored growth.

3.2.2 Demethylation of Free and Ester-linked Ferulic Acid by Sonicated Cellular Extracts of SR3. The O-demethylase activity of the SR3 strain was not excreted in the culture medium. To demethylate ferulic acid esterified to WEAX, which cannot penetrate the cells, a cell extract of SR3 strain was prepared by anaerobic sonication (MSON 05 Bioblock, 20 Hz, 3 cycles of sonication of 2 min; 2 s pulses separated by 2 s lag phase). The sonicated cellular extract of SR3 was tested in anaerobiose on free ferulic acid and esterified ferulic acid at 37 °C with a bacteria protein/ferulic acid ratio of 100 mg/μmol. Composition of reaction medium has been chosen in agreement with previous works on the demethylation of methoxylated phenolic compounds by other strains.[15,20–22] In order to limit the viscosity of the medium, a 0.5% arabinoxylan concentration was used, corresponding to a 50 μM ferulic acid concentration.

3.2.3 Analytical Methods. Free and total phenolic acids were determined by RP-HPLC, with 3,4,5-trimethoxy-*trans*-cinnamic acid as internal standard as described by Figueroa-Espinoza and Rouau.[3] For the determination of total phenolic acid, samples were first de-esterified with 2.0 M NaOH during 1 h at 20 °C under argon prior to ether extraction. A Waters 996 photodiode array detector (Millipore Co., Milford, MA) was used to record ferulic and caffeic acid spectra. The phenolic acid content was determined in duplicates. WEAX content was determined according to the semi-automated method of Rouau and Sur-get[23] as described by Figueroa-Espinoza and Rouau.[3] SE-HPLC of WEAX was performed as described by Figueroa-Espinoza and Rouau.[3] Xylanase activity of SR3 cellular extract was determined with the Xylazyme AX Test Tablet procedure (Megazyme International, Ireland) using azurine-crosslinked wheat arabinoxylan as substrate. Proteins of the sonicated cellular extract were determined by the method of Bradford.[24]

4 Results

4.1 Strain Growth on Free Ferulic Acid Enriched Medium

The growth of three anaerobic strains (WB1, FD and SR3) and one facultative aerobic strain (DG6) has been studied on ferulic acid-enriched mediums during 70 h and 25 h in anaerobiose and aerobiose, respectively. The generation time and the production of caffeic acid from 5 mM free ferulic acid obtained for each of the four strains are presented in Table 1. The SR3 strain, giving the highest production of caffeic acid and the shortest generation time among the three anaerobic strains, was selected for further study. The DG6 strain gave much lower generation times than the anaerobic strains. However, no caffeic acid was produced while free ferulic acid was totally consumed after 10 h. According to Garbic-Galic,[11] caffeic acid could have been a transient product, immediately degraded by side-chain transformation and ring cleavage in the aerobic pathway. However, it was not detected in our study.

Table 1   *Generation time and caffeic acid production of the SR3, WB1, FD and DG6 strains*

	SR3	WB1	FD	DG6
Generation time (h)	6	29	31	2
Caffeic acid production (mM)*	2.6	1.5	1.9	0

*from 5 mM initial ferulic acid concentration

4.2 Production of Caffeic Acid from Free Ferulic Acid by Sonicated Cellular Extracts of SR3 Strain

In order to *O*-demethylate ferulic acid linked to arabinoxylans, the cellular extract of SR3 was sonicated to release its demethylase activity. The consumption of free ferulic acid and the production of caffeic acid by the sonicated SR3 cellular extract are presented in Table 2. After 18 h of reaction, the totality of ferulic acid has been used. Caffeic acid production was proportional to the consumption of ferulic acid during the first 5 h then caffeic acid concentration remained constant while ferulic acid was still consumed. Caffeic acid was probably degraded by oxidation phenomena. Indeed, the use of a reducing agent (cysteine) can reduce the caffeic acid loss from 40% to 15%. With cysteine (ferulic acid/cysteine = 0.01), the final amount of caffeic acid yielded 84% of the initial free ferulic acid content after 18 h of reaction.

The sonicated SR3 cellular extract (100 mg of protein/μmol of ferulic acid) has been applied to a 0.5% arabinoxylan solution corresponding to a 50 μM ferulic

Table 2 *Ferulic and caffeic acids consumption and production by the sonicated cellular extract of SR3 (100 mg protein extract/μmol ferulic acid; with cysteine)*

Time (h)	Ferulic Acid Consumption (% of initial ferulic acid)	Caffeic Acid Production (% of initial ferulic acid)
0.25	5 ± 1	4 ± 1
2	68 ± 5	69 ± 4
18	100 ± 0	84 ± 3

Nb: all values are means of two cellular extract actions

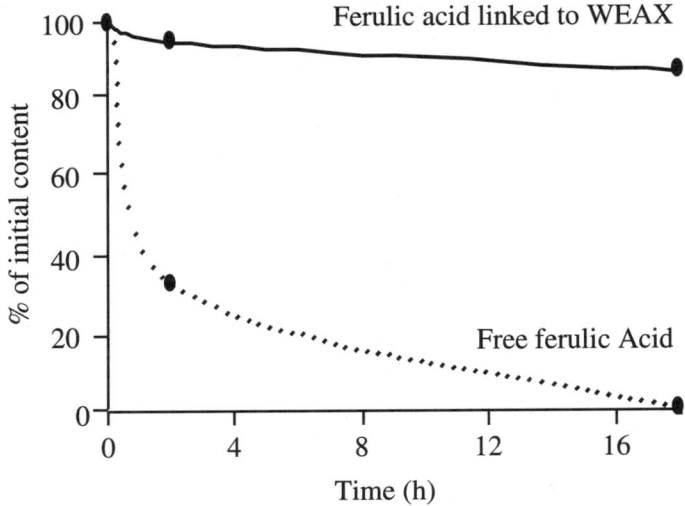

Figure 1 *Consumption of free and WEAX esterified ferulic acid by the sonicated cellular extract of SR3 (100 mg of protein extract/μmol ferulic acid)*

acid concentration. At the opposite to what happened with free ferulic acid, no free neither ester-linked caffeic acid was produced from ferulic acid esterified to WEAX, although esterified ferulic acid slightly decreased (Figure 1). SE-HPLC of WEAX, carried out after 0 h, 2 h and 18 h of action of the sonicated SR3 cellular extract, is presented in Figure 2. A depolymerisation of WEAX was observed with time, due to a slight endoxylanase activity of the sonicated SR3 cellular extract (111 units/g *versus* 8800 units/g of standard xylanase used as reference). Neither oligomer nor monomer was produced by SR3 cellular extract (ion exchange chromatography; results not shown).

5 Conclusion

Free ferulic acid was used by all the three *O*-demethylating anaerobic bacteria, *Acetobacterium woodii* (strain WB1), *Eubacterium callanderi* (strain FD) and

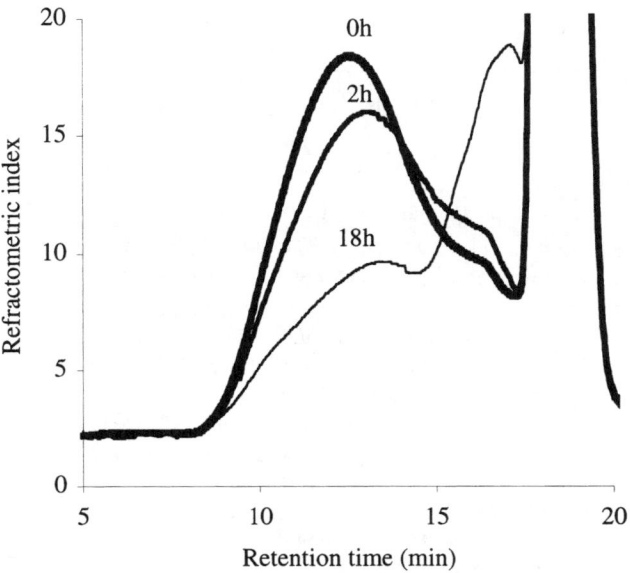

Figure 2 *Size-exclusion chromatography profile of WEAX after the action of the SR3 cellular extract*

Clostridium methoxybenzovorans (strain SR3) and the facultative aerobic bacteria *Enterobacter cloaceae* (strain DG6) tested. However, only anaerobic bacteria gave caffeic acid as an end-product of the reaction. The SR3 strain gave the highest production of caffeic acid and the shortest generation time. Its demethylase activity was not secreted into the culture medium but can be released by sonication of the cells. The sonicated cellular extract produced caffeic acid from free ferulic acid with a 84% yield. Caffeic acid, being unstable, is stabilised by the use of cysteine as reducing agent during the *O*-demethylation reaction. No activity of the sonicated cellular extract was registered on ferulic acid when it was esterified to arabinoxylans. This could be explained by the inhibition of the *O*-demethylase activity when carboxyl group of ferulic acid is engaged in the ester-link with arabinoxylans or by the steric hindrance of WEAX avoiding enzyme–substrate contact. Further studies on methylferulate or feruloylated oligomers are necessary to verify this hypothesis.

Acknowledgements

This work was finacially supported by the Transformation des Produits Végétaux department of INRA (Action Nouvelle Soutenue).

References

1. M. S. Izydorczyk and C. G. Biliaderis, *Carbohydr. Polym.*, 1995, **28**, 33.
2. M. M. Smith and R. D. Hartley, *Carbohydr. Res.*, 1983, **118**, 65.
3. M. C. Figueroa-Espinoza and X. Rouau, *Cereal Chem.*, 1998, **75**, 259.

4. M. C. Figueroa-Espinoza, M.-H. Morel, A. Surget and X. Rouau, *J. Sci. Food Agric.*, 1999, **79**, 460.

5. M. S. Izydorczyk, C. G. Biliaderis and W. Bushuk, *J. Cereal Sci.*, 1990, **11**, 153.

6. A. Kato, T. Wada, K. Kobayashi, K. Seguro and M. Motoki, *Agric. Biol. Chem.*, 1991, 55, 1027.

7. M. C. Figueroa-Espinoza, M.-H. Morel and X. Rouau, *Res. Adv. Agric. & Food Chem.*, 2000, **1**, 73.

8. G. Oudgenoeg, R. Hilhorst, S. R. Piersma, C. G. Boeriu, H. Gruppen, M. Hessing, A. G. J. Voragen and C. Laane, *J. Agric. Food Chem.*, 2001, **49**, 2503.

9. W. S. Pierpoint, *Biochem. J.*, 1969, **112**, 609.

10. R. Bache and N. Pfenning, *Arch. Microbiol.*, 1981, **130**, 255.

11. D. Garbic-Galic, *Appl. Environ. Microbiol.*, 1985, **50**, 1052.

12. T. Mechichi, M. Labat, B. K. C. Patel, T. H. S. Woo, P. Thomas and J.-L. Garcia, *Int. J. System. Bacteriol.*, 1999, 49, 1201.

13. D. O. Mountford and R. A. Asher, *Arch. Microbiol.*, 1986, **145**, 55.

14. D. O. Mountford, W. D. Grant, R. Clarke and R. A. Asher, *Int. J. System. Bacteriol.*, 1988, **38**, 254.

15. E. Stupperich and R. Konle, *Appl. Environ. Microbiol.*, 1993, **59**, 3110.

16. A. Tschech, and N. Pfennig, *Arch. Microbiol.*, 1984, **137**, 163.

17. G. B. Fincher and B. A. Stone, *Adv. Cereal Sci. Technol.*, 1986, **8**, 207.

18. X. Rouau and D. Moreau, *Cereal Chem.*, 1993, **70**, 626.

19. R. E. Hungate, *Methods Microbiol.*, 1969, **136**, 194.

20. M. H. Berman and A. C. Frazer, *Appl. Environ. Microbiol.*, 1992, **58**, 925.

21. A. El Kasmi, S.Rajasekharan and S. W. Ragsdale, *Biochemistry*, 1994, **33**, 11217.

22. F. Kaufmann, G. Wohlfarth and G. Diekert, *Arch. Microbiol.*, 1997, **168**, 136.

23. X. Rouau and A. Surget, *Carbohydr. Polym.*, 1994, **24**, 123.

24. M. M. Bradford, *Anal. Biochem.*, 1976, **72**, 248.

Enzymatically and Chemically De-esterified Lime Pectins: Physico-chemical Characterisation, Polyelectrolyte Behaviour and Calcium Binding Properties

Marie-Christine Ralet, Vincent Dronnet, Jean-François Thibault

UNITÉ DE RECHERCHE SUR LES POLYSACCHARIDES, LEURS ORGANISATIONS ET INTERACTIONS, INSTITUT NATIONAL DE LA RECHERCHE AGRONOMIQUE, RUE DE LA GÉRAUDIÈRE, BP 71627, F-44316 NANTES CEDEX 3, FRANCE

1 Abstract

A series of pectins with different levels and patterns of methyl esterification was produced by de-esterification of a very highly methylated lime pectin with acid, base, plant pectin methyl esterase and fungus pectin methyl esterase. The intrinsic pK values and the calcium binding properties were determined in dilute salt-free solutions. The variations of pK_a *versus* the ionisation degree were found to depend on the de-esterification process but a unique value of 2.90 ± 0.15 was estimated for the intrinsic pK value. A sudden dimerisation through calcium ions, around a degree of methylation of 35%, of pectins de-esterified chemically or by fungus pectin methyl esterase was shown. A progressive dimerisation process was hypothesised for plant pectin methyl esterase de-esterified pectins.

2 Introduction

Pectins are a complex family of heterogeneous branched polysaccharides arising from the primary cell walls and intercellular regions of higher plants.[1,2] The term pectin points out a group of polysaccharides in which the presence of partly methyl-esterified galacturonic acid and of rhamnose are two distinctive features.[1] Pectins are classified as high methoxyl (HM) and low methoxyl (LM), depending on their degree of esterification and are used industrially for their gelling properties.[1,3] Native pectins are usually HM; they can be de-esterified, usually by controlled acid treatment, to give LM pectins. Other de-esterification means, namely alkali, enzymes and ammonia, can be used.[4–9] LM pectins are able to strongly react with calcium ions and, in defined conditions, to form gels. In the present study, plant and fungal pectin methyl esterases and chemical treatments (acid or base) have been used to generate series of chemically- and enzymatically-de-esterified pectins. The dissociation of these well-characterised pectin samples was followed by potentiometric titration and the degree of binding of calcium ions was determined by conductimetry.

3 Experimental

3.1 Synthesis of Model Pectins

A commercial pectin (L72) from Mexican lime peel (*Citrus aurantifolia*), with a degree of methylation (DM) of 72% was esterified in acid-methanol medium to give a pectin (E81) of a DM of 81%. A series of pectins with defined DM were prepared by enzymatic or chemical treatment of E81.[10] B15-B71, A61-A75, F11-F76 and P16-P76 were generated by treating E81 with base, acid, fungus pectin methyl esterase (f-PME) or by plant pectin methyl esterase (p-PME), respectively.

3.2 Potentiometric Measurements

pH measurements were performed at $25.0 \pm 0.2\,°C$ with a pH-meter LPH 430 T (Radiometer Analytical SA) fitted out with a combined pH-electrode Ingold (type U 402-S7/120) and a temperature probe (XT 130, Radiometer). Titrations were performed on salt-free solutions of pectin samples in the acidic form $(C_p \sim 1\ \text{meq/L})$ with freshly prepared 10 meq/L NaOH solution.

3.3 Conductimetric Measurements

Transport parameters were determined using conductimetric measurements as already described.[7] All conductimetric measurements were carried out at $25.0 \pm 0.2\,°C$ with a CDM 83 conductimeter (Radiometer Analytical SA) equipped with a double platinum electrode CDC 241U (Radiometer Analytical SA). Titrations were performed on salt-free solutions of pectin samples in the acidic form $(C_p \sim 1\ \text{meq/L})$ with freshly prepared 10 meq/L solutions of KOH, LiOH and $Ca(OH)_2$.

4 Results and Discussion

4.1 Potentiometric Titrations

The apparent dissociation constant (pK_a) was calculated as a function of the degree of dissociation (α) according to Equation (1):

$$pK_a = pH + \log\left(\frac{1-\alpha}{\alpha}\right) \tag{1}$$

The treatment of experimental data was deduced from the model of Lifson and Katchalsky.[11] The structural charge density ($\bar{\xi}$) was calculated from Equation (2):

$$\bar{\xi} = \frac{\varepsilon^2}{bDkT} \times \frac{100 - DM}{100} = 1.61 \times \frac{100 - DM}{100} \tag{2}$$

where ε is the electron charge, kT the Boltzmann term, b the length of the monomeric unit (0.435 nm)[12] and D the dielectric constant of the solvent. The effective charge density (ξ) can be calculated from Equation (3):

$$\xi = \alpha\bar{\xi} \tag{3}$$

The intrinsic dissociation constant (pK_0) can be obtained by superimposing the experimental curves $pK_a = f(\alpha)$ on the theoretical ones $\Delta pK = f(\alpha)$, as described by Rinaudo and Milas.[13]

The relationships between pK_a and the degree of dissociation α of polygalacturonic acid and pectins from the B-series are presented in Figure 1 together with the Lifson and Katchalsky[11] theoretical curves $\Delta pK = f(\alpha)$. As already pointed out,[9,14] experimental values were lower than theoretical ones for degrees of dissociation above 0.5 and DM < 50, phenomenon which can be ascribed to a condensation of the monovalent counterions.[15] A pK_0 value of ~ 2.80 could be

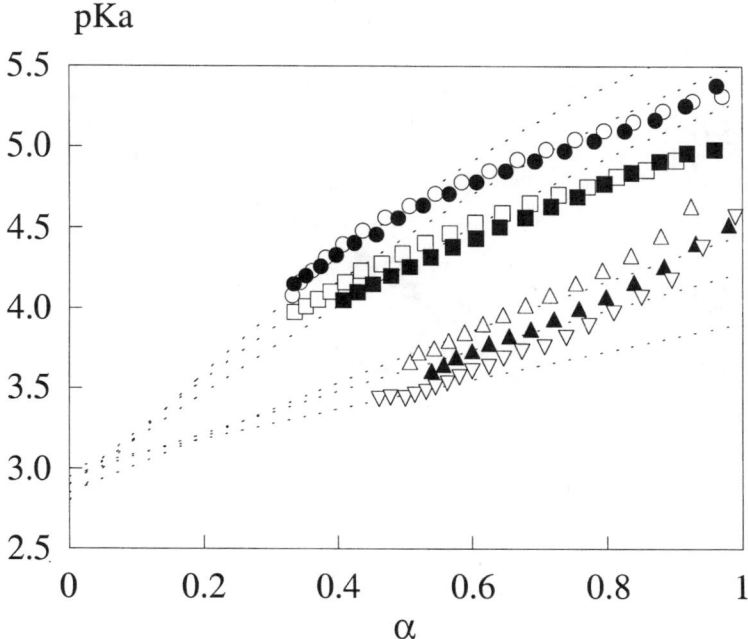

Figure 1 *Variations in pK_a with the degree of dissociation for base-de-esterified pectins (B-series). The lines are the corresponding theoretical $\Delta pK_a = f(\alpha)$ functions. ($\bigcirc$) polygalacturonic acid; ($\bullet$) B15; ($\square$) B34; ($\blacksquare$) B43; ($\triangle$) B64; ($\blacktriangle$) B71; ($\triangledown$) E81*

extrapolated for polygalacturonic acid, in agreement with previously published data.[9] pK_a curves revealed a concave curvature for high DMs, a monotonic quasi-linear curve for intermediate DMs and a convex curvature for lower DMs. For a given degree of dissociation, the pK_a decreased with an increase of DM. Except for intermediate DMs (45% < DM < 70%), the agreement between experimental and theoretical curves is quite poor. Lowly methoxylated pectins (DM < 45%) exhibited the same global behaviour as polygalacturonic acid, the experimental values being lower than the theoretical ones for high degrees of dissociation. Condensation is known to occur for a structural density value above one (*i.e.* DM < 38%) but probably takes place progressively.[16] In agreement with previously reported data,[9,17] a pK_0 value of 2.90 ± 0.15 was determined for all pectic samples from the B-series, confirming that pK_0 is independent of the DM for pectins with randomly distributed charges.[13,14] Pectins from the A- and F-series exhibited the same behaviour as pectins from the B-series (data not shown) and pK_0 values of 2.9 ± 0.1 were also extrapolated.

As for pectins from the B-series, for a given degree of dissociation, the pK_a values of pectins from the P-series decreased with an increase of DM (Figure 2). Concave, quasi-linear and convex curvatures were observed for pK_a curves but the convex curvatures appeared for higher DMs (60–65%) than those observed for pectins from the B-series. The initial slope of the pK_a curves remained

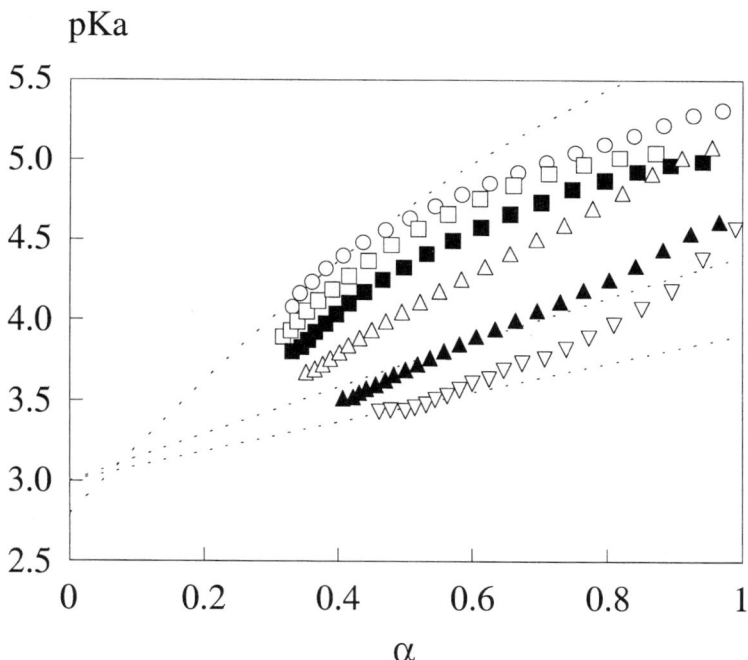

Figure 2 *Variations in* pK_a *with the degree of dissociation for plant PME-de-esterified pectins (P-series). The lines are the corresponding theoretical* $\Delta pK_a = f(\alpha)$ *functions:* (○) *polygalacturonic acid;* (□) *P32;* (■) *P41;* (△) *P60;* (▲) *P70;* (▽) *E81*

roughly constant at ~ 3.3 from P16 to P66. Such high values were observed for polygalacturonic acid and low DM samples from the B- and F-series (F11, F19, F31, B15). Above DM ~ 30, slope values were shown to decrease regularly with an increase of DM for pectins from the B- and F-series. pK_0 values of 2.9 ± 0.1 were extrapolated for pectins from the P-series by plotting the effective charge density *versus* the degree of dissociation (figure not shown). The excess of condensation observed for pectins from the P-series is likely to be due to the blockwise arrangement of free carboxyl groups, even for high DM samples.

4.2 Interaction with Calcium Ions

The limiting law for the equivalent conductivity of polyelectrolyte without external salts is given by:

$$\Lambda = f(\lambda_c + \lambda_p) \tag{4}$$

where Λ is the equivalent conductivity (S.cm^2/eq) of the salts in solution, λ_p, the equivalent conductivity of the active monomer carried by the polyelectrolyte, λ_c, the equivalent conductivity of the isolated counterion in pure solvent at infinite dilution at 25°C and f, the transport parameter.

By measuring the conductivity of three ionic forms of the polyelectrolyte (Li-, K- and Ca-forms) and by considering transport parameter independent of the nature of the monovalent counterion, the transport parameters for monovalent (f_{Li}^+, $_K^+$) and calcium (f_{Ca}^{++}) cations, and the equivalent conductivity of the polyelectrolyte (λ_p), can be calculated.

The calcium transport parameter values were compared with theoretical predictions from Manning's model:[15]

$$z\bar{\xi} < 1 \qquad f = 1 - \frac{0.55(|z|\bar{\xi})^2}{|z|\bar{\xi} + \pi} \tag{5}$$

$$z\bar{\xi} \geq 1 \qquad f = \frac{0.87}{|z|\bar{\xi}} \tag{6}$$

where z is the charge of the counterion.

Calcium transport parameter decreased with decreasing DM for pectins from the B-, A-, F- and P-series. In the range 20% < DM < 75%, pectins from the P-series exhibited significantly lower calcium transport parameter values than pectins from the B-, A-, and F-series of similar DM, indicating a stronger binding of calcium ions (Table 1). Figure 3 shows the ratio of experimental to theoretical values of the calcium transport parameter *versus* the degree of esterification. In the range 35% < DM < 80%, experimental to theoretical ratio remained roughly constant for B-, A- and F-series around 0.65, in agreement with previously reported data on alkali-de-esterified pectins.[6] The sudden drop in the ratio of experimental over theoretical values around DM 35%, in agreement with previous findings on alkali-de-esterified pectins,[6] can be explained by an inter-molecular binding of the calcium ions to carboxyl groups of two molecules leading to the formation of dimers.[6,14] Pectins from the P-series exhibited a

Table 1 *Calcium transport parameter: experimental and theoretical values*

DM	ξ	f_{Ca} (exp)	f_{Ca} (theo)
Mother pectin GRINSTED ^{TM} *Pectin URS 1200*			
E81	81.1	0.305	0.615	0.945
Samples from alkali de-esterification			
B71	72.7	0.440	0.575	0.894
B64	64.6	0.570	0.485	0.763
B43	41.2	0.947	0.315	0.460
B34	32.4	1.088	0.240	0.400
B15	15.6	1.359	0.130	0.320
Samples from acid de-esterification			
A75	73.4	0.428	0.575	0.899
A72	69.0	0.499	0.535	0.868
A61	57.7	0.681	0.427	0.639
Samples from de-esterification with f-PME			
F76	75.7	0.391	0.580	0.914
F69	66.5	0.539	0.468	0.807
F58	56.6	0.699	0.400	0.623
F43	42.3	0.929	0.280	0.468
F31	31.6	1.101	0.247	0.395
F19	17.6	1.327	0.145	0.328
F11	10.0	1.449	0.135	0.300
Samples from de-esterification with p-PME			
P76	75.6	0.394	0.535	0.913
P73	72.8	0.438	0.510	0.895
P70	69.9	0.485	0.475	0.874
P66	66.0	0.547	0.440	0.795
P60	59.4	0.654	0.340	0.665
P53	52.7	0.762	0.300	0.571
P46	46.1	0.869	0.255	0.501
P41	39.8	0.969	0.210	0.449
P32	32.9	1.080	0.175	0.403
P24	20.8	1.275	0.145	0.341
P16	14.2	1.381	0.125	0.315
Polygalacturonic acid			
PGA	2.0	1.578	0.107	0.276

radically different behaviour and a roughly continuous decrease in the ratio of experimental over theoretical calcium transport parameter values with decreasing DM was observed. The de-esterification by p-PME is known to lead to a blockwise arrangement of carboxyl groups and segments long enough to form 'egg-boxes' might be rapidly generated during the de-esterification kinetics. Manning's theoretical values calculated for pectins from the P-series must however be considered cautiously as this theory uses a model of uniform charge density along the polyion.

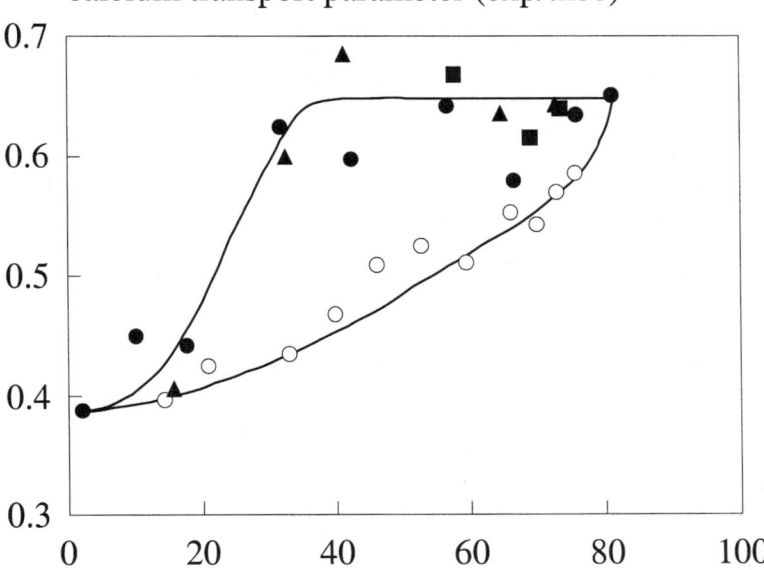

calcium transport parameter (exp/theo)

Figure 3 *Variations of the ratio of experimental to theoretical values of calcium transport parameter with the degree of methylation:* (▲) *base-de-esterified pectins;* (■) *acid-de-esterified pectins;* (●) *fungus PME-de-esterified pectins;* (○) *plant PME-de-esterified pectins*

5 Conclusion

The polyelectrolyte behaviour and calcium binding properties of lime pectins with different levels and pattern of methyl-esterification were studied. The variations of pK_a *versus* the ionisation degree and the calcium transport parameters were found to depend on the de-esterification mode of the pectic molecules but a constant pK_0 value of ~ 2.9 was extrapolated whatever the DM and the de-esterification mode. A transition in calcium binding properties was evidenced around DM 35% for both chemically and f-PME de-esterified pectins. This fact was interpreted as a dimerisation process. After some initial decrease in the methylation degree (DM < 65–70%), p-PME de-esterified pectins exhibited peculiar polyelectrolyte behaviour and calcium binding property, which can be ascribed to the presence of sequences of contiguous free carboxyl groups. An excess of condensation was revealed for DM < 60–65% and dimerisation process could take place progressively with decreasing DM.

Acknowledgements

We thank Danisco-Cultor (Denmark) for providing the pectin samples. Financial support from the European Community within the EU-Biotechnology Program contract number ERBIO4CT960685 is gratefully acknowledged.

References

1. A. G. J. Voragen, W. Pilnik, J.-F. Thibault, M. A. V. Axelos and C. M. G. C. Renard. 'Food Polysaccharides and their Applications', Food Science and Technology Series, Marcel Dekker, New York, 1995, Chapter 10, p. 287.
2. P. Albersheim, A. G. Darvill, M. A. O'Neill, H. A. Schols and A. G. J. Voragen. 'Pectins and Pectinases', Progress in Biotechnology, Elsevier Science BV, Amsterdam, 1996, p. 47.
3. C. D. May, *Carbohydr. Polym.*, 1990, **12**, 79.
4. R. Kohn and O. Luknar, *Collec. Czech. Chem. Commun.*, 1975, **40**, 959.
5. R. Kohn, O. Markovic and E. Machova, *Collec. Czech. Chem. Commun.*, 1983, **48**, 790.
6. J.-F. Thibault and M. Rinaudo, *Biopolymers*, 1985, **24**, 2131.
7. J.-F. Thibault and M. Rinaudo, *British Polym. J.*, 1985, **17**, 181.
8. J.-M. Denès, A. Baron, C. M. G. C. Renard, C. Péan and J.-F. Drilleau, *Carbohydr. Res.*, 2000, **327**, 385.
9. E. Racapé, J.-F. Thibault, J. C. E. Reitsma and W. Pilnik, *Biopolymers*, 1989, **28**, 1435.
10. G. Limberg, R. Körner, H. C. Buchholt, T. M. I. E. Christensen, P. Roepstorff and J. D. Mikkelsen, *Carbohydr. Res.*, 2000, **327**, 293.
11. S. Lifson and A. Katchalsky, *J. Polym. Sci.*, 1954, **13**, 43.
12. D. A. Rees and A. W. Wight, *J. Chem. Soc. (B)*, 1971, 1366.
13. M. Rinaudo and M. Milas, *J. Polym. Sci.*, 1974, **12**, 2073.
14. G. Ravanat and M. Rinaudo, *Biopolymers*, 1980, **19**, 2209.
15. G. S. Manning, *J. Chem. Phys.*, 1969, **51**, 924.
16. N. D. Truong, G. Medjahd, D. Sarazin and J. François, *Polymer Bull.*, 1990, **24**, 101.
17. F. Michel, J.-F. Thibault and J.-L. Doublier, *Carbohydr. Polym.*, 1984, **4**, 283.

Incorporation of Unsaturated Isoleucine Analogues into Proteins *In Vivo*

Thierry Michon,[a] Francis Barbot[b] and David Tirrell[c]

[a]UNITE DE BIOCHIMIE ET DE TECHNOLOGIE DES PROTEINES, RUE DE LA GERAUDIERE, INSTITUT NATIONAL DE LA RECHERCHE AGRONOMIQUE, 44316 NANTES CEDEX 03, FRANCE
[b]UNIVERSITE DE POITIERS, BAT. GON AV. DU RECTEUR PINEAU, 86022 POITIERS CEDEX, FRANCE
[c]DIVISION OF CHEMISTRY AND CHEMICAL ENGINEERING, 210–41 CALIFORNIA INSTITUTE OF TECHNOLOGY, PASADENA, CALIFORNIA 91125, USA

1 Abstract

The translational activity of various unsaturated analogs of L-isoleucine was evaluated using an *Escherichia coli* strain auxotrophic for isoleucine. It was observed that the alkene [2-amino-3-methyl-4-pentenoic acid (**2**)] and alkyne [2-amino-3-methyl-4-pentynoic acid (**3**)] derivatives of L-isoleucine can support protein synthesis at levels approximately 50% of that observed in cultures supplemented with isoleucine. However, no incorporation of the αC or βC methylated derivatives could be detected. In order to examine the stereoselectivity of incorporation, the (2*S*, 3*S*) and (2*S*, 3*R*) diastereomers of **2** and **3** were prepared. The extents of isoleucine substitution *in vivo* were 80% and 70% for (2*S*, 3*S*)-**2** and (2*S*, 3*S*)-**3**, respectively, under the conditions examined in this study.

2 Introduction

The *in vivo* incorporation into proteins of amino acids analogs bearing non-biological chemical reactivity within their side chain would allow a completely new chemistry of proteins. For instance, this would have applications in the design of new materials by combining proteins to synthetic polymers, nucleic acids or carbon hydrates. In the past it was shown that *E. coli* is extremely permissive for the incorporation of artificial amino acids (see ref. 1 for review).

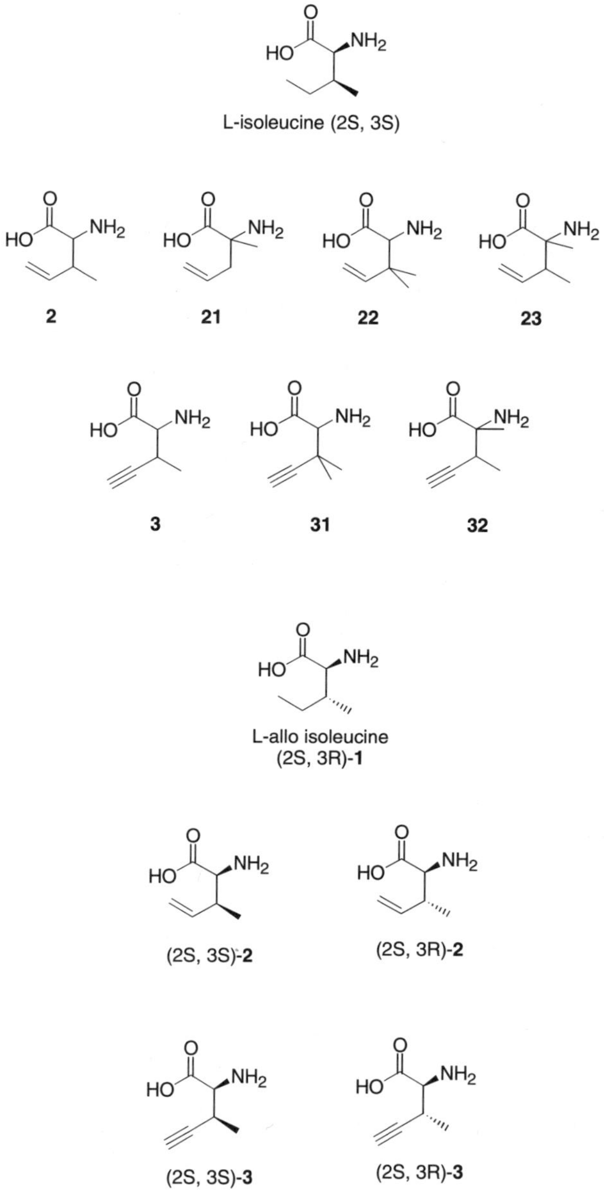

L-isoleucine (2S, 3S)

2   **21**   **22**   **23**

3   **31**   **32**

L-allo isoleucine
(2S, 3R)-**1**

(2S, 3S)-**2**   (2S, 3R)-**2**

(2S, 3S)-**3**   (2S, 3R)-**3**

Figure 1 *Structures of the isoleucine analogs used in this study*

One of the reasons probably lies within the fact that in the course of evolution the selection pressure never applied to these amino acid analogs. As the laws governing protein structure and function emerge, it is becoming increasingly productive to design 'artificial proteins' as building blocks for new kinds of supra-molecular chemistry.[2,3] Optimizing such building blocks requires control of the driving forces that direct protein folding and assembly. This is achieved largely by controlling the physical and chemical properties of the amino acid side chains of the protein of interest.[4] In an effort to increase the range of chemical function that can be incorporated into proteins, we have focused our investigations on the translational activity of unsaturated amino acid analogs because of the versatile chemistry of alkenes and alkynes.[5] For example, supramolecular structures made up of weakly hydrogen bonded cyclic peptides can be stabilized through inter-peptide cross-linking[6] utilizing ruthenium-catalyzed ring-closing metathesis of pendant alkene functions.[7] We have recently shown that homoallylglycine and homopropargylglycine can be incorporated *in vivo* into recombinant proteins.[8–10] In this paper we examine the incorporation of unsaturated isoleucine analogs (Figure 1) into a heterologous protein, mouse dihydrofolate reductase (mDHFR), over-expressed in an *Escherichia coli* host. Analogues **2** and **3** can be efficiently incorporated into *E. coli* proteins, and incorporation is stereospecific, favoring the (2*S*, 3*S*) diastereomers. The results are discussed from the standpoint of the insensitivity of the isoleucyl-tRNA synthetase (IleRS) editing mechanism to non-canonical amino acids.

3 Experimental Section

3.1 Amino Acid Synthesis

Diastereomers mixtures (Figure 1) of 2-amino-3-methyl-4-pentenoic acid (**2**), 2-amino-2-methyl-4-pentynoic acid (**21**), 2-amino-3-dimethyl-4-pentenoic acid (**22**), 2-amino-2,3-dimethyl-4-pentenoic acid (**23**), 2-amino-3-methyl-4-pentynoic acid (**3**), 2-amino-3-dimethyl-4-pentynoic acid (**31**), 2-amino-2,3-dimethyl-4-pentynoic acid (**32**) were prepared according to Aidene *et al.*[11] The regioselective reaction between α-unsaturated organozincs and *N*-(phenylsulfanyl)iminoesters was used as a starting step to obtain the α-aminoesters.[11]

Detailed procedures for the preparation of (**2**) and (**3**) pure diastereomers will be described elsewhere (Michon *et al.*, manuscript in preparation).

For all the compounds used, ^{1}H NMR spectra were in agreement with the expected structures.

3.2 Construction of an *E. coli* Expression Strain

An *E. coli* strain (AIV$_s$) that is auxotrophic for valine and isoleucine was prepared. The procedure will be described elsewhere (Michon *et al.* manuscript in preparation). An *in vivo* test system was designed in order to determine the extent of incorporation of isoleucine analogs into an over expressed reporter protein. Plasmid pQE15 (Qiagen) carries a gene encoding mouse dihydrofolate reductase (mDHFR) under control of a strong bacteriophage T5 promotor that is recognized by *E. coli* RNA polymerase. A repressor-binding site has been introduced downstream from the T5 promotor on pQE15 that allows a programmed induction of DHFR expression by addition of isopropyl-β-D-thiogalactopyranoside (IPTG) to the medium. The gene encoding mDHFR also encodes an N-terminal hexahistidine sequence to permit protein purification by immobilized metal affinity chromatography. pQE15 confers ampiciline resistance. The auxotrophic *E. coli* strain AIV_s was transformed with pQE15 to give AIV_s/pQE15.[12] The repressor plasmid pLys-IQ modified from pLys-S (Novagen, Madison, WI, USA) to contain the $LacI^q$ gene encoding the lac repressor was then introduced into the AIV_s/pQE15 strain to give the AIV_s-IQ/pQE15 strain. This strain was made in an attempt to prevent 'leaky' expression of the DHFR gene before induction (see results). This strain was used for all the assays.

3.3 Protein Expression

To test for analog incorporation 5 mL of M9AA medium supplemented with ampicillin (200 μg/mL), chloramphenicol (35 μg/mL), 2 mM $MgCl_2$, 100 μM $CaCl_2$, 0.2% glucose and 0.5 mg thiamin chloride were inoculated with a single colony of AIV_s-IQ/pQE15. After overnight growth at 37°C the culture was diluted with fresh M9 medium in order to obtain an OD_{600} of 0.1. When the OD_{600} reached 1 (after about 3.5 hours) the cells were sedimented (5000 g, 10 min, 4°C) and washed twice with 0.9% NaCl. The cells were resuspended in 50 mL fresh M9 medium supplemented with the 19 amino acids (16 mg/L) but lacking isoleucine. Tests tubes containing 10 mL aliquots of this culture were prepared and supplemented with 250 μL water (negative control), 40 mg/L L-isoleucine (2S, 3S/2R, 3R) (positive control), or 80 mg/L of each of the analogs. After 10 min of growth DHFR expression was induced by addition of IPTG at a final concentration of 0.4 mM. The culture were grown at 37°C for 4 hours and 1 mL aliquots were spun down. Pellets were suspended in 50 μL of a 10 mM $mgCl^2$ solution containing 5 μg/mL DNAse and 10 μg/mL RNAse. The suspensions were frozen, thawed and sonicated prior to electrophoresis. In each case the remaining 9 mL were centrifuged and the pellets stored at -20°C overnight before DHFR purification. Cellular proteins were resolved by SDS PAGE, and mDHFR was detected by western blotting with antibodies raised against the histidine tag (Qiagen, Inc., Santa Clarita, CA, USA).[12]

3.4 DHFR Purification

Pellets were thawed for 30 min, and resuspended in 600 μL of buffer (6 M guanidine-HCl, 0.1 M NaH_2PO_4, 0.01 M Tris/HCl, pH 8). The mixture was vigorously shaken at room temperature for 1 h. The cell debris was sedimented (10 000 g, 30 min at room temperature) and the lysate was submitted to affinity chromatography on pre-packed Nickel columns according to the Ni-NTA spin column procedure described by Qiagen. The DHFR fractions were recovered in a buffer containing 8 M urea, 0.1 M NaH_2PO_4 and 0.01 M Tris/HCl pH 4.5. UV spectra of the samples were taken and the amount of DHFR obtained was quantified using $\varepsilon = 30500$ M^{-1} cm^{-1} at 280 nm.[13]

3.5 Estimation of the Level of Isoleucine Replacement into DHFR

Amino acid analysis were obtained from the Beckman Research Institute, Division of Immunology, City of Hope, Duarte CA. The extent of replacement of isoleucine was estimated based on the diminution of isoleucine from its expected value. MALDI-TOF spectra were performed on a Voyager-DE STR spectrometer using 3,5-dimethoxy–4-hydroxy-cinnamoic acid as matrix (Mass Spectrometry Center of the Beckman Institute, California Institute of Technology).

4 Results

4.1 Analog Incorporation

Incorporation of selected amino acid analogs into proteins made *in vivo* has been known for many years.[1,14–16] In our study an *E. coli* strain auxotrophic for isoleucine was used to assay the extent of *in vivo* incorporation of isoleucine analogues into mouse dihydrofolate reductase (mDHFR), a test protein readily over expressed in bacterial cultures. It was expected that the *in vivo* incorporation of analogs into the proteins would kill the cells (cell lysis). In order to estimate the permissivity of *E. coli* protein biosynthesis, we fed the cells with diastereomer mixtures of isoleucine analogs. As shown in Figure 2, mDHFR was detected only when the culture medium was supplemented with **2** and **3**. It was shown earlier that amino acids with bulkier side chains than isoleucine could be accommodated in the binding site of the synthetase,[17,18] but none of the analogs bearing a double methylation on the βC carbon gave detectable amounts of mDHFR.

Among the 20 proteinogenic amino acids, only L-isoleucine and L-threonine carry a symmetry center in their side chain and only (2*S*, 3*S*)-isoleucine and (2*S*, 3*R*)-threonine are found in proteins. When L-isoleucine was replaced in the culture medium by L-alloisoleucine, the (2*S*, 3*R*)-**1** diastereomer of isoleucine, the cell growth slowed dramatically, without evidence of cell lysis. As western blotting did not show detectable DHFR expression in such cultures, we conclude that (2*S*, 3*R*)-**1** is incorporated into protein slowly if at all under our experimental conditions (Figure 3B, lane 5). However, small amounts of DHFR could be isolated from such culture (*ca.* 8% of the level of expression obtained in media supplemented with L-isoleucine, see Table 1). DHFR synthesis was not due to leaky expression before induction, as uninduced cells did not produce DHFR in media supplemented with L-isoleucine (Figure 3B, lane 3). Instead it appears that

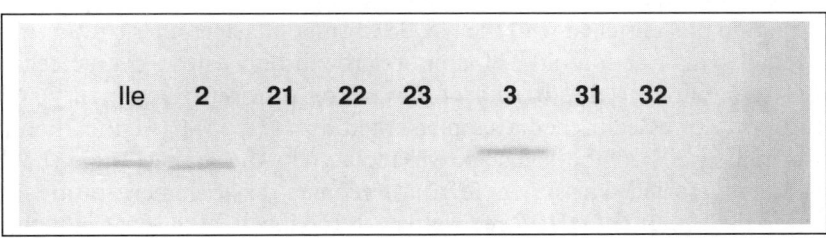

Figure 2 *Western blot obtained from a SDS PAGE of* E. coli *proteins 4 hours after medium shift. Amino acids added as indicated at the top of the figure. DHFR was detected with antibodies raised against the amino terminal hexahistidine tag*

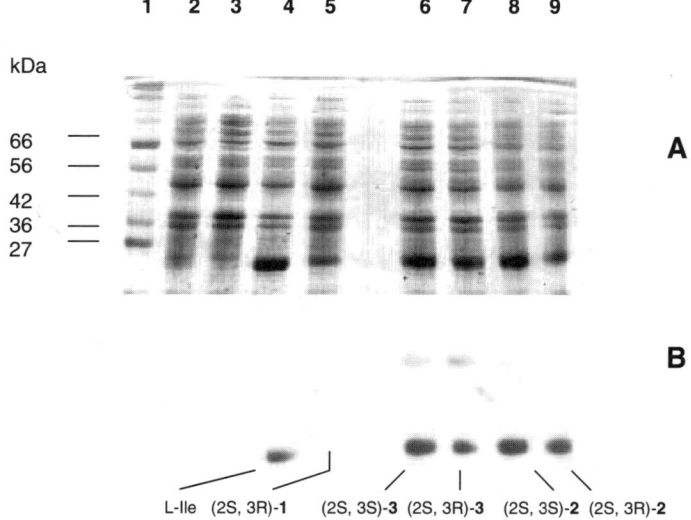

Figure 3 *(A) SDS PAGE of* E. coli *proteins 4 hours after medium shift. Lane 1: molecular weight standards; 2: no isoleucine added, induction; 3: isoleucine added, no induction. Lanes 4 to 9: analogs added as indicated at the bottom of the figure. (B) Autoradiogram of a western blot obtained from the gel above. DHFR was detected with antibodies raised against the amino terminal hexahistidine tag*

the host can provide a small pool of L-isoleucine, probably by means of intracellular proteolysis.

In order to compare the efficiency of the methyl position carried by βC on the incorporation, we prepared pure diastereomers of **2** and **3**.

(2S, 3S)-**2** did not support cell growth and was toxic to the host as shown by a drop in OD_{600} signaling cell lysis. Most importantly, (2S, 3S)-**2** supported synthesis of DHFR (Figure 3B, lane 8). Approximately 80% of the L-isoleucine in DHFR was replaced by the analog, as determined by amino acid analysis (Table 1). In contrast to (2S, 3S) the (2S, 3R) analog neither supported cell growth nor caused measurable lysis. DHFR could be isolated after induction of cultures supplemented with (2S, 3R)-**2** (Figure 3B lane 9), but in amounts too low to permit estimation of replacement efficiency (Table 1).

A fraction of **3** enriched with the (2S, 3S) isomer (diastereomeric ratio 8.6/1.4) was tested. This fraction did not support growth but instead caused cell lysis (Table 1), confirming the toxicity of the analog previously reported.[19] When DHFR was expressed in media supplemented with (2S, 3S)-**3** (Figure 2B lane 6) approximately 70% of the isoleucine was replaced by the analog (Table 1). When a fraction of **3** enriched in the (2S, 3R) diastereomer (diastereomeric ratio 8.3/1.7) was used the amount of DHFR recovered was too low to allow determination of the extent of incorporation.

MALDI-TOF mass spectra were recorded on the purified DHFR samples. The difference in mass obtained between native DHFR and its modified forms

Table 1 *Incorporation of isoleucine analogs into DHFR*

amino acid	DHFR Yield[a] μg/mL	$\overline{M}_n$[b]	% replacement AAA[c]	MALDI-TOF[d]
L-isoleucine (2S, 3S)	29.5	24044		
(2S, 3R)-**1**	2.4	ND	ND	ND
(2S, 3S)-**2**	16.4	24018	81	86
(2S, 3R)-**2**	1.35	ND	ND	ND
(2S, 3S)-**3**	11	24007	72	70
(2S, 3R)-**3**	ND	ND	ND	ND

[a]Yield of DHFR after purification from 10 mL culture, as determined by absorption at 280 nm.
[b]When mass spectra (MALDI-TOF) were recorded, all samples gave a main peak (85 to 95% of the signal intensity). The m/z value of this main peak was used to estimate $\overline{M}_n$, the average mass of DHFR.
[c]Extent of replacement of isoleucine as determined by amino acid analysis.
[d]Extent of replacement of isoleucine as estimated from the difference in $\overline{M}_n$ obtained between native DHFR and its modified forms.

were consistent with the % replacement determined by amino acid composition (data not shown).

5 Discussion

The first critical step in the incorporation of amino acids analogs is their uptake from the culture medium; the analog must be transported across the cytoplasmic membrane either by the machinery used for the uptake of its natural counterpart or by other import machinery. In the case of non polar isoleucine analogs this step might not be limiting as isoleucine and other non polar amino acids are likely to cross the phospholipids bilayer by simple diffusion. However, the leucine-isoleucine-valine (LIV)-binding proteins present at the surface of *E. coli* reveals great sequence similarity.[20] It was recently showed that the specificity of these systems might be weak enough to tolerate closely structurally related amino acids.[21]

In a second step the analog has to be coupled to a tRNA species by an aminoacyl-tRNA synthetase and must circumvent the editing pathways that normally limit misacylation of tRNAs. The selectivity (*s*) of an aminoacyl-tRNA synthetase toward an amino acid is defined as the ratio of the rate of editing to the rate of activation. It is noteworthy that the Met-tRNA synthetase exhibits a very high selectivity towards homocysteine ($s = 11000$) and norleucine ($s = 1000$) which are both biological amino acids present in the cell.[22] In contrast it was previously demonstrated in our group that the synthetase is able to mis-charge Met-tRNA with at least three unsaturated methionine analogues.[8,9] The editing mechanism of *E.coli* isoleucyl-tRNA synthetase (IleRS) has been exten-sively studied.[23,24] Its selectivity for natural amino acids is high, ranging from $s = 6000$ for valine to $s = 8.5 \times 10^6$ for alanine.[22] IleRS possesses two sites: one for the binding of the amino acid prior to its activation through the formation of the AMP–AA phosphoester bond; the other, the editing site, for the hydrolysis of this bond when amino acids smaller than isoleucine (which easily fit into the binding pocket) have been inappropriately acylated.[25,26] Analogs **2** and **3** tested in this study appear to circumvent the editing mechanism of IleRS because they are too large to fit into the editing site. Our results show that IleRS is stereoselective as only (2S, 3S) isoleucine analogs are incorporated into protein at measurable rate. This is in agreement with previous binding studies which demonstrated that L-2 amino-3S-methylhexanoic acid binds to IleRS (K_a = 20 mM^{-1}) with a stronger affinity than its diastereomer L-2 amino-3R-methylhexanoic acid (K_a = 0.6 mM^{-1}).[17]

Finally, the analog may be edited at the ribosome level. If the editing mechanism of the aa-tRNA synthetase seems to insure a weak discrimination between natural and artificial amino acids one should expect a more efficient discrimination at the ribosome level. Studies performed *in vitro* show that this step is probably permissive enough.[27–30] The misscharged tRNA must avoid discrimination by elongation factor Tu (EFTu) (Figure 4). According to this scheme both the rate of GTP hydrolysis (k_2) and the rate of the EFTu.GDP complex dissociation with the ribosome (k_3) are defined as internal kinetic standard constants which do not depend on the presence of any cognate or non-cognate amino acid-tRNA. By contrast the dissociation constants between the ribosome and either the ternary complex EFTu.GTP.aa-tRNA (governed by k_{-1}) or the aa-tRNA after GTP conversion to GDP (governed by k_4) depend on the strength of the binding between the ribosome and the aa-tRNA (see ref. 31 for review). It is likely that the efficiency of transfer of the non-cognate amino acid to the on-growing peptidic chain is much lower than in the case of the natural amino acid. In the experiments reported here, the medium shift method prevents

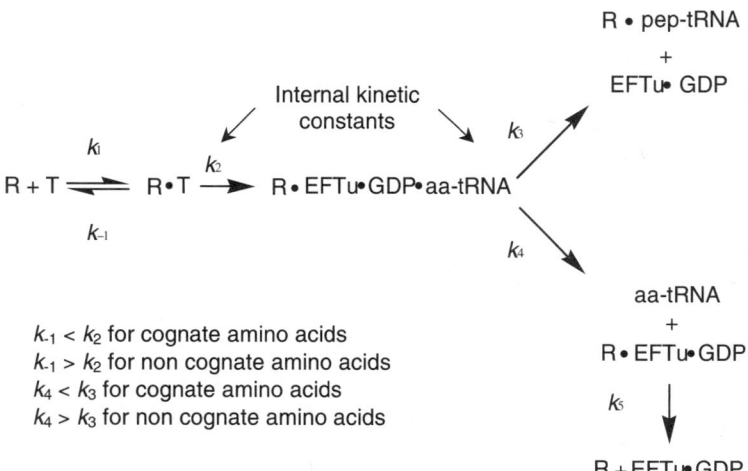

Figure 4   *Minimal mechanism by which aa-RNs are incorporated into nascent protein (see discussion for comments). AatRNA, aminoacyl tRNA; EFTu, elongation factor; GTP, guanosine triphosphate; GDP, guanosine diphosphate; R, ribosome; T, ternary complex (aa-tRNA•GTP•EFTu). Redrawn from Thompson 1988*[31]

competition by Ile and high level of incorporation of the analogs were obtained. However, it has been reported that furanomycin, an analogue of isoleucine, is readily incorporated into proteins.[18] Interestingly, in spite of large structural discrepancies between furanomycin and isoleucine, the equilibrium constant of the ternary complex formation (furanomycyl ~ tRNA•EFTu•GTP) equals that for isoleucyl ~ tRNA•EFTu•GTP formation. In the course of evolution nature selected editing processes on the basis of naturally occurring amino acids, which are present in the cell. This seems to open relatively large possibilities for the incorporation of artificial amino acids making use of the natural machinery of the cell.

Acknowledgment

The authors thank NATO for its financial support to T. M.'s year-long sabbatical. We are grateful to Kristi Kiick for fruitful discussions.

References

1. G. Hortin and I. Boime, *Methods in Enzymology*, 1983, **96**, 777.
2. M. Krejchi, E. Atkins, A. Waddon, M. Fournier, T. Mason and D. Tirrell, *Science*, 1994, **265**, 1427.
3. S. Sakamoto, I. Obataya, A. Ueno and H. Mihara, *J. Chem. Soc., Perkin Trans.*, 1999, **2**, 2059.
4. M. R. Ghadiri and C. Choi, *J. Am. Chem. Soc,*. 1990, **112**, 1630.
5. T. Michon and D. A. Tirrell, *Biofutur*, 2000, **197**, 34.
6. T. Clark and M. R. Ghadiri, *J. Am. Chem. Soc.*, 1995, **117**, 12364.
7. S. B. Nguyen, R. H. Grubbs and J. W. Ziller, *J. Am. Chem. Soc.*, 1993, **115**, 9858.
8. J. van Hest and D. Tirrell, *FEBS Lett.*, 1998, **428**, 68.
9. J. van Hest, K. Kiick and D. Tirrell, *J. Am. Chem. Soc.*, 2000, **122**, 1282.
10. K. L. Kiick, J. C. van Hest and D. A. Tirrell, *Angew. Chem. Int. Ed. Engl.*, 2000, **39**, 2148.
11. M. Aidene, F. Barbot and L. Miginiac, *J. Organometallic Chem.*, 1997, **534**, 117.
12. J. Sambrook, E. F. Fritsch and T. Maniatis, 'Molecular Cloning: A Laboratory Manual', 2nd Ed., Cold Spring Harbor Laboratory Press, Cold Spring Harbor, NY.
13. J. Andrews, C. A. Fierke, B. Birdsall, G. Ostler, J. Feeney, G. C. Roberts and S. J. Benkovic, *Biochemistry*, 1989, **28**, 5743.
14. W. Hendrickson, J. Horton and D. Lemaster, *EMBO J.*, 1990, **9**, 1665.
15. E. D. Fenster and H. S. Anker, *Biochemistry*, 1969, **8**, 269.
16. T. W. Tuve, and H. H. Williams, *J. Am. Chem. Soc.*, 1957, **79**, 5830.
17. J. Flossdorf, H.-J. Pratorius and M.-R. Kula, *Eur. J. Biochem.*, 1976, **66**, 147.
18. T. Kohno, D. Kohda, M. Haruki, S. Yokoyama and T. Miyazawa, *J. Biol. Chem.*, 1990, **265**, 6931.
19. H. Gershon, J. Shapira, J. S. Meek, and K. Dittmer, *J. Am. Chem. Soc.*, 1954, **46**, 3484.
20. R. Tam and M. Saier, *Microbiol. Rev.*, 1993, **57**, 320.
21. D. R. Liu and P. G. Schultz, *Proc. Natl. Acad. Sci. USA*, 1993, **96**, 4780.
22. H. Jakubowski and E. Goldman, *Microbiol. Rev.*, 1993, **56**, 412.
23. A. R. Fersht, *Biochemistry*, 1977, **16**, 1025.
24. A. R. Fersht and C. Dingwall, *Biochemistry*, 1979, **18**, 2627.

25. O. Nureki, D. Vassylyev, M. Tateno, A. Shimada, T. Nakama, S. Fukai, M.Konno, T. Hendrickson, P. Schimmel and S. Yokoyama, *Science*, 1998, **280**, 578.
26. L. Silvian, J. Wang and T. Steitz, *Science*, 1999, **285**, 1074.
27. J. M. Pezzuto and S. M. Hecht, *J. Biol. Chem.*, 1980, **255**, 865.
28. G. Baldini, B. Martoglio, A. Schachenmann, C. Zugliani and J. Brunner, *Biochemistry*, 1988, **27**, 7951.
29. J. D. Bain, C. G. Glabe, T. A. Dix, and A. R. Chamberlin, *J. Am. Chem. Soc.*, 1989, **111**, 8013.
30. J. Ellman, D. Mendel, S. Anthony-Cahill, C. J. Noren and P. G. Schultz, *Methods Enzymol.*, 1991, **202**, 301.
31. R. Thompson, *TIBS*, 1988, **13**, 91.

Binding of Two Lipid Monomers by Plant Lipid Transfer Proteins, LTP1

Jean-Paul Douliez and Didier Marion

LABORATOIRE DE BIOCHIMIE ET TECHNOLOGIE DES PROTÉINES, INRA, RUE DE LA GÉRAUDIÈRE, BP 71627, 44316 NANTES, FRANCE

1 Summary

The binding of two single chained lipid monomers by plant lipid transfer proteins, LTP1s, becomes an attractive field of research which could help with our understanding of the functional role of this protein family. This has been investigated in the case of wheat, barley and maize LTP1. The titration with myristoyl-lysophosphatidylcholine could be followed either by LTP1 intrinsic fluorescence or isothermal titration calorimetry, ITC. The fluorescence titration exhibited a behaviour different in the case of maize LTP1 compared to that of barley and wheat, while ITC returned analogous patterns. Those experiments returned a dissociation constant of about 1 μM and showed that these proteins could indeed bind two monomers of a single chained lipid. This result was corroborated by molecular modelling where the structure of the complex between LTP1 and two lipid monomers could be derived. Finally, by using several double chained lipids, we showed by fluorescence, ITC and molecular modelling that wheat LTP1 is also capable of binding two monomers of such lipids.

2 Introduction

Plant non specific lipid transfer proteins (ns-LTP) are well known for their ability to bind and transfer lipids.[1,2] They exhibit a basic pI and a 9 kDa molecular weight[3] while eight cysteines all involved in disulphide bridges help in maintaining the structure of the protein. The three dimensional structure of ns-LTP1 has been determined by ^{1}H NMR and crystallography and reveals an hydrophobic cavity within the protein.[4-7] Kader et al.,[8] reported binding of lipids by maize LTP1 by using displacement fluorescence methods which involve labelled lipids. The binding of two lipid monomers was suspected from those experiments.

73

However, it was not possible to determine any binding constant from these data.[8] More recently, we analysed the binding constant of wheat LTP1 complexed with several lipids as obtained by intrinsic tyrosine fluorescence.[9] Our data confirmed the lack of specificity for fatty acids and phospholipids with various chain length with Kd of about 10^{-6} M. Since the finding that wheat LTP1 is capable of binding two lipids,[7] it becomes a new goal to determine whether it can be a general feature of the LTP1 family. This task is reviewed in the present paper where two monomers of a single chained lipid were shown to be bound by wheat, maize and barley LTP1. Interestingly, wheat LTP1 was also shown to bind two monomers of a double chained lipid. This work should help in our understanding of the physiological role of this protein family and the mechanism of lipid transfer.

3 Materials and Methods

3.1 Fluorescence Titration

The purification of LTP1 is performed according to ref. 10. Fluorescence intensity is measured at 25°C with a Fluoromax-Spex (Jobin et Yvon, France). Excitation is set at 275 nm while emission spectra are recorded from 280 to 340 nm. 1 mL of a 50 μM LTP1 is poured into the cuvette and titration is performed by adding, in a stepwise manner, the lipid solution. Titration curves report maximum fluorescence intensity at 305 nm *vs* the molar ratio lipid/protein, Ri.

3.2 Isothermal Titration Calorimetry

ITC is performed using a MicroCal titration microcalorimeter (Northampton, MA). Solutions are degassed under vacuum prior to use. Protein at a concentration of 1 mg/mL is poured in the calorimeter cell and lipid (2.5 mg/mL) is added automatically by aliquots of 7 μL. Thermogram data are integrated using the Origin software supplied by MicroCal Inc.

4 Results and Discussion

4.1 Binding of Myristoyl-lysophosphatidylcholine by LTP1s

Tyrosine intrinsic fluorescence has been shown to be a powerful tool for studying the binding of LTP1 to various lipids.[9] In that case, the interaction is followed by monitoring the increase of fluorescence emission upon addition of lipids. Figure 1 depicts a typical titration behaviour for the binding of lipids by wheat and barley LTP1.[9,11] The fluorescence intensity increases upon addition of lipids until a plateau is reached. The saturation occurs at around a molar ratio lipid/protein, Ri, around two. The data can be fitted, allowing to determine both the stoichiometry and the affinity of the lipid for the protein. Wheat LTP1 is known to bind several lipids with an affinity in the range of the micromolar.[9,11,12] The stoichiometry is between one and two, showing that two mono-

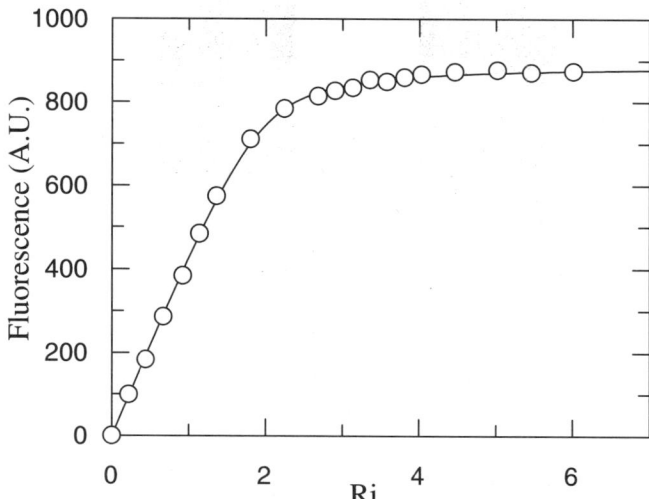

Figure 1 *Typical fluorescence titration behaviour for the binding of lipids by wheat and barley LTP1. In the case of maize, the fluorescence starts to increase at around Ri = 0.5*

mers of these lipids can be bound by the protein. This has been well established by the structure of the complex between wheat LTP1 and two monomers of myristoyl-lysophosphatodylcholine.[7] Recently, this feature was also shown in the case of barley LTP1.[11]

Although three dimensional crystal structures of maize LTP1 with various lipids[13] showed only one monomer embedded in the protein cavity, we recently demonstrated that this protein is also capable of binding two monomers of double chained lipids (Douliez *et al.*, in preparation). In that case, the fluorescence behaviour was slightly different than for wheat or barley LTP1 since the fluorescence intensity only increases after Ri = 0.5. This revealed a potential different mode of binding in case of this protein.

Isothermal titration calorimetry (ITC) that has already been used for probing the binding of lipids to proteins[14,15] can also be used to probe the binding of lipids by plant LTP1. In that case, the titration is followed by measuring the heat variation per second as a function of time after each injection. Upon addition of LMPC, exothermic peaks whose intensity decreased with the injection time are recorded (Figure 2). At around Ri = 1.4, endothermic peaks appeared, forming a bell curve to vanish at approximately Ri = 2. This behaviour is analogous for barley,[11] wheat and maize LTP1 (Douliez *et al.*, in preparation).

4.2 Molecular Modelling of the LMPC–Barley LTP1 Complex

Because the structure of the native barley LTP1 (1LIP.pdb)[6] maize[5] and that of the wheat LTP1 complexed with two LMPC (1BWO.pdb)[7] are known, it appears interesting to make use of molecular modelling to derive the structure of barley and maize LTP1 complexed with two LMPC. After superimposition of

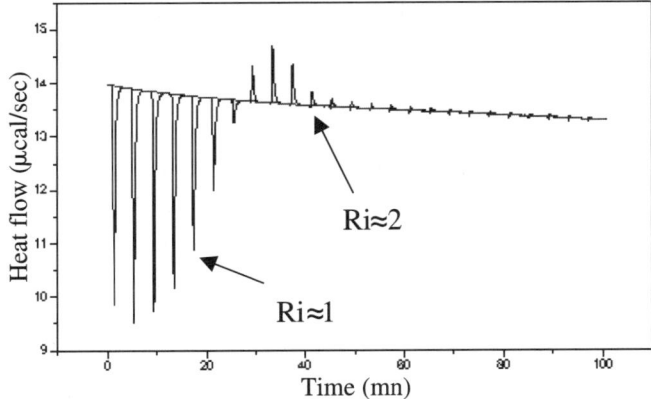

Figure 2 *Typical isothermal titration calorimetry behaviour for the binding of lipids by plant LTP1. The exothermic/endothermic behaviour reveals that two lipid monomers are bound by the protein*

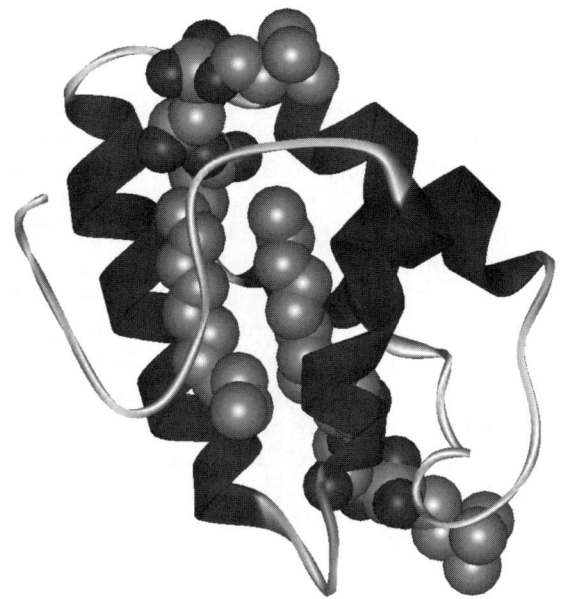

Figure 3 *Three dimensional structure of wheat LTP1 complexed with two monomers of myristoyl-lysophosphatidylcholine.[7] That of barley or maize can also be derived by using molecular modelling[11]*

barley or maize LTP1 on the wheat protein, a template procedure is used, both lipids are re-introduced in the protein on the basis of the structure of wheat LTP1 (1BWO), and a minimisation performed without any constraints. The interaction energy between lipids and protein are analogous in the case of wheat, maize and

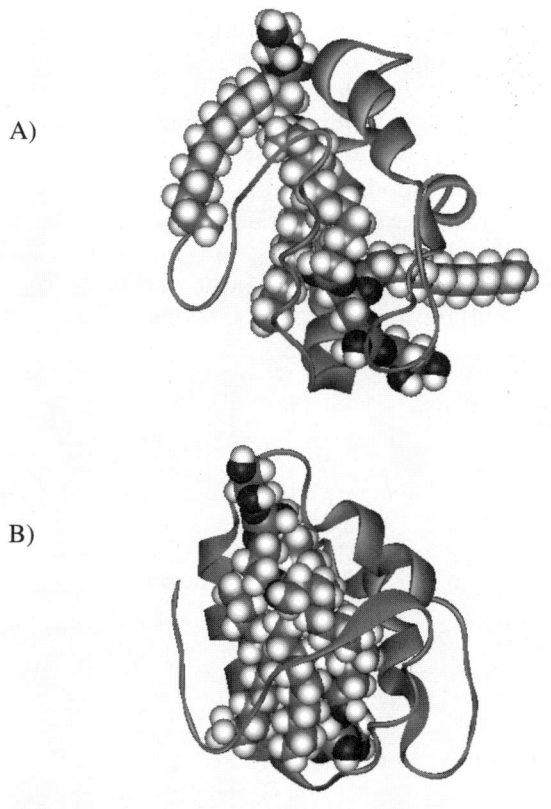

Figure 4 *Three dimensional structure of the wheat LTP1 with two monomers of a double chained lipid as derived by using molecular modelling. (A) the first model in which only one chain of each of the two lipids fits within the cavity; (B) the second model in which both chains of the two lipids are loaded*

barley (Douliez *et al.*, 2001; in preparation). This indicates that on the basis of the structure of wheat LTP1, 1BWO, barley and maize LTP1 can also bind two monomers.

4.3 Binding of Double Chained Lipids by Wheat LTP1

We clearly showed that the binding of two monomers of single chained lipids is a general feature of plant LTP1. It then appeared of interest to investigate more carefully the binding of double chained lipids. This was previously done in the case of various natural lipids.[9] The binding of two monomers could be suspect from that data however, because these lipids form multilamellar vesicles in solution, and the stoichiometry could have been overestimated. More recently, we used double chained lipids as a surfactant and a polymer grafted lipid. Such lipid-like molecules are convenient because they form micelles rather than large vesicles in solution. Fluorescence titration returned a typical pattern as in the case of single chained lipids (Figure 1). ITC revealed only exothermic peaks with a saturation at around $R_i = 2$. In both cases, fitting the data returned analogous values for the affinity and the stoichiometry (Douliez *et al.*, submitted). It was concluded that wheat LTP1 can also bind two monomers of double chained lipids. Interestingly, molecular modelling could also be used to derive the structure of such a complex. In fact, two different models could be proposed starting from the structure with two LMPC. In the first, only one chain of each of the two lipids fits within the hydrophobic cavity of the protein (Figure 4A). In the second, both chains of each lipid are loaded inside, showing a new insight into the swelling capacity of the cavity (Figure 4B).

5 Conclusion

We presented various works dealing with the binding of two lipid monomers by plant LTP1. We concluded that in case of single chained lipids, this could be a general feature of the LTP1 family. This also happens in the case of the binding of double chained lipids by the wheat LTP1 and two different models are proposed to illustrate such a feature.

References

1. J. C. Kader, *Ann. Rev. Plant Physiol. Plant Mol. Biol.*, 1996, **47**, 627.
2. J. P. Douliez, T. Michon, K. Elmorjani and D. Marion, *Cereal Chem.*, 2000, **32**, 1.
3 J. C. Kader, *Biochim. Biophys. Acta*, 1975, **380**, 31.
4. E. Gincel, J. P. Simorre, A. Caille, D. Marion, M. Ptak and F. Vovelle, *Eur. J. Biochem.*, 1994, **226**, 413.
5. D. H. Shin, J. Y. Lee, K. Y. Qwang, K. K. Kim and S. W. Suh, *Structure*, 1995, **3**, 189.
6. B. Heinemann, K. Andersen, P. Nielsen, L. Bech and F. Poulsen, *Prot. Science*, 1996, **5**, 13.
7. D. Charvolin, J. P. Douliez, D. Marion, C. Cohen-Addad and E. Pebay-Peyroula, *Eur. J. Biochem.*, 1999, **264**, 562.
8. A. Zachowski, F. Guerbette, M. Grobois, A. Jolliot-Croquin and J. C. Kader, *Eur. J. Biochem.*, 1998, **257**, 443.
9. J. P. Douliez, T. Michon and D. Marion, *Biochim. Biophys. Acta*, 2000, **1467**, 65.
10. J. P. Douliez, S. Jégou C. Pato, C. Larré, D, Mollé and D. J. Marion, *J. Agric. Food Chem.*, 2000, **49**, 1805.
11. J. P. Douliez, S. Jégou, C. Pato, D. Mollé, V. Tran and D. Marion, *Eur. J. Biochem.*, 2001, **268**, 384.
12. C. Pato, M. LeBorgne, G. LeBaut, P. LePape, D. Marion and J. P. Douliez, *Biochem. Pharm.*, 2001, **62**, 555.
13. G. W. Han, J. Y. Lee, H. K. Song, C. Chang, K. Min, J. Moon, D. H. Shin, M. L. Kopka, M. R. Sawaya, H. S. Yuan, T. D. Kim, J. Choe, D. Lim, H. J. Moon and S. W. Suh, *J. Mol Biol.*, 2001, **308**, 263.
14. T. Wiseman, S. Williston, J. F. Brandts and L. N. Lin, *Anal. Biochem.*, 1989, **17**, 131.
15. K. Miller and D. P. Cistola, *Mol. Cell. Biochem.*, 1993, **123**, 29.

Topographical Comparisons of Family 13 α-Amylases using Molecular Modelling Techniques

Gabriel Paës, Gwénaëlle André and Vinh Tran

UNITÉ DE PHYSICO-CHIMIE DES MACROMOLÉCULES, INRA,
BP 71627 – 44316 NANTES CEDEX 03, FRANCE

1 Introduction

α-Amylases (EC 3.2.1.1) are widespread endo-enzymes responsible for the internal hydrolysis of $\alpha(1-4)$ glycosidic linkages of amylose, amylopectin or glycogen. Among the numerous industrial applications, α-amylases are highly represented in glucose and dextrin processing, brewery and detergent industries. A molecular identification of functional and/or structural fragments (composed of segments of consecutive residues) of these enzymes would lead to an improvement of their properties through rational design approaches. For this purpose, a general strategy consists in combining algorithms of sequence alignment for molecules belonging to the same family with those of secondary structure predictions. The first goal is to detect consensus regions with hypothetical functional and structural roles that could be respectively correlated to the activity and the stability. A subsequent goal is to understand the 3D folding of these molecules and their catalytic action from a mechanistic point of view. But it must be underlined that only starting from the sequence of residues (1D) could be misleading due to the absence of global 3D vision of the molecule and potent superposition of similar enzymes with same functionality and architecture.

In our laboratory, we are exploring the concept that the essential characteristic of a class of molecules with the same functionality could rely in specific motifs (identified as a small ensemble for residues or even a fragment of residues). This ensemble of motifs and their location in 3D space (topography) should define a pattern belonging to the same functional module as defined by Coutinho and Henrissat[1] or more probably shared between several modules. Such a pattern could be the essential print in the folding and/or the functionality of these molecules. To get to this identification of superposed motifs (often far from any consensus region detected by classical sequence alignment methods), we shall point out the importance of a relevant 3D superposition of the molecules with similar functionality as the preliminary stage to locate the motifs with visual tools of molecular modelling.

2 Examples of α-Amylases

We shall illustrate our approach with three enzymes of the super-family 13 of hydrolases.[2,3] Numerous α-amylases have been catalogued, crystallized and modified according to their catalytic specificity.[4] In the IUBMB* classification, enzymes are only classified according to their catalytic action (α-amylases belong to class 2 of hydrolases) and the natural substrate; unfortunately their structures are not considered. Then, a new classification has been developed by Jespersen *et al.*[5] and Henrissat[2] where glycosyl-hydrolases and glycosyl-transferases belong to families 13, 70 and 77. Later, Takada *et al.*[6] gave more refined criteria for family 13 which still contain *stricto sensu* α-amylases but some of them could also act as transferases. Among known enzymes[†] of family 13, seven have accessible structure from the PDB.[‡] From a structural point of view, they all have the widespread TIM barrel folding.[7] This $(\beta/\alpha)_8$ architecture corresponds to domain A except the extended loop between $\beta3$ and $\alpha3$ delimiting domain B. For each enzyme, it is generally assumed that the ensemble of eight $\beta\alpha$ loops (longer than the $\alpha\beta$ ones) is mainly responsible for functional specificity.[8] Therefore, we shall compare both sequence alignments and 3D superpositions between two enzymes of the same sub-family (strict hydrolase): barley α-amylase,[9] G4-amylase,[10] then between two enzymes of each sub-family (strict hydrolase, additional transferase activity): barley α-amylase, taka-amylase.[11] In the following text, barley α-amylase, taka-amylase and G4-amylase will be called AMY, TAKA and AMG respectively. Furthermore, for sake of clarity, we shall only focus on the location of the backbone of their loops. Information about these enzymes is summarized in Table 1.

3 Protocol

3.1 Sequence Alignments

The primary sequence of an enzyme is the first set of available structural information. Therefore, the three sequences were first classically aligned with the INRA server MultiAlin§ based on the algorithm of Corpet.[12]

3.2 Initial Superposition

In order to identify the conserved residues in 3D space whose topology could be related to the functionality of the family, a global superposition is necessary.

---

* International Union of Biochemistry and Molecular Biology (www.iubmb.unibe.ch)
† http://afmb.cnrs-mrs.fr/~cazy/CAZY
‡ www.rcsb.org/pdb
§ //prodes.toulouse.inra.fr /multialin

Table 1 *Main characteristics of the enzymes*

Name		barley α-amylase	G4-amylase	TAKA-amylase
Abbreviation		AMY2	G4-AMY	TAKA
EC		3.2.1.1	3.2.1.60	3.2.1.1
PDB code		1AMY	2AMG	7TAA
Origin		*Hordeum vulgare*	*Pseudomonas stutzeri*	*Aspergillus oryzae*
Res. numb.		403	418	476
Catalytic triad		D179-E204-D289	D193-E219-D294	D206-E230-E297
Reaction		A : endo-α–(1,4)	A : exo-α–(1,4)	A : endo-α–(1,4)
				B : α–(1,4)
βα loop	1	5–19	21–36	13–42
	2	39–67	56–92	62–97
	3	88–154	113–168	118–181
	4	178–186	192–200	205–216
	5	203–229	218–243	229–243
	6	243–261	253–268	249–270
	7	284–306	289–317	292–313
	8	321–333	332–344	328–357

Several approaches have been proposed in the literature, *e.g.* based on disulphide bridge[13] or on catalytic triad[14] but selected residues are not well dispatched along the enzymes, or belong to mobile segments liable to vary. An original superposition method has been proposed here as the starting point of the comparison between these enzyme topographies. This superposition is based on the core of the enzyme architecture with the conserved $(\beta/\alpha)_8$ barrel. Because of our interest on the eight $\beta\alpha$ loops in 3D space, the terminal residues of each β sheet segment were selected and their backbones were taken for the global pair–pair superimposition of enzymes. For this family 13, the topography of the TIM barrel has been very well kept because all starting residues α helix segments were also superposed. With this initial superposition it is possible to compare loops of different length.

3.3 Identification of Superimposed Residues in 3D Space

At the bottom of Table 1, we have reported the numbering of the eight $\beta\alpha$ loops to be taken into account. Then, only considering the backbone of these segments, each loop, for a given enzyme, has been analysed with reference to all other loops of the compared enzyme. With the initial superposition previously described, all $\beta\alpha_n$ loops (with $n = 1$–8) were at least superimposed with their end residues but all other residues superimposed with their backbone are also looked for with a basic visual criterion.

4 Results and Discussions

4.1 Superposed Segments

Tables 2a and 2b show the superposed residues when referring to loop 7 of AMY

Table 2   *Topographical superposition of loop 7 of barley α-amylase and TAKA amylase*

(a)

AMY	TAKA
F284	F292
V285	V293
D286	E294
N287	N295
H288	H296
D289	D297
T290	
G291	
S292	
T293	A342
Q294	P341 L8
H295	D340
M296	
W297	
P298	P299
F299	R300
P300	F301
S301	
D302	
R303	
V304	A311
M305	K312
Q306	N313

(b)

TAKA	AMY
F292	F284
V293	V285
E294	D286
N295	N287
H296	H288
D297	D289
N298	
P299	P298
R300	F299
F301	P300
A302	
S303	
Y304	
T305	
N306	
D307	
I308	
A309	
L310	
A311	V304
K312	M305
N313	Q306

and TAKA respectively. In these tables, residues in grey are those located at the catalytic face of the enzyme and therefore likely to be in contact with the substrate in a preliminary docking phase. Residues in black are much more interesting because they delimit the catalytic cleft and certainly act in the docking process. In this case, the backbones of following segments of loop 7 are superposed in 3D space: {F284 to D289}, {P298-F299-P300} and {V304-M305-Q306} of AMY to {F292 to D297}, {P299-R300-F301} and {A311-K312-N313} of TAKA respectively. Moreover, the trace of segment {T293-Q294-H295} of AMY loop 7 is also superposed to that of segment {A342-P341-D340} of TAKA loop 8. The latter case clearly demonstrates that this typical feature could not be obtained from any classical sequence alignment since the propagation direction of the two superposed segments are opposite. The structural features of loop 7 of AMY and TAKA are also shown on Figure 1.

All such conserved segments in 3D space of all $\beta\alpha$ loops should further be analysed carefully as possible motifs related to enzyme functionality.

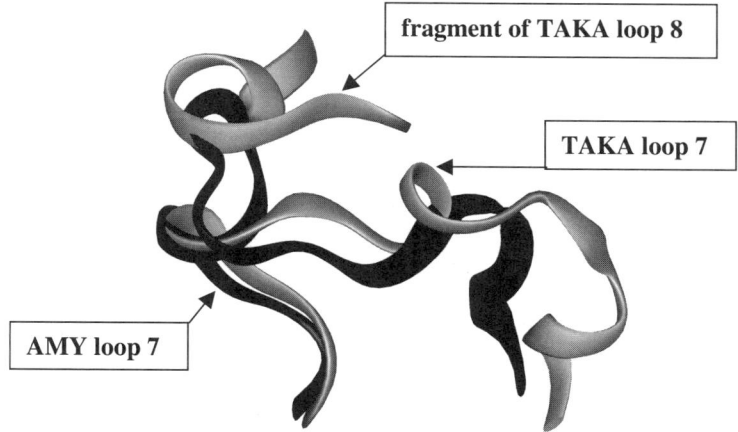

fragment of TAKA loop 8

TAKA loop 7

AMY loop 7

Figure 1 *Superposition of AMY and TAKA loop 7*

4.2 Comparison with Sequence Alignments

From a sequences multiple alignment of these three enzymes, Figures 2a and 2b only show the alignments of $\beta\alpha$ loops of AMY/AMG and AMY/TAKA respectively. On these figures, for each block of alignment, the rectangle on the first line corresponds to the size of the loop of enzyme 1 (extreme values are starting and ending residues for this loop), the second line is the sequence of enzyme 1. In the third line, grey and black patterns mean that residues are respectively located at the enzyme surface and the catalytic cleft. The thick dash on the fourth line corresponds to superposed residues in 3D space according to our protocol with the number of superposed residues above. More interestingly, vertical or oblique lines delimit the superposed segments between the compared enzymes. The three next lines correspond to symmetric information concerning enzyme 2. Finally, the last line corresponds to consensus symbols of the 1D sequence alignment.

The first comparison (Figure 2a) between two similar enzymes of this family 13 (strict hydrolases) shows a good agreement between 1D alignment and the location of these $\beta\alpha$ loops because no significant shifts occur for the end residues of these loops. But 3D superposition give additional functional or structural information because these loops could be fully superposed in 3D space (*e.g.* loops 4 and 8) or lowly superposed (*e.g.* loops 3 and 5). The analysis of loop 3 shall give basic schemes to interpret these enzyme matches. The importance of loop 3 must be first underlined as it describes completely domain B. In the present comparison, only half the residues are 3D superposed which means that this loop should contain key information about similarities between enzymes of family 13 and enzyme specificities as well. Then, one must carefully look at residues delimiting the catalytic cleft (in black) and check if they belong or not to a 3D superposed segment. In the former case, it would mean that the location of these residues could be part of a specific pattern related to the functionality of the considered family or sub-family. In the latter case, it would mean that this single residue

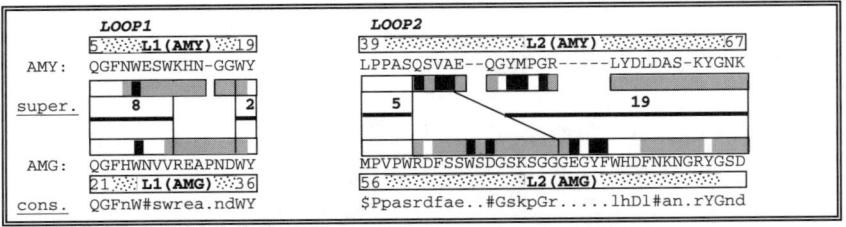

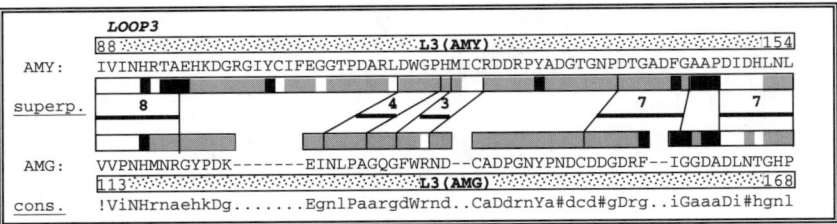

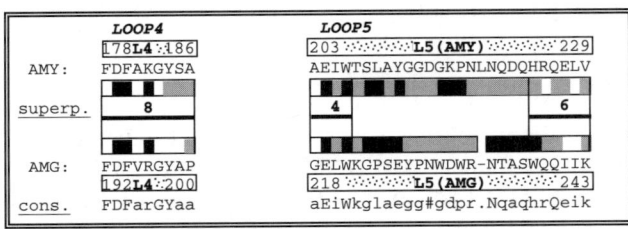

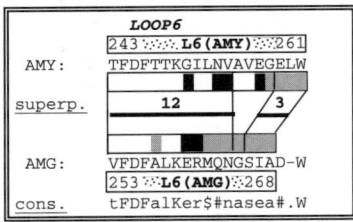

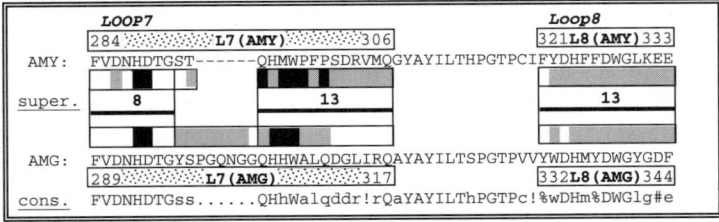

(a)

Figure 2 *Comparison between 1D alignment and 3D superposition: (a) between AMY and AMY amylases; (b) between AMY and TAKA amylases*

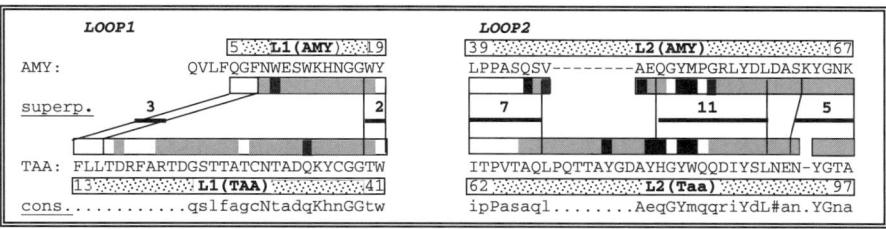

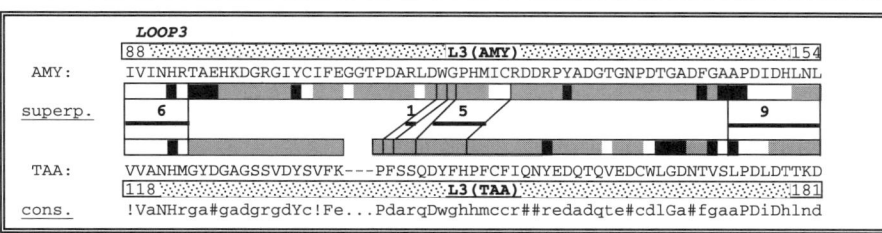

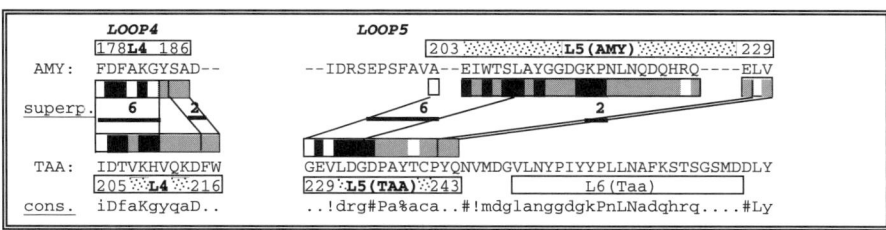

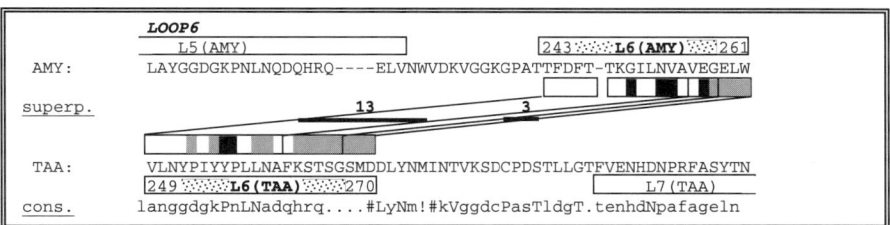

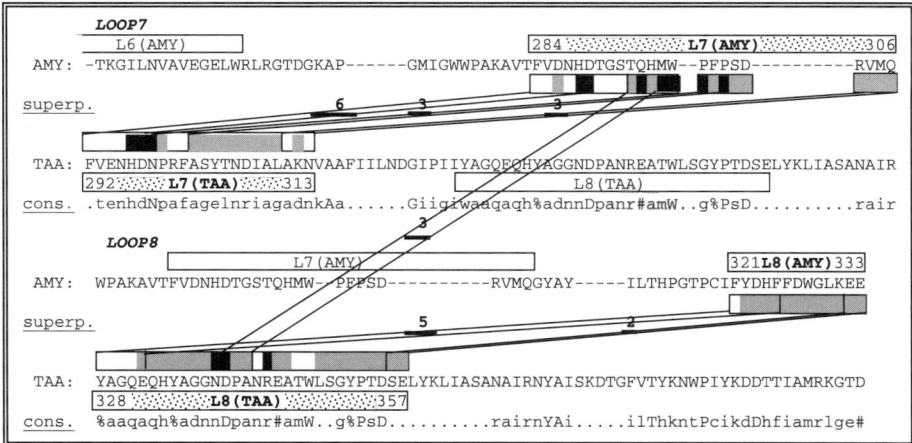

(b)

arrangement is more specific of the enzyme. Therefore, H92 and F143 (AMY numbering) would probably be part of a pattern characteristic to family 13 or more strictly to sub-families of pure hydrolases. On the contrary, Y104 and Y130 should be more specific to barley α-amylase docking with amylose substrate.

The second comparison (Figure 2b) between a pure hydrolase (AMY) and one also capable to act as transferase (TAKA) clearly demonstrates that sequence alignment could be in contradiction with a 3D superposition of the loops. As in the former comparison, the location of the four first loops can be satisfactorily obtained by sequence alignment and 3D superposition as well. But on the contrary, sequence alignment techniques cannot predict the 3D superposition of the four last loops which can only be related to enzyme functionality. As a demonstration of the assumption, the catalytic residues are well 3D superposed and follow the loop shifts. Although from loop 4 to the end of the sequence, the 1D alignment is still acceptable (*cf.* consensus line), the necessary shift to match these last loops in 3D space cannot be predicted. More surprisingly, loop 5, 6 and 7 of AMY grossly correspond to loops 6, 7 and 8 of TAKA, on the basis of the 3D superposition. For example, $D289_{AMY}$ (loop 7) corresponds to $D294_{TAKA}$ (loop 7) instead of $D340_{TAKA}$ (loop 8) as predicted by 1D alignment. Therefore, this 3D superposition can 'correct' some matches found with the sequence alignment. For example, the segment $\{H288\text{-}D289\}_{AMY/loop7}$ should not be compared to $\{N339\text{-}D340\}_{TAKA/loop8}$ as suggested by the sequence alignment (both segments delimit the catalytic clef) but to $\{H296\text{-}D297\}_{TAKA/loop7}$, also delimiting the catalytic cleft.

5 Conclusions

The global superposition of enzymes of family 13 on the common $(\beta/\alpha)_8$ barrel is remarkably efficient to start $\beta\alpha$ loops comparison in the 3D space. This 3D superposition can focus directly to the new structural information for similar enzymes in term of functionality when compared to predictions based on sequence alignment. Interestingly, it has been seen in some comparisons, between less similar enzymes, that our 3D superposition is more powerful and more accurate to match segments of residues whose location is tightly related to functionality. In the shown examples of three enzymes of family 13, this 3D superposition is necessary to detect similarities and discrepancies between sub-families of strict hydrolase and hydrolase-transferase.

References

1. P. M. Coutinho and B. Henrissat. 'Carbohydrate-active Enzymes: An Integrated Database Approach', in *Recent Advances in Carbohydrate Bioengineering*, H. J. Gilbert, G. J. Davies, B. Henrissat and B. Svensson (eds.), Royal Society of Chemistry, Cambridge, 1999, p. 3–12.
2. B. Henrissat, *Biochem. J.*, 1991, **280**, 309.
3. B. Henrissat and A. Bairoch, *Biochem. J.*, 1993, **293**, 781.
4. B. Svensson, *Plant. Mol. Biol.*, 1994, **25**, 141.

5. H. M. Jespersen, E. A. MacGregor, M. R. Sierks and B. Svensson, *Biochem. J.*, 1991, **280**, 51.

6. H. Takata, T. Kuriki, S. Okada, Y. Takesada, M. Iizuka, N. Minamiura and T. Imanaka, *J. Biol. Chem.*, 1992, **267**(26), 18447.

7. G. K. Farber and G. A. Petsko, *Trends Biochem. Sci.*, 1990, **15**(6), 228.

8. J.-P. Y. Scheerlinck, I. Lasters, M. Claessens, M. De Maeyer, F. Pio, P. Delhaise and S. J. Wodak, *Proteins*, 1992, **12**, 299.

9. A. Kadziola, J.-I. Abe, B. Svensson and R. Haser, *J. Mol. Biol.*, 1994, **239**, 104.

10. Y. Morishita, K. Hasegawa, Y. Matsuura, Y. Katsube, M. Kubota and S. Sakai, *J. Mol. Biol.*, 1997, **267**, 661.

11. A. M. Brzozowski and G. J. Davies, *Biochem.*, 1997, **36**, 10837.

12. F. Corpet, *Nucl. Acids Res.*, 1988, **16**(22), 10881.

13. J. M. Mas, P. Aloy, M. A. Mart-Renom, B. Oliva, C. Blanco-Aparicio, M. A. Molina, R. de Llorens, E. Querol and F. X. Aviles, *J. Mol. Biol.*, 1998, **284**, 541.

14. G. Andre, PhD Thesis, Université de Nantes, 1998, Chapter 9.

Biopolymer Assemblies

Glutenina Macropolymer: A Gel Formed by Glutenin Particles

Clyde Don,[1] Wim Lichtendonk,[1] Johan Plijter[1] and Rob J. Hamer[1,2]

[1]TNO NUTRITION AND FOOD RESEARCH, PO BOX 360, 3700 AJ ZEIST, THE NETHERLANDS
[2]CENTRE FOR PROTEIN TECHNOLOGY TNO-WUR, PO BOX 8129, 6700 EV WAGENINGEN, THE NETHERLANDS

1 Abstract

The quality of wheat-based foods and the processing properties of wheat flour dough are strongly related to the presence and properties of very large glutenin protein aggregates. These very large aggregates are insoluble in 1.5% (w/v) SDS and can be recovered as a gel, the so-called Glutenin Macro Polymer (GMP) fraction, after ultracentrifugation. GMP quantity and gel properties strongly correlate with flour technological quality. We therefore studied factors governing GMP formation. Flour from four wheat varieties: Galahad-7, Caprimus, Soissons and Classic were used in this study. GMP was isolated from each of the flour samples and characterised. Plateau values of G' paralleled differences in wheat quality. Further detail was obtained by studying GMP dispersed in 1.5% SDS. Re-aggregation, viscometry and Confocal Scanning Laser Microscopy confirmed that GMP consists of very large particles, able to form a gel. Clear differences in average particle size and re-aggregation (gel-forming) properties could be observed. The size and shape of the particles point at a possible origin in the protein bodies from immature wheat endosperm.

2 Introduction

The glutenin fraction of wheat gluten has long been considered to have a prominent role in the strengthening of dough.[1] This fraction can be isolated as an SDS insoluble gel-layer,[2] named Glutenin Macro Polymer (GMP). GMP is one of the best flour quality indicators today. GMP quantity correlates strongly with dough properties and loaf volume.[3] Pritchard *et al.* have demonstrated that

GMP rheological properties are also very relevant, due to their strong correlation with loaf volume and dough extensibility.[4–7] GMP consists of very large structures of both high molecular and low molecular weight glutenin subunits (HMWGS and LMWGS respectively). Several models, mainly based on the ability of the subunits to form disulfide bonds, have been proposed for the structure of the glutenin network,[8–10] but no consensus has been reached. A link is suspected between HMWGS subunit composition and glutenin network properties. Payne has demonstrated a positive correlation with loaf volume if HMWGS 5 + 10 are present.[11] HMWGS are without doubt important in determining dough properties. Varieties lacking HMWGS (triple null varieties) are unable to form a visco-elastic gluten. Dough from a variety like Galahad-7, only having HMWGS-7, is extremely extensible and hardly exhibits elastic properties. In contrast, a variety like Glenlea (having a double set of HMWGS) is overstrong.[12] Both Popineau and Lafiandra and recently, Lefebvre and Popineau[12,13] have demonstrated that HMWGS-5 is necessary to form elastic properties. There must be a link between HMWGS composition, GMP and gluten properties. We have recently proposed[14] that GMP is a gel formed by both physical and chemical interactions. In this we also have indicated the importance of particle–particle interactions. In the present study we set out to understand the factors determining the formation and properties of GMP, thus unravelling underlying quality factors.

3 Materials and Methods

3.1 Wheat Flour

Flour was obtained from the following varieties: Galahad-7, Caprimus, Soissons and Classic. The protein content was respectively: 11.5%, 10.3%, 10.6% and 13.7%.

Soissons and Classic are so-called '5 + 10' varieties, Caprimus is a '2 + 12' variety. Galahad-7 only has HMWGS-7.

3.2 Isolation of GMP

Flour (1.4 g) was suspended in 1.5% SDS (28 mL) and centrifuged at 80.000 g for 30 minutes at 20°C in a Kontron Ultracentrifuge.[2] The supernatant was decanted and the gel-layer collected as GMP.

3.3 Rheology of GMP Gels

1 g of material was carefully taken from the top of the gel and transferred into the measuring cell of a Bohlin VOR rheometer. The cell consisted of two parallel plates with a gap of 1 mm. Measurements were performed at 20°C in a strain sweep mode at amplitudes ranging from 1% to 100%. Data were expressed as G' *vs* strain or delta (δ) *vs* strain.

3.4 GMP Dispersion

GMP dispersions were prepared by transferring *ca.* 1 g of GMP gel to a tube containing 10 mL 1.5% SDS solution. The gel was mixed with the solvent by briefly stirring with a spatula. Then the tube was sealed and placed on a roller-bank for 3 h at ambient temperature. This produced a visually homogenous dispersion that was used for further characterisation.

3.5 Protein Analysis

Protein content was measured using an UV absorption method described in ref. 15. The UV method was successfully calibrated with a set of Kjeldahl protein values of GMP samples.

3.6 Re-aggregation Experiments

Dispersions of GMP were diluted with 1.5% SDS solution to a concentration of 0.5, 1 or 2 mg/mL respectively. Each dispersion was then centrifuged for 30 min at 80 000 g at 20 °C. Each supernatant or newly formed gel was weighed and analysed for protein content. Results were expressed as % protein recovered in the gel phase.

3.7 Rheological Characterisation of GMP Dispersions

The viscosity of GMP dispersions could be measured using a Ubbelohde capillary viscometer. The pass through time was in the order of 300–400 seconds to optimise accuracy. Data were expressed as reduced viscosity L/g.

3.8 Confocal Laser Microscopy

Freshly prepared samples of GMP dispersions were stained for protein with rhodamine and observed using a Leica Confocal Laser Scanning Microscope.

4 Results and Discussion

GMP was isolated from each of the flour samples. Data reflecting flour technological quality, the GMP quantity and protein concentration composition of each GMP isolate are presented in Table 1.

GMP protein varies between varieties: Galahad-7 < Caprimus < Soissons < Classic. Each GMP demonstrates typical gel like behaviour (results not shown). Plateau values of G' in the strain sweep were: 6, 13, 38 and 100 Pa for Galahad-7, Caprimus, Soissons and Classic respectively. The ratio of viscous to elastic behaviour is reflected in δ, for δ lower than 45°, elastic behaviour predominates in the structure. The δ values in the plateau region for Galahad-7, Caprimus, Soissons and Classic are 25°, 14°, 9° and 9°. Galahad-7 has the weakest GMP (G'), also its δ increases the most (> 45°) when the strain is increased and passes the plateau region. This is indicative of a weakly stabilised network. It is the G' of GMP that parallels best the variation in technological classification. When GMP is a particle network stabilised by a combination of covalent and non-covalent interactions, polymer physical theory points at the importance of both concentration and properties of particles.[16] The re-aggregation experiments shown in Figure 1 demonstrates a different behaviour of the varieties at three protein concentrations. With Galahad-7 and Caprimus lowering the protein concentration leads to a decrease in gel recovered. In contrast, with Soissons and Classic the relative recovery is not affected by protein concentration. This points at the importance of not only concentration but also of particle size and associative properties.

The poor recovery at lower protein concentration for Galahad-7 and Ca-primus is likely due to an increased solubility at lower protein concentration. The firmer and more elastic GMP from '5 + 10' varieties Classic and Soissons appear to have comparatively stable aggregates.

Viscometric characterisation of the dispersions was carried out to further reveal differences in particle properties. Figure 2 shows the typical concentration *vs* reduced viscosity plots for the four varieties. The intercept represents the intrinsic viscosity that is related to the average size of the particle.

The intrinsic viscosity is increased from Galahad-7 to Classic. Assuming a spherical shape the relative sizes for Galahad-7, Caprimus, Soissons and Classic are: 1:8:12:15. This again points at considerable differences between the four varieties. Physical theory states that the rheological properties of a gel are related to the size and concentration of the contained particles. Although it is not yet possible to quantify this, the differences in G' plateau values are clearly reflected

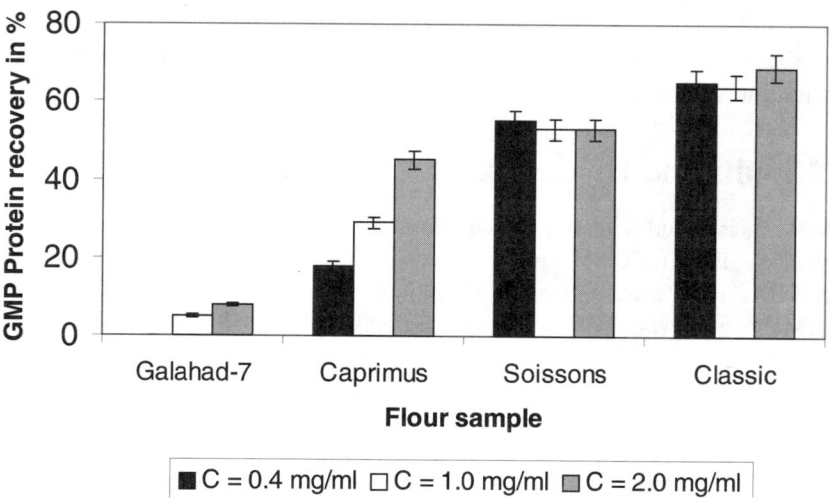

Figure 1   *Re-aggregation experiment*

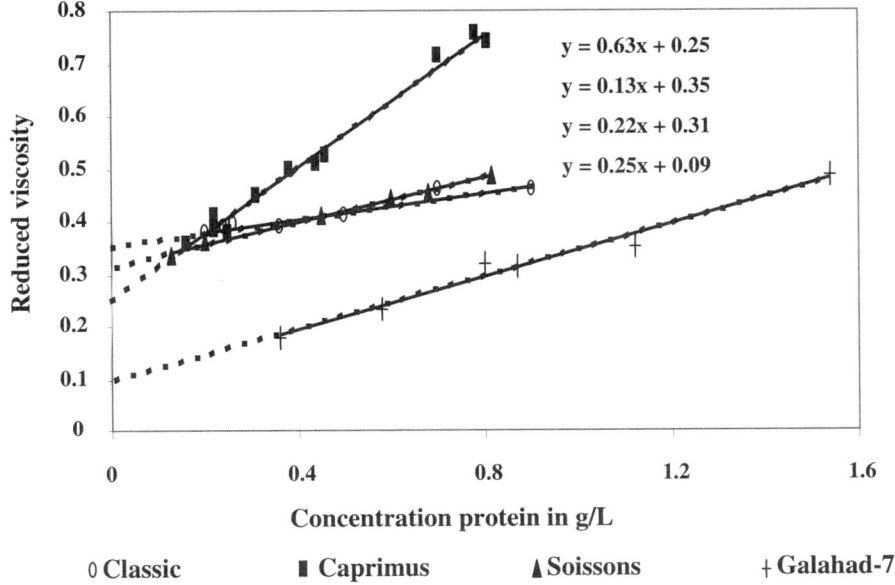

Figure 2 *Reduced viscosity* versus *protein concentration for GMP dispersions*

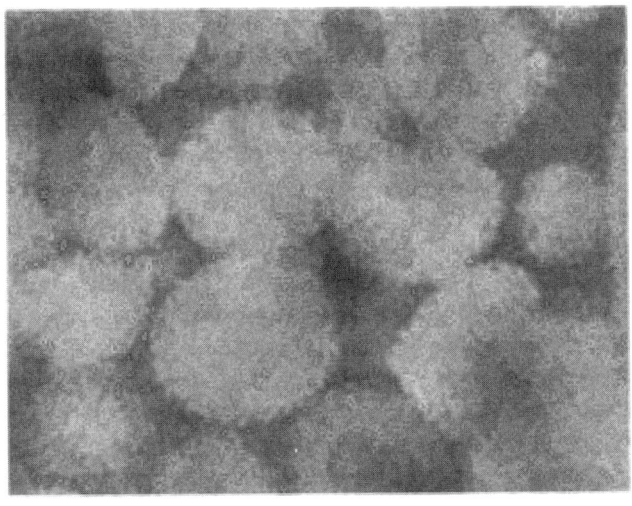

$Bar \approx 10\,\mu m$

Figure 3 *A CSLM image of dispersed protein particles from GMP*

in differences in size. The intrinsic viscosity values found point at a very large size of the particles.

We used CSLM to confirm this finding. CSLM revealed the presence of large $(10-30 \mu m)$ particles with Classic, Soissons and Caprimus. Due to a lack of contrast no particles could be observed with Galahad-7. Figure 3 shows a typical example of particles of *ca. 25 μm* diameter.

Both the size and shape of these particles reminded us of earlier TEM photographs of glutenin particles by Graveland and – more importantly – of protein bodies in the immature wheat endosperm.[17,18] The latter are *ca. 3–5 μm* diameter. The presence of SDS in our solvent however leads to swelling, which could explain the larger sizes found in this study.

This remarkable resemblance suggests a tentative link: the particles in GMP are pre-formed and originate from the protein bodies observed in the immature wheat endosperm.

5　Conclusions

In this paper we have confirmed the importance of GMP and GMP rheological values in explaining the variation in technological quality of wheat.

Our hypothesis that GMP is a particle network was confirmed by re-dispersion and re-aggregation experiments. Clear differences between particles could be observed. These differences are possibly related to difference in HMWGS composition and help explain differences in G′ plateau values for GMP.

The average size differences of the particles by viscometry prompted examination of GMP by CSLM. With this technique we were able to confirm the particle nature of GMP. We propose that these particles are pre-formed and originate from protein bodies of immature wheat endosperm.

Acknowledgements

The assistance of J. van Riel and M. Paques in performing the CSLM experiments is gratefully acknowledged.

References

1. F. MacRitchie, *Cereal Foods World*, 1980, **25**, 382.
2. A. Graveland, P. Bosveld, W. J. Lichtendonk and J. H. E. Moonen, *J. Sci. Food Agric.*, 1982, **33**, 1982, 1117.
3. P. L. Weegels, A. M. van de Pijpekamp, A. Graveland, R. J. Hamer and J. D. Schofield. *J. of Cereal Sci.*, 1996, **23**, 103.
4. P. E. Pritchard, *Aspects of Applied Biology (UK)*, 1993 (no. 36), 75.
5. J. H. E. Moonen, A. Scheepstra and A. Graveland. *Euphytica*, 1986, **31**, 677.
6. H. D. Sapirstein and J. Suchy, *Cereal Chemistry*, 1999, **76**, 164.
7. M. Kelfkens and W. J. Lichtendonk. *Getreide Mehl und Brot*, 2000, 54.
8. J. A. D. Ewart, *J. Sci. Food Agric.*, 1979, **30**, 482.
9. A. Graveland, P. Bosveld, W. J. Lichtendonk, J. P. Marseille, J. H. E. Moonen and A.

Scheepstra, *J. Cereal Sci.*, 1985, **21**, 117.

10. L. Gao, P. K. W. Ng and W. Bushuk, *Cereal Chem.*, 1992, **69**, 452.
11. P. I. Payne, M. A. Nightingale, A. F. Krattiger and L. M. Holt, *J. Sci. Food Agric.*, 1987, **40**, 51.
12. D. Lafiandra, R. Ovidio, E. Porcedu, B. Margiotta and G. Colaprico, *J. of Cereal Sc.*, 1993, **18**, 197.
13. Y. Popineau, M. Cornec, J. Lefebvre and B. Marchylo., *J. of Cereal Sci.*, 1994, **19**, 231.
14. R. J. Hamer and T. van Vliet, in *Wheat Gluten*, eds. P. R. Shewry and A. S. Tatham, Royal Society of Chemistry, Cambridge, 2000, p. 125.
15. G. M. Hall. *Methods of Testing Protein Functionality*, Blackie Academic & Professional, 1996, 34.
16. P. J. Flory, *Principles of Polymer Chemistry*, Cornell University Press, Ithaca, 1953.
17. A. Graveland and M. H. Henderson, in Proceedings of the 3rd International Gluten Workshop – Budapest, eds. R. Lasztity and F. Bekes, 1987, 238.
18. A. D. Evers and D. B. Bechtel, *Wheat: Chemistry and Technology I.*, *3 ed.*; Eagan Press, Minneapolis, St Paul, 1988.

Swelling and Hydration of the Pectin Network of the Tomato Cell Wall

A. J. MacDougall and S. G. Ring

DIVISION OF FOOD MATERIALS SCIENCE, INSTITUTE OF FOOD
RESEARCH, NORWICH RESEARCH PARK, COLNEY, NORWICH,
NR4 7UA, UK

1 Introduction

The primary cell wall of dicotyledonous plants consists of cellulose microfibrils dispersed within a matrix of predominantly non-cellulosic polysaccharides, including xyloglucans and pectic polysaccharides. The xyloglucans are neutral polysaccharides which bind to the cellulose microfibrils through secondary interactions, and have the ability to crosslink the fibrillar cellulose network. This fibrillar network is then dispersed in a network of the pectic polysaccharides.[1] The pectic polysaccharide network also forms the middle lamella in dicotyledons and is responsible for cell–cell adhesion.

The pectic polysaccharides are structurally complex.[2,3] They consist of a backbone of $(1\rightarrow4)$ α-D-galacturonosyl residues interrupted with typically a 10% substitution of $(1\rightarrow2)$-α-L-rhamnopyranosyl residues. A portion of the rhamnosyl residues are branch points for neutral sugar side-chains which contain L-arabinose and D-galactose. The rhamnosyl substitution is thought to cluster in 'hairy' regions leaving 'smooth' sequences of the galacturonan backbone. The backbone may be partially acetylated and may be further substituted with terminal xylose. A portion of the galacturonosyl residues of extracted pectins are partially methyl esterified. The swelling and hydration of the pectin network in the plant cell wall and middle lamella will depend on the extent of crosslinking and the affinity of the polymer for water. In pectin networks there is the potential for both covalent and non-covalent crosslinks, and the greater the extent of crosslinking, the greater the potential restorative force resisting swelling.

The characterisation of the crosslinking of the pectic network is still a matter of research. In some plant families, *e.g.* the *Chenopodiaceae*, of which the most investigated is the sugar beet (*Beta vulgaris* L.), the potential for covalent crosslinking through phenolic residues, such as diferulic acid, has been demonstrated.[4] There is also the possibility of crosslinking of the pectic network through ester crosslinks involving the D-galacturonic acid of the pectic polysaccharide backbone.[5] As yet, the crosslink has not been isolated. The crosslink for which there is a more detailed structural characterisation is a 1:2 borate ester diol of a pectic polysaccharide fragment (rhamnogalacturonan II) released from the primary cell wall of plants by treatment with *endo*-α-1,4-polygalacturonase.[6] In addition to these covalent crosslinks, there is also the potential for secondary interactions to contribute to network formation. Of these, the most studied is the interaction of pectic polysaccharides with calcium ions[7,8] which can result in network formation and gelation of moderately concentrated solutions of the pectic polysaccharides.[9,10] The main requirement for gelation is the presence of stretches of unsubstituted D-galacturonosyl residues in the pectic polysaccharide backbone, and a sufficient number of such regions, to form an interconnected network. From studies on the solution behaviour of pectic polysaccharides it is proposed that the conformation of pectic polysaccharides in the junction zone or crosslink is that of an 'egg box',[9] although other types of association are also possible.[11]

The properties of the pectic network will have a major influence on the properties of the cell wall and middle lamella. The extent of crosslinking will contribute to the stiffness of the cell wall, and influence the porosity of the cell wall to macromolecular species such as enzymes. A highly crosslinked network could have a reduced porosity with a result that enzymes would be excluded and could only act at the network surface.

2 Network Crosslinking

2.1 Cell Wall Studies

The pectin network of the middle lamella is potentially involved in cell-cell adhesion. Using the Ca^{2+} chelating agent cyclohexane diamine tetraacetic acid (CDTA) as an extractant it is possible to probe some of the properties of this network. In Figure 1 is shown the effect of increasing CDTA concentration on cell separation in tomato pericarp at pH 6.8. As the CDTA concentration is increased to 30 mM there is a marked increase in the number of free cells; as the concentration is further increased to 90 mM CDTA there is a further, less marked increase. In another experiment the effect of pH on the ability of 100 mM CDTA to separate cells was examined (Figure 2). There was an increase in cell separation as the pH was increased from 5 to 7, *i.e.* in the pH range where CDTA becomes a more effective chelator of Ca^{2+}. Both sets of data suggest that the cells of tomato pericarp are held together by a pectin network which is crosslinked by Ca^{2+} ions, and furthermore that the pectin, extracted at neutral pH from isolated CWM, is primarily responsible for forming the pectin network involved in the cellular adhesion of this tissue.

2.2 Calcium Mediated Crosslinking of Tomato Pectin

In previous reports we have demonstrated that CDTA, used to extract pectin

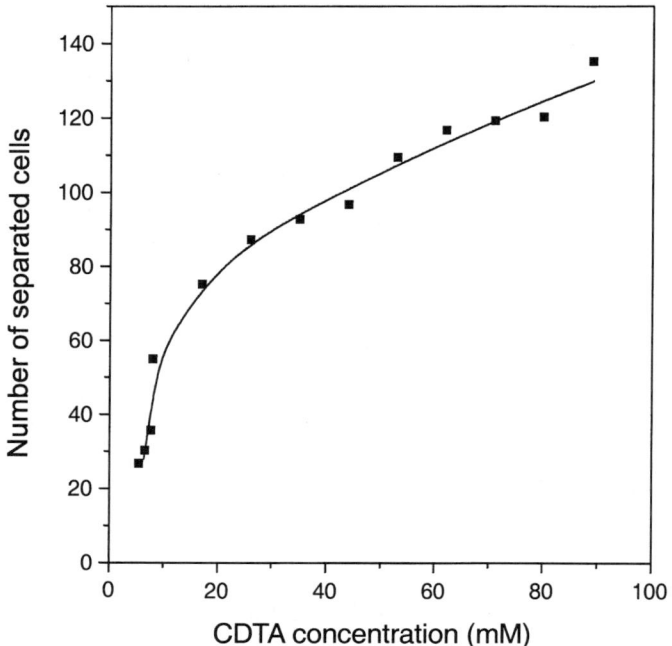

Figure 1 *The effect of CDTA concentration on cell separation in mature green tomato pericarp tissue. Cell separation assays were carried on pericarp fragments (5 to 10 cells) extracted with buffered phenol*

from cell walls, can be removed quantitatively[12] and that the purified pectin forms elastic gels on addition of calcium ions.[10] The Ca^{2+} binding and swelling behaviour of this pectin have been characterised. For Ca^{2+} ions interacting with a pectin, this equilibrium can be quantitatively described by a stability constant, K:

$$K = [Ca^{2+}2COO^-]/[Ca^{2+}][2COO^-]$$

where following convention, and the observed stoichiometry of binding, one Ca^{2+} ion is considered to bind to a fragment of pectin containing two carboxyl functions. The change in stability constant with both degree of methyl esterification and galacturonate chain length has been determined. The stability constant increases with decreasing degree of methyl esterification. For a pectin with a degree of methyl esterification of 65% the interpolated value of log K is ~2.4.[7] This constant describes the binding in solution, which could involve both intra and inter molecular associations. For the tomato pectin of carbohydrate composition, D-galactose 10%; L-arabinose 3%; L-rhamnose 0.7%; D-galacturonic acid 71% w/w, and degree of methyl esterification 68%, the Ca^{2+} binding behaviour in aqueous solution was comparable to that of other pectins (log K ~2.7) with a similar degree of methyl esterification.[13]

The binding of Ca^{2+} in the tomato pectin *network* was determined by follo-

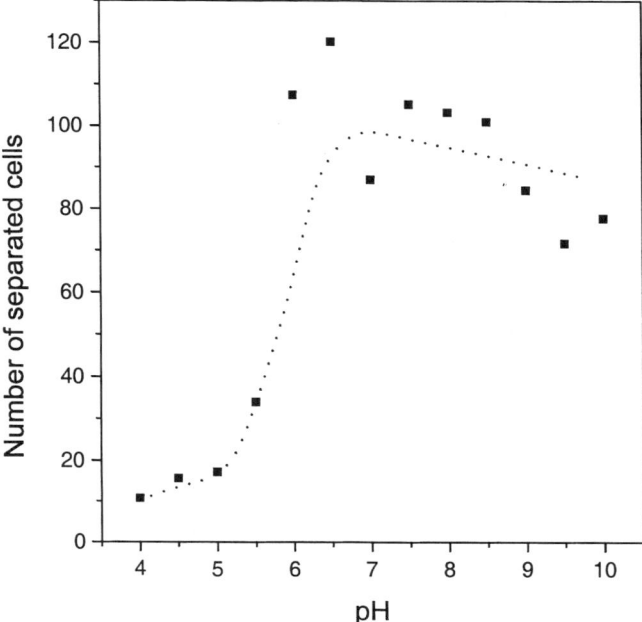

Figure 2 *The effect of pH on cell separation in mature green tomato pericarp tissue. Cell separation assays were carried out on pericarp fragments (5 to 10 cells) extracted with buffered phenol*

wing the dissolution of a pectin gel in water, and a value of log K of ~ 3.9 was found.[13] This is more than an order of magnitude greater than the comparable solution case, indicating that a crosslink of sufficient permanence to create an elastic gel has a higher affinity for Ca^{2+} than the chain in solution.

2.3 Cell Wall Speciation of Ca^{2+}

The apoplast is a complex ionic environment with a range of species being present including pectin, inorganic and organic cations, organic acids such as citrate and cell wall proteins. There is a range of potential ionic interactions, with the various ionic species being involved in different complex equilibria. Identification of the potential factors influencing these equilibria could provide insight into the biochemical regulation of cell wall behaviour. To attempt to calculate speciation within the cell wall, information is required on the components present, their concentration, and the stability constants describing the various ionic equilibria.

Pectin concentration can be estimated from the *in vitro* experiments on cell wall swelling (qv). For the determination of ion and organic acid content, two complementary approaches may be used. Apoplastic sap may be expressed after application of pressure and its ion and solute content determined.[14] This will give an indication of the free cation concentration within the cell wall. The main cations found in tomato fruit apoplastic sap (24–26 days after anthesis) were K^+, Na^+, NH_4^+, Mg^{2+} and Ca^{2+}, at concentrations of $\sim 20, 0.5, 0.5, 5.2$ and 6.3 mM respectively.[15] The ion and organic acid content of the apoplast may also be obtained using non-aqueous methods involving the isolation of a cell wall fraction. After correction for cytoplasmic contamination, it is then possible to obtain the total ion content of the apoplast, which will include free and bound forms.[16] In both ripe and unripe tomato cell wall, the Mg^{2+} and Ca^{2+} levels were ~ 18 and ~ 60 μmol g^{-1}, respectively. At a swelling ratio of 3.0 g/g this would lead to a concentration of Ca^{2+} of ~ 36 mM. The large difference between the total and free Ca^{2+} levels, presumably reflects the extent of binding of Ca^{2+} by the pectic polysaccharides in the wall. To test this proposition information is required on the affinity of the various species within the cell wall, including pectic polysaccharides and organic acids, for the different ions. Having estimates of pectin concentration in the cell wall, its affinity for Ca^{2+} ions, and the ionic composition of the apoplast, it is possible to predict the speciation behaviour as a function of pH. The predicted concentrations of Ca^{2+}, pectate and calcium pectate as a function of pH in the range 3–7 is shown in Figure 3 (based on a log stability constant for the formation of calcium pectate of 3.87). There is an excess of pectate available for interaction with Ca^{2+}. As pH increases, free Ca^{2+} levels fall and there is a corresponding increase in formation of Ca^{2+} pectate. The free

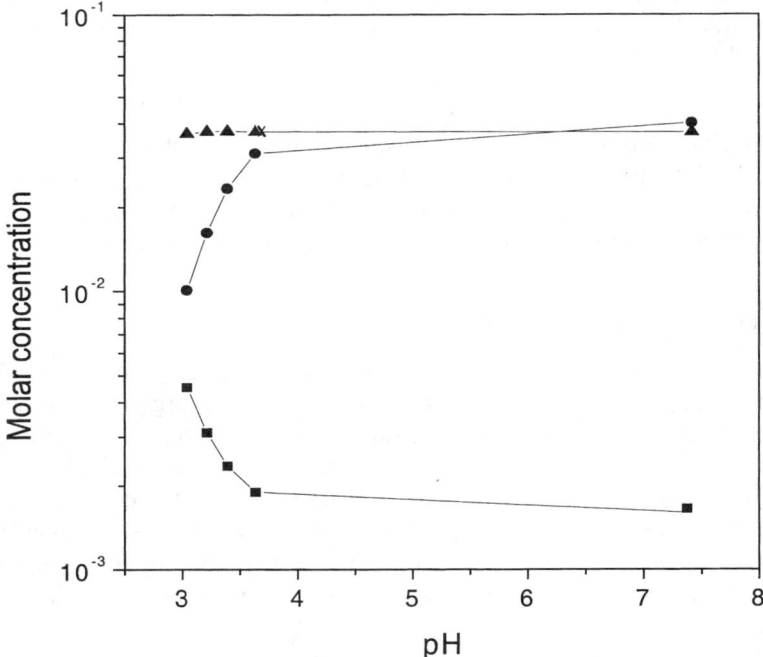

Figure 3 *Predicted speciation of pectate at ionic concentrations estimated for the plant cell wall. Pectate (●), Ca^{2+} (■), Ca^{2+} pectate (▲)*

Ca^{2+} level is predicted to be 1.6 mM at pH 7, compared with a measured value of 6.3 mM. Repeating the calculation with a log stability constant, K, of 2.7 (the affinity of Ca^{2+} for the pectin chain in solution rather than gel) gives a value of ~ 14 mM for the level of free Ca^{2+} at pH 7. These calculated values of free Ca^{2+} span the reported values of Ca^{2+}. From the reported compositions of the tomato cell wall (both Ca^{2+} and pectate), only a fraction ($\sim 50\%$) of the free anhydrogalacturonate of the pectin chain could be involved in Ca^{2+} mediated crosslinking. The other cations of the apoplast balance the remaining excess charge of the pectate in the cell wall. It is these more weakly interacting cations which can contribute to the Donnan effect of the cell wall network. The magnitude of the resulting swelling pressure would depend on the valency of the inorganic and organic counterions present.

There has been speculation that organic acids (malic, citric and oxalic acids), within the cell wall could significantly complex Ca^{2+} and effect Ca^{2+} crosslinking of the pectin. Of these, citric acid forms the strongest interaction with Ca^{2+}. From published values of the respective stability constants[17] it is possible to estimate the effect of addition of citrate on calcium binding behaviour. In Figure 4 is shown the effect of including citrate, at the levels estimated to be in the cell wall of ripe tomato fruit, with the cell wall hydrated at 3 g/g. The speciation behaviour of Ca^{2+} in the presence of pectate (log K Ca pectate 3.9) is presented. Over the pH range 3–7 the effect on Ca^{2+} pectate formation is very small. The

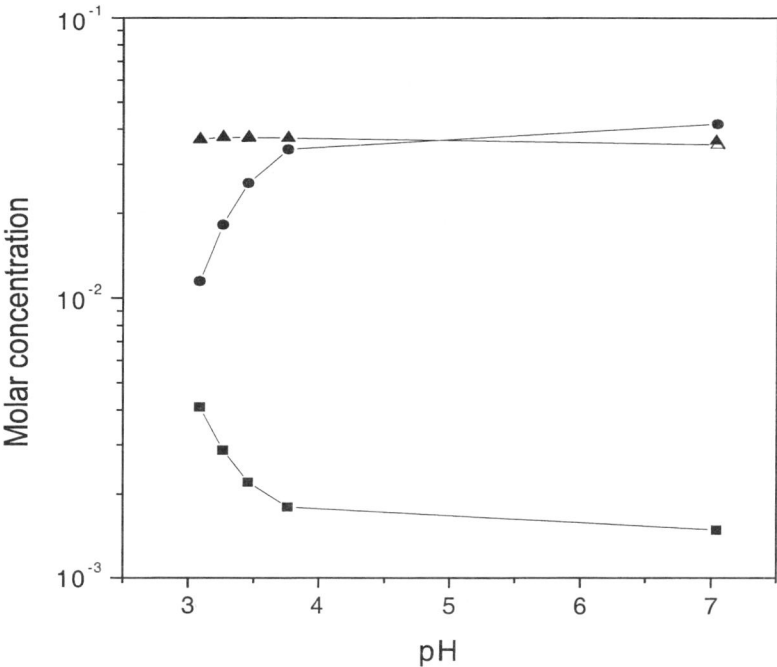

Figure 4 *Predicted speciation of pectate for cell wall concentrations of ions with the addition of citrate at 70 mM. Pectate ($\bullet$), Ca^{2+} ($\blacksquare$), Ca^{2+} pectate ($\blacktriangle$).*

organic acids, at the levels found, are thus predicted to have little effect on the stability of the Ca^{2+} crosslinks at cell wall concentrations of pectin. In Figure 5 is shown corresponding calculation for Ca^{2+} and pectate in the presence of 100 mM CDTA. Over the pH range 3–7 there is an increasing complexation of the Ca^{2+} by the CDTA with very low levels of Ca^{2+} pectate formation at neutral pH. These calculations are consistent with the known behaviour of CDTA as an extractant, including its effectiveness as a function of pH, and as an agent which causes cell separation in tomato.

2.4 Other Crosslinking Agents

Although Ca^{2+} mediated crosslinking of pectins and pectic polysaccharides has been extensively studied, other cationic species have the potential to crosslink these networks.[18] Several different classes of structural proteins have been identified in plant cell walls, and prominent among these are the hydroxyproline-rich plant glycoproteins (HRGPs).[19] HRGPs are basic proteins; they are rich in hydroxyproline and serine, and are glycosylated mainly with arabinose and galactose. Early hypotheses concerning the structural role of HRGPs concentrated on the possibility that covalent crosslinking through tyrosine residues could lead to the formation of a protein network. There is evidence also for covalent linkage between HRGPs and pectin.[20,21] At the same time there has been consistent speculation that HRGPs form ionic complexes with pectins.[19,22,23]

It was possible to form elastic gels between tomato pectin and a variety of different basic compounds including the highly basic peptides, poly-L-lysine (average dp 6 or 19), poly-L-arginine (average dp 56), and a synthetic extensin peptide (a seven amino acid fragment of carrot extensin)[24] containing His-His-Tyr-Lys-Tyr-Lys. All the gels were free standing. At initial levels of addition, the resulting gel was essentially clear and has the ability to recover from small static deformations. At higher levels, (peptide:pectin charge ratio up to 1.37) a marked opacity, indicative of substantial polymer aggregation was apparent, coupled with a marked shrinkage and syneresis of the gel. In all cases gelation could be reversed by leaching the gels under acidic conditions, demonstrating that the interaction was non-covalent, and reversible. The shear modulus (G') of the gels formed with peptides or Ca^{2+} ions was examined as a function of the molar concentration of crosslinking agent (Figure 6). The most effective crosslinking agent (on a molar basis) for inducing gelation was poly-L-arginine, followed by poly-L-lysine, extensin peptide and finally Ca^{2+} ions. These data reflect the different affinities of the crosslinking agent for the pectin chain and suggest that while a calcium crosslink is a cooperative association involving a number of calcium ions, a single multiply charged peptide can also function as a crosslink. However, the maximum level of crosslinking in the peptide/pectin gels, as indicated by the shear modulus ($\sim 1100 \, Nm^{-2}$, Figure 6), is low compared to the level found in Ca^{2+}/pectin gels prepared under identical conditions at saturating levels of calcium ions ($2500 \, Nm^{-2}$). This suggests that for the peptides there are a limited number of potential crosslinking sites on the pectin backbone. The

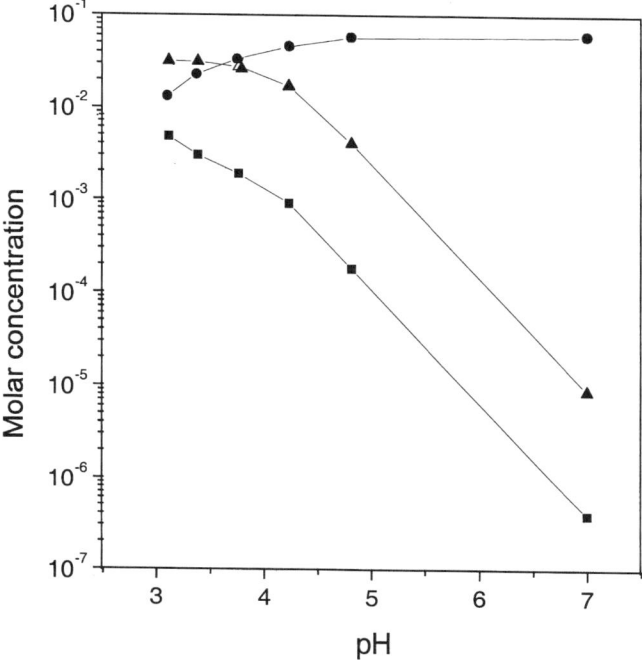

Figure 5 *Predicted speciation of pectate for cell wall concentrations of ions with the addition of 100 mM CDTA. Pectate (●), Ca^{2+} (■), Ca^{2+} pectate (▲)*

extensin peptide induced a higher degree of crosslinking than poly-L-lysine dp 6, despite having roughly half the charge. These experiments demonstrate that basic peptides can act as crosslinking agents for the pectin network. These peptides have a sufficiently high affinity for the charged pectin backbone that moderately concentrated solutions of the pectin (1 to 2% w/w) can be induced to form gels. At much higher pectin concentrations there is the potential for more weakly interacting cations to act as crosslinkers. A key parameter in the description of the behaviour of the pectin network *in vivo*, is the control of the hydration and swelling of the pectin network.

3 Network Swelling

The physicochemical basis of the hydration of polymer networks has been the subject of continuing study.[25-27] For neutral polymer networks hydration can be characterised in terms of a single parameter describing the affinity of polymer and solvent.[28] At high polymer concentrations this contribution to network hydration can be large. However, the available literature on the hydration behaviour of synthetic polymers suggests that in the plant cell wall a more important potential contribution to hydration comes from the polyelectrolyte

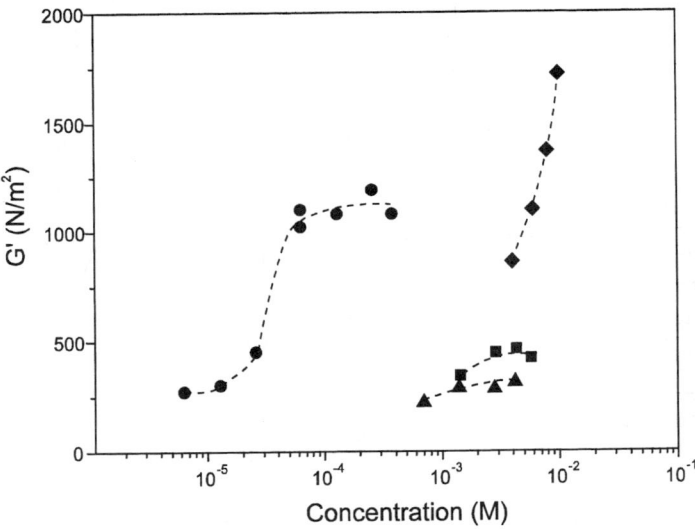

Figure 6 *Shear modulus as a function of the molar concentration of added compound for gels formed with pectin (1.3% w/w) and different cations: (●) poly L-arginine dp 56; (▲) poly L-lysine dp 6; (■) extensin peptide; (♦) Ca²⁺*

characteristic of the pectin network. For polyelectrolyte gels, the requirement for electrical neutrality leads to an excess of counterion within the gel compared to the external medium.[28] This excess generates an osmotic pressure difference between the gel and external medium, which increases with decreasing ionic strength. The excess osmotic pressure leads to the swelling of the gel until a balance is achieved between the osmotic pressure which drives swelling and the restorative force arising from the deformation of the crosslinked network.[28] At intermediate salt concentrations an estimate of the contribution to osmotic pressure, π, due to a polyelectrolyte may be obtained from[26,28,29]

$$\pi \approx \frac{RTc^2}{A(c + 4Ac_s)} \tag{1}$$

for univalent electrolytes, where c and c_s are the molar concentrations of polymer segment and salt, and A is the number of monomers between effective charges. The greater the charge on the polymer, and the lower the ionic strength, the greater the osmotic pressure generated. However, at high charge densities, the phenomenon of counterion condensation can reduce the counterion fraction which can contribute to this Donnan effect.[30,31] In pectin networks, where the crosslinks are formed by specific ionic complexation, a further proportion of the ionisable residues on the polyelectrolyte is excluded from contributing to the generation of the hydration force. Where pectin networks are crosslinked by interaction with Ca²⁺ ions, the galacturonate sequences play a dual role. They contribute to swelling when ionised, but when involved in Ca²⁺ mediated crosslinking they no longer contribute to swelling and instead play a role in

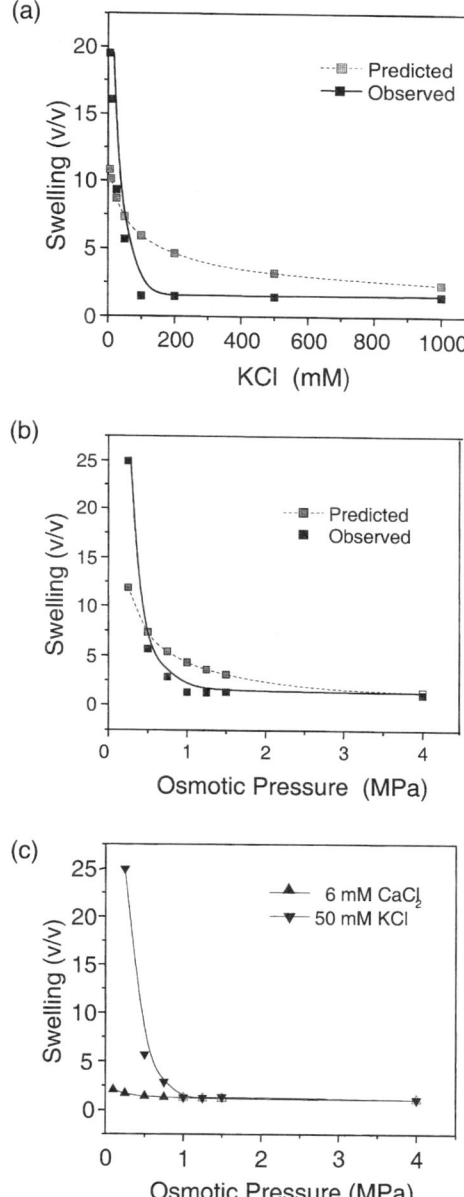

Figure 7 *Swelling of a tomato pectin film (a) as a function of ionic strength at an osmotic stress of 0.5 MPa; (b) as a function of osmotic stress in 50 mM KCl; (c) as a function of osmotic stress comparing K^+ and Ca^{2+} counterions*

resisting network expansion. These effects are illustrated in Figure 7 which shows the effects of salt concentration, cation and osmotic stress[32] on the swelling behaviour of an isolated tomato pectin film. In Figure 7a is shown the increased swelling which is observed on decreasing ionic strength. For comparison, the behaviour predicted by the application of Equation (1) is also shown. In Figure 7b the salt concentration is held constant at 50 mM and an increased swelling is observed on decreasing osmotic stress. As before, the predicted behaviour is shown for comparison. Both of these figures are illustrative of a general polyelectrolyte contribution to swelling, with Equation (1) correctly predicting the form of the increase in swelling although not the detail. The effect of changing the counterion is illustrated in Figure 7c where the swelling of the pectin as a function of osmotic stress in 50 mM KCl and 6 mM $CaCl_2$ is compared. The calcium crosslinked film is much more resistant to swelling at low osmotic stresses. Mg^{2+} counterions similarly inhibit swelling at low osmotic stresses at a concentration of 50 mM.

These data on isolated tomato pectin films show that the osmotic stress that the cell wall and middle lamella is exposed to can have a potentially large impact on cell wall swelling.[33] To test this proposal, the swelling of isolated cell wall material in 50 mM KCl was examined as a function of osmotic stress.[34] The cell wall swelling falls from 8 g/g to 3 g/g as the osmotic stress is increased to 0.15 MPa (Figure 8). The osmotic stress being exerted on the cell wall by solutes contained within plant cells can be determined from microprobe measurements of the hydrostatic (turgor) pressure in individual cells.[35] Reported values for the expanding cells of higher plants are generally in the range 0.1 to 1 MPa. For mature green tomato fruit pericarp cells the turgor pressure has been estimated at 0.14 MPa, falling to 0.03 MPa[36] as the fruit ripen – the decrease being associated with failure of the cell membranes to continue to act as an effective barrier to the movement of solutes. These observations suggest that during

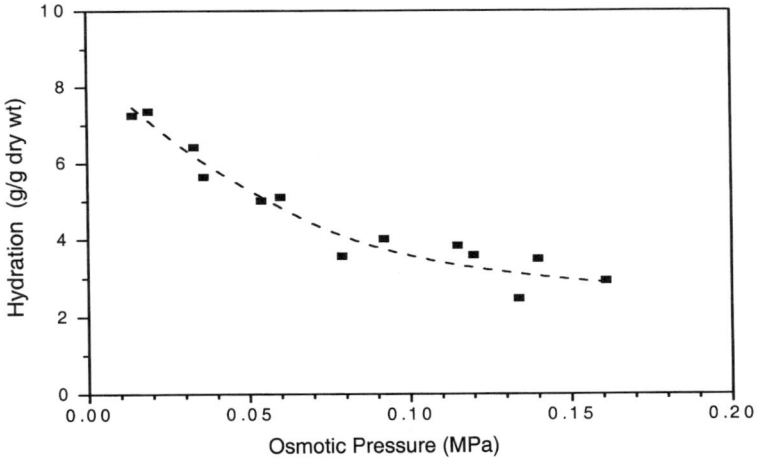

Figure 8 *The effect of externally applied osmotic pressure on the hydration of tomato cell walls*

ripening the osmotic stress that the cell wall is exposed to will fall with a consequent potential increase in cell wall swelling and change in cell wall properties.

4 Conclusions

In unripe tomato fruit the pectin network of the middle lamella is held together by ionic interactions involving Ca^{2+} ions. If it is assumed that the interaction of Ca^{2+} with pectin in the network can be adequately described by a single stability constant the ionic equilibria in the apoplast can be calculated using a speciation approach. From the known composition of the apoplast and tomato pectin it is predicted that a fraction of the charged galacturonosyl residues in pectin are involved in crosslinking, while another fraction has more loosely associated cations which can contribute to the hydration of the pectin network through a Donnan effect. In addition to Ca^{2+} crosslinking, the pectin network can be crosslinked by basic peptides including a fragment of the cell wall protein extensin. This suggests that *in vivo* the extensin may exert a structural effect through this interaction. The hydration of the pectin network is strongly influenced by the osmotic stress to which it is exposed. At the concentrations of pectin found in the plant cell wall under osmotic stresses found *in vivo*, weaker ionic interactions such as those involving Mg^{2+} counterions have the potential to crosslink the network.

References

1. N. C. Carpita and D. M. Gibeaut, *Plant J.*, 1993, **3**, 1.
2. H. A. Schols and A. G. J. Voragen, *Carbohydr. Res.*, 1994, **256**, 83.
3. H. A. Schols, A. G. J. Voragen and I. J. Colquhoun, *Carbohydr. Res.*, 1994, **256**, 97.
4. L. Saulnier and J.-F. Thibault, *J. Sci. Food Agric.*, 1999, **79**, 396.
5. J. A. Brown and S. C. Fry, *Plant Physiol.*, 1993, **103**, 993.
6. T. Ishii, T. Matsunaga, P. Pellerin, M. A. O'Neill, A. Darvill and P. Albersheim, *J. Biol. Chem.*, 1999, **274**, 13098.
7. R. Kohn, *Pure Appl. Chem.*, 1975, **42**, 371.
8. C. Garnier, M. A. V. Axelos and J.-F. Thibault, *Carbohydr. Res.*, 1994, **256**, 71.
9. E. R. Morris, D. A. Powell, M. J. Gidley and D. A. Rees, *J. Mol. Biol.*, 1982, **155**, 507.
10. A. J. MacDougall, P. W. Needs, N. M. Rigby and S. G. Ring, *Carbohydr. Res.*, 1996, **293**, 235.
11. M. C. Jarvis and D. C. Apperley, *Carbohydr. Res.*, 1995, **275**, 131.
12. A. J. MacDougall, N. M. Rigby and S. G. Ring, *Plant Physiol.*, 1997, **114**, 353.
13. C. W. Tibbits, A. J. MacDougall and S. G. Ring, *Carbohydr. Res.*, 1998, **310**, 101.
14. C. Grignon and H. Sentenac, *Ann. Rev. Plant Physiol. Plant Mol. Biol.*, 1991, **42**, 102.
15. Y.-L. Ruan, J. W. Patrick and C. J. Brady, *Aust. J. Plant Physiol.*, 1996, **23**, 9.
16. A. J. MacDougall, R. Parker and R. R. Selvendran, *Plant Physiol.*, 1995, **108**, 1679.
17. R. M. Smith and A. E. Martell, *Critical Stability Constants*, Plenum Press, New York, 1976.
18. S. Bystrický, A. Malovíková and T. Sticzay, *Carbohydr. Polym.*, 1990, **13**, 283.
19. A. M. Showalter, *Plant Cell*, 1993, **5**, 9.

20. X. Qi, B. X. Behrens, P. R. West and A. E. Mort, *Plant Physiol.*, 1995, **108**, 1691.
21. J. D. Brady, I. H. Sadler and S. C. Fry, *Biochem. J.*, 1996, **315**, 323.
22. J. Sommer-Knudsen, A. Bacic and A. E. Clarke, *Phytochemistry*, 1998, **47**, 483.
23. M. J. Kieliszewski and D. T. A. Lamport, *Plant J.*, 1994, **5**, 157.
24. J. Chen and J. E. Varner, *EMBO J.*, 1985, **4**, 2145.
25. A. V. Dobrynin, R. H. Colby and M. Rubinstein, *Macromolecules*, 1995, **28**, 1859.
26. R. Skouri, F. Schosseler, J. P. Munch and S. J. Candau, *Macromolecules*, 1995, **28**, 197.
27. M. Rubinstein, R. H. Colby, A. V. Dobrynin and J.-F. Joanny, *Macromolecules*, 1996, **29**, 398.
28. P. J. Flory, *Principles of Polymer Chemistry*, Cornell University Press, 1953.
29. J.-L. Barrat and J.-F. Joanny, *Adv. Chem. Phys.*, 1996, **94**, 1.
30. G. S. Manning, *Berenges Phys. Chem.*, 1996, **100**, 923.
31. G. S. Manning and J. Ray, *J. Biomol. Struct. Dynamics*, 1998, **16**, 461.
32. V. A. Parsegian, R. P. Rand, N. L. Fuller and D. C. Rau, *Methods Enzymol.*, 1986, **127**, 400.
33. P. Ryden, A. J. MacDougall, C. W. Tibbits and S. G. Ring, *Biopolymers*, 2000, **54**, 398.
34. A. J. MacDougall, N. M. Rigby, P. Ryden, C. W. Tibbits and S. G. Ring, *Biomacromolecules*, 2001, **2**, 450.
35. A. D. Tomos and R. A. Leigh, *Ann. Rev. Plant Physiol. Plant Mol. Biol.*, 1999, **50**, 447.
36. K. A. Shackel, C. Greve, J. M. Labavitch and H. Ahmadi, *Plant Physiol.*, 1991, **97**, 814.

Self-assembly of Acacia Gum and β-Lactoglobulin in Aqueous Dispersion

C. Sanchez,[1] C. Schmitt,[1,2] G. Mekhloufi,[1] J. Hardy,[1] D. Renard[3] and P. Robert[3]

[1]LABORATOIRE DE PHYSICO-CHIMIE ET GÉNIE ALIMENTAIRES, ENSAIA-INPL, BP172, 54505 VANDOEUVRE-LÈS-NANCY CEDEX, FRANCE
[2]PRESENT ADRESS: NESTLÉ RESEARCH CENTER, DEPARTMENT OF FOOD SCIENCE AND PROCESS RESEARCH, VERS-CHEZ-LES-BLANC, CH–1000 LAUSANNE 26, SWITZERLAND
[3]UNITÉ DE PHYSICO-CHIMIE DES MACROMOLÉCULES, INRA, BP 71627, 44316 NANTES CEDEX 3, FRANCE

1 Introduction

Complex coacervation is a liquid–liquid or, in some cases, liquid–solid phase separation which may be described basically by the formation of primary soluble macromolecular complexes that interact to form electrically neutralised aggregates, then unstable liquid droplets and/or precipitates that ultimately sediment to form the coacervated phase containing both biopolymers.[1–5] From a practical point of view, complex coacervation may be used in microencapsulation processes, in the formation of extracellular matrices for biomaterials design, in biotechnology (treatment of food processing waste and activated sludge, protein purification, biosensor design) and in the creation of new multifunctional food ingredients. A better knowledge of complex coacervation could also provide fundamental informations on important biological self-assembly processes such as DNA–protein interactions, cell cytosol organisation or tissue formation.

A survey of the abundant literature shows clearly that pH, ionic strength, type of ions, protein to polysaccharide ratio, size, shape, charge density and flexibility of macromolecules are important parameters controlling the extent of phase separation as determined at equilibrium.[3,4,6–10] On the contrary, out of equilibrium conditions have attracted much less attention. Some important questions such as the transition from macromolecular complexes/aggregates to the appearance of coacervates, the structure of coacervates and the phase ordering kinetics

111

from their appearance to the equilibrium remain to be elucidated.

A preliminary kinetic approach showed that a complex interplay between growth, coalescence/flocculation and sedimentation of coacervates occurred in the β-lactoglobulin acacia gum system.[11,12] More detailed structural and kinetic informations on this model system are reported in the following using confocal scanning laser microscopy and time resolved small angle static laser light scattering.

2 Materials and Methods

2.1 Materials

Acacia gum sample from *Acacia senegal* trees (lot 97J716) was a gift from the CNI company (Rouen, France). Acid processed β-lactoglobulin (lot 818) was a gift from Lactalis Research Center (Retiers, France). The chemical composition of powders and physico-chemical properties of acacia gum (AG) and β-lactoglobulin (BLG) macromolecules were reported previously.[13–16]

2.2 Preparation of BLG/AG Mixed Dispersions

AG and BLG aqueous stock dispersions at 0.1 wt% or 1 wt% total biopolymer concentration were prepared in deionized water (MilliQ, MilliPore, USA) or bidistilled water, respectively.[12] The pH of the resulting dispersions was adjusted to 3.6 or 4.2 using HCl or NaOH. BLG/AG dispersions at protein to polysaccharide weight ratios (Pr:Ps) of 1:1 or 2:1 were obtained by gently mixing stock dispersions. For light scattering experiments, stock dispersions were filtered through 0.22 μm Millex microfilters (MilliPore, USA) before mixing.

2.3 Confocal Scanning Laser Microscopy (CSLM)

The structure of BLG/AG mixtures at 1 wt% total biopolymer concentration was determined by CSLM. The procedure for BLG and AG labelling, respectively by FITC and RITC, and the technical set-up were described previously.[12] The focal plane of observation was about 30 μm from the inverted objective from the microscope. Pictures were processed ∼ 160 s after mixing stock dispersions using the Laser Sharp MRC–1024 software version 3.2 (Bio-Rad, Germany). In order to better visualize microstructures, pictures were transformed in negative images and brightness and contrast were optimized using the imagoWeb freeware version 1.1.[17] All experiments were duplicated using freshly prepared dispersions.

2.4 Small Angle Static Laser Light Scattering (SALS)

Phase ordering kinetics of BLG/AG mixtures at 0.1 wt% total biopolymer concentration, 2:1 Pr:Ps weight ratio and pH 4.2 were followed during 90 min using SALS. Light scattering experiments were carried out using a Mastersizer S long bench (Malvern Ltd, UK). A He–Ne laser light ($\lambda = 0.6334$ μm) was passed through a 0.5 mm width measurement cell (Malvern Ltd) in which the BLG/AG mixture was injected using a syringe. The beam was converged by a Reverse Fourier 300 mm focussing lens. 42 Detectors, covering a scattering wave vector Q range of 8×10^{-3} to 10 μm^{-1}, collected the scattered light. Experimental scattering intensity was obtained every 30 s (sample time: 10 s) by substracting the background intensity (cell + filtered MilliQ water) from the raw scattered intensity of the sample. In addition, corrections were applied to take into account the geometry of the detectors. The resulting scattered intensity was further corrected for turbidity.[18] Experiments were duplicated using freshly prepared dispersions.

3 Results and Discussion

3.1 Confocal Scanning Laser Microscopy (CSLM)

The microstructure of BLG/AG dispersions at 1:1 or 2:1 Pr:Ps weight ratios, pH 3.6 or 4.2 is shown Figure 1. A first look at the micrographs clearly shows that both pH and Pr:Ps weight ratio have a pronounced influence on the initial microstructure of mixed dispersions. At 1:1 Pr:Ps, a great number of polydispersed coacervates appeared. The apparent diameters (d_{app}) of coacervates ranged

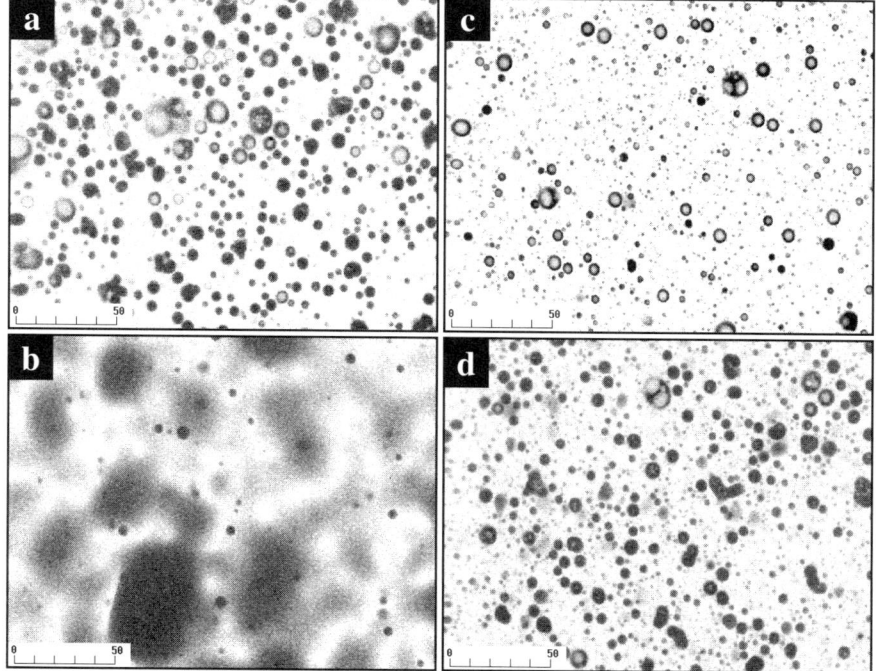

Figure 1 *CSLM micrographs obtained on BLG/AG dispersions at 1 wt% total biopolymer concentration, pH 3.6 (a, b) or 4.2 (c, d) and Pr:Ps weight ratios of 1:1 (a, c), 2:1 (b,d). Scale bar refers to 50 μm*

from 1 to 30 μm at pH 3.6 and from 1 to 20 μm at pH 4.2 (Figure 1a, c). Most coacervates displayed a d_{app} in the range 5–10 μm and 1–5 μm at pH 3.6 and 4.2, respectively. This is a clear indication that BLG/AG dispersions at pH 3.6 were less stable than at pH 4.2. A likely reason is that coacervates in the former case were electrically neutral, which favours flocculation/coalescence phenomena, whereas they were negatively charged in the latter case, which favours particle stabilization through charge repulsion.[14] An interesting feature was the presence of vesicular or multivesicular coacervates for both pH (Figure 1a, c). Similar structures have been observed in a preliminary CSLM study.[12] The origin of such structures remains unclear. However, very similar giant vesicles induced by self-assembly of polymers have been shown by CSLM.[19] One hypothesis would be the micellization of a macrosurfactant formed by interactions between AG and BLG. It is possible that the high molecular mass surface-active Arabinogalactan-Protein (AGP) component of AG plays a key role in the appareance of vesicles. We plan in the future to verify the latter hypothesis by separating the different molecular fractions of AG.

At 2:1 Pr:Ps ratio, BLG/AG dispersions at pH 3.6 were less stable than at pH 4.2 (Figure 1b, d). A rapid sedimentation of insoluble coacervates onto the observation slide was the result of this instability (dark areas in Figure 1b). At pH 4.2, vesicular coacervates were present and a number of flocculation/coalescence phenomena between coacervates were visible. As a general trend, BLG/AG dispersions at 2:1 Pr:Ps ratio were less stable than at ratio 1:1. For instance, size distribution of coacervates was in the range 1–5 μm at ratio 1:1 and rather in the range 5–10 μm at ratio 2:1. This may be explained by the better charge neutralization of coacervates in the latter case.[14]

3.2 Small Angle Static Light Scattering (SALS)

When a binary mixture is quenched in the unstable coexistence region, by thermal treatment or pressure changes or simply by mixing the two macromolecules, fluctuations in density grow with time, and finally result in a complete phase separation. The dynamics of phase separation are generally divided into early, intermediate and late stages. The different stages can be described by means of the temporal evolution of the scattered intensity function.[20–22]

The scattered intensity function initially displayed a correlation peak at a correlation length Λ ($\Lambda = 2\pi/Q_{max}$ where Q_{max} is the scattering wave vector corresponding to the maximum scattered intensity I_{max}) of 19 μm (Figures 2, 3). In the time interval between 0 and 1350 s, Q_{max} shifted towards smaller Q values indicating the growth of structural domains. The presence of a correlation peak is generally ascribed to the spinodal decomposition phase separation mechanism.[20,21] This feature can be observed for the nucleation and growth mechanism when numerous particles or when depleted areas around growing particles are present.[22,23] The evolution of the correlation length Λ is shown in Figure 3. Following a short lag time, Λ increased from 15–19 μm (t = 0 s) up to 45 μm (t = $\sim$ 700–800 s). After $\sim$ 700–800 s, the correlation length reached a constant value but the system coarsening continued, in agreement with the increase of

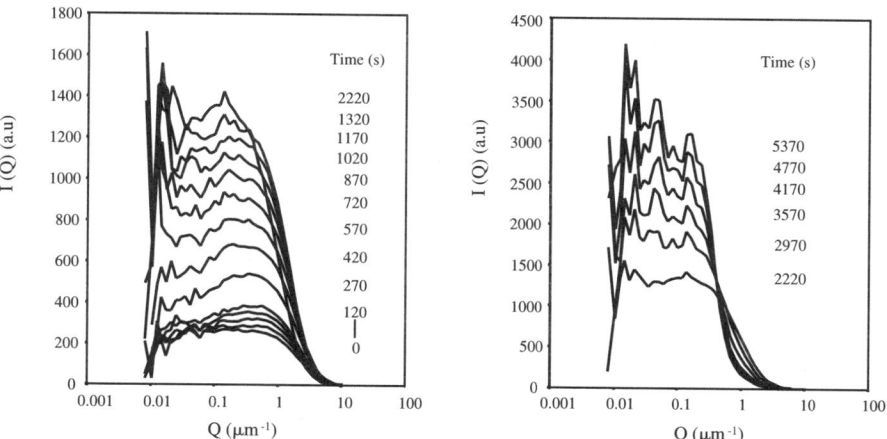

Figure 2 *Temporal evolution of scattered intensity functions [I(Q) vs Q] as determined by SALS on BLG/AG dispersions at 0.1 wt% total biopolymer concentration, pH 4.2 and Pr:Ps weight ratio of 2:1*

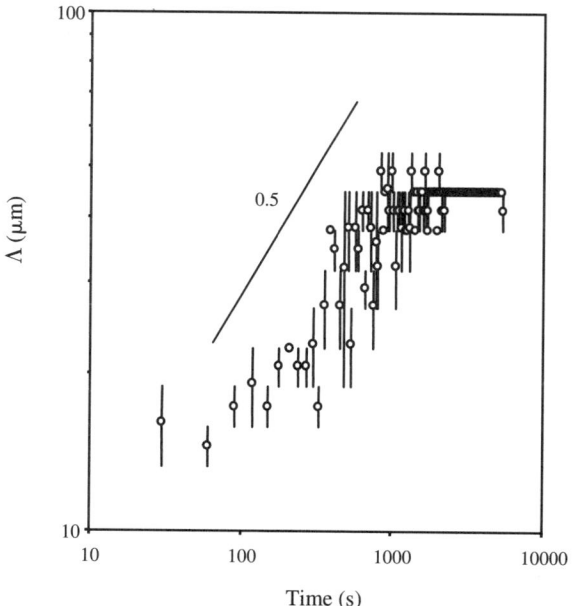

Figure 3 *Temporal evolution of the correlation length Λ (μm) as determined by SALS on BLG/AG dispersions at 0.1 wt% total biopolymer concentration, pH 4.2 and Pr:Ps weight ratio of 2:1. Bars are standard deviation based on duplicate experiments. Drawn line is a power-law function with an exponent of 0.5*

scattered intensity at small Q (Figure 2). The increase of Λ as a function of time followed a power-law relationship $Q_{max} \sim t^{-\alpha}$ with α equal to ~ 0.4–0.5 (Figure 3). This value is intermediate between that obtained for diffusion-induced coalescence ($\alpha = 0.33$) and that obtained for hydrodynamics-induced coarsening ($\alpha =$

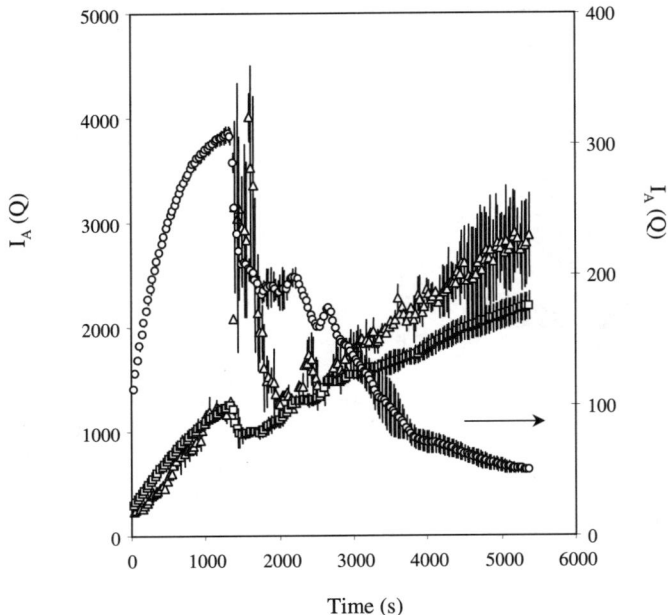

Time (s)

Figure 4 *Temporal evolution of scattered intensities $I_A(Q)$ averaged according to three Q ranges to distinguish large, medium and small sized particles as determined by SALS on BLG/AG dispersions at 0.1 wt% total biopolymer concentration, pH 4.2 and Pr:Ps weight ration of 2:1. [△]: large particles (Q: 0.008–0.042 μm^{-1}), [□]: medium particles (Q: 0.05–0.67 μm^{-1}), [○]: small particles (Q: 0.8–10.4 μm^{-1}). Bars are standard deviation based on duplicate experiments*

1), indicating a correlation between growth mechanisms.[24]

Coarsening kinetics can be better visualized by averaging the scattered intensities according to three ranges of Q. The three ranges were assigned to large, medium and small sized particles (Figure 4). Following a first phase (0–1320 s) where an increasing number of polydispersed droplets appeared, a sudden increase of large particles was observed in parallel with a decrease of medium and small particles (1350–1380 s). This large-scale coalescence/flocculation process could be at the origin of the formation of multivesicular structures. In a second step, large particles disappeared due to gravity effects (sedimentation) or adsorption onto glass walls of optical cell or collapse of structures. A transition regime appeared between ∼ 1800 and ∼ 2300 s where the number of large particles still decreased but the number of medium and small particles increased again. The observation was important since it suggests that different coarsening rates exist in the system, probably stemming from the different molecular masses or charge densities of AG molecular fractions. It is known, for instance, that large polymers induced faster phase separation than smaller ones.[3] Finally, small particles decreased to a large extent to the benefit of medium and large particles that grew continuously.

As the shape of the different scattered intensity functions and the power-law growth of structural domains suggested that the system was located in the late stage regime of the phase separation, SALS data were scaled by plotting I/I_{max} *vs* Q/Q_{max}.[23] Unfortunately, scaled data did not overlap into a single master curve (results not shown). Possible explanations would be that more than one length scale occurred in the phase separating system (effect of polydispersity) or that the final stage characterized by sharp interfaces was not reached. According to the Porod law, scattering arising from sharp interfaces should produce a high-Q tail decreasing as Q^{-4}.[24] In our system, the scaling law exponent varied with time from 2.9 to 3.6, indicating that coarsening of droplets with time is coupled with a reorganisation of droplets interfaces.

4 Conclusion

Complex coacervation between BLG and AG induced the formation of supra-molecular structures such as droplets, vesicles or multihollow spheres. SALS experiments demonstrated the presence of a decomposition process with correlated growth mechanisms. Coarsening kinetics were mainly characterized by an interplay between initial particles growth, large-scale flocculation/coalescence and transient growth of small particles. Polydispersity of particles size and roughness of interfaces precluded from applying dynamic scaling of the structure function.

Acknowledgements

Many thanks are due to Prof. C. G. de Kruif (NIZO, The Netherlands), Prof. C.-M. Lehr and Dr A. Lamprecht (Sarrebrücken University, Germany) for their direct or indirect contributions to this study.

References

1. H. G. Bungenberg de Jong, *Colloid Science*, H. R. Kruyt (ed.), Elsevier, Amsterdam, 1949, Vol. 2, p. 233.
2. V. B. Tolstoguzov, *Food Proteins and their Applications*, S. Damodaran and A. Paraf (eds.), Marcel Dekker, New York, 1997, Chapter 6, p. 171.
3. C. Schmitt, C. Sanchez, S. Desobry-Banon and J. Hardy, *Crit. Rev. Food Sci. Nutr.*, 1998, **38**, 689.
4. K. W. Mattison, Y. Wang, K. Grymonpré and P. L. Dubin, *Macromol. Symp.*, 1999, **140**, 53.
5. K. Kaibara, T. Okazaki, H. B. Bohidar and P. L. Dubin, *Biomacromolecules*, 2000, **1**, 100.
6. V. B. Tolstoguzov, *Food Hydrocoll.*, 1991, **4**, 429.
7. D. J. Burgess, *Macromolecular Complexes in Chemistry and Biology*, P. L. Dubin, J. Bock, R. Davis, D. N. Schulz and C. Thies (eds.), Springer Verlag, Berlin, 1994, Chapter 17, p. 281.
8. J. Xia and P. L. Dubin, *Macromolecular Complexes in Chemistry and Biology*, P. L. Dubin, J. Bock, R. Davis, D. N. Schulz and C. Thies (eds.), Springer Verlag, Berlin, 1994, Chapter 15, p. 247.
9. L. Piculell, K. Bergfeldt and S. Nilsson, *Biopolymer Mixtures*, S. E. Harding, S. E. Hill

and J. R. Mitchell (eds.), Nottingham University Press, Nottingham, 1995, Chapter 2, p. 13.

10. J.-L. Doublier, C. Garnier, D. Renard and C. Sanchez, *Curr. Opin. Colloid Interf. Sci.*, 2000, **5**, 184.

11. C. Sanchez, S. Despond, C. Schmitt and J. Hardy, *Food Colloids: Fundamentals of Formulation*, E. Dickinson and R. Miller (eds.), Royal Society of Chemistry, Cambridge, 2001, p. 332.

12. C. Schmitt, C. Sanchez, A. Lamprecht, D. Renard, C. M. Lehr, C. G. de Kruif and J. Hardy, *Coll. Surf. B: Biointerf.*, 2001, **20**, 267.

13. C. Schmitt, PhD Thesis, INPL, Vandœuvre-lès-Nancy, France, 2000.

14. C. Schmitt, C. Sanchez, F. Thomas and J. Hardy, *Food Hydrocoll.*, 1999, **13**, 483.

15. C. Schmitt, C. Sanchez, S. Despond, D. Renard, F. Thomas and J. Hardy, *Food Hydrocoll.*, 2000, **14**, 403.

16. C. Sanchez, D. Renard, P. Robert, C. Schmitt and J. Lefebvre, *Food Hydrocoll.*, 2001, (submitted).

17. http://fabrizio.jth.it

18. T. Hashimoto, M. Itakura and H. Hasegawa, *J. Chem. Phys.*, 1986, **85**, 6118.

19. F. Ilhan, T. H. Galow, M. Gray, G. Clavier and V. M. Rotello, *J. Am. Chem. Soc.*, 2000, **122**, 5895.

20. F. Mallamace and N. Micali, *Light Scattering – Principles and Development*, W. Brown (ed.), Clarendon Press, Oxford, 1996, Chapter 12, p. 381.

21. J. K. G. Dhont, *J. Chem. Phys.*, 1996, **105**, 5112.

22. K. Binder, *Materials Science and Technology – A Comprehensive Treatment*, R. W. Cahn, P. Haasen and E. J. Kramer (eds.), VCH, Weinhem, 1991, Vol. 5, Chapter 7, p. 405.

23. J. Maugey, T. van Nuland and P. Navard, *Polymer*, 2001, **42**, 4353.

24. H. Furukawa, *Adv. Phys.*, 1985, **34**, 703.

Creation of Biopolymeric Colloidal Carriers Dedicated to Controlled Release Applications

Denis Renard,[1] Paul Robert,[1] Laurence Lavenant,[1] Dominique Melcion,[1] Yves Popineau,[1] Jacques Guéguen,[1] Cécile Duclairoir,[2] Evelyne Nakache,[2] Christian Sanchez[3] and Christophe Schmitt[4]

[1]INRA CENTRE DE RECHERCHES DE NANTES, RUE DE LA GERAUDIERE, BP 71627 44316 NANTES CEDEX 3, FRANCE
[2]EQUIPE POLYMERES INTERFACES LCMT, UMR 6507 ISMRA, 14050 CAEN CEDEX, FRANCE
[3]LPCGA ENSAIA/INPL, 54505 VANDOEUVRE-LES-NANCY, FRANCE
[4]NESTLE RESEARCH CENTER, CH–1000 LAUSANNE 26, SWITZERLAND

1 Introduction

In the last decade, micro and nanosized colloidal carriers have received a growing scientific and industrial interest.[1] These vectors may be capsules (with liquid core surrounded by a solid shell), particles (polymeric matrices), vesicles or liposomes, or multiple or single emulsions, and have found a wide range of applications. They may be loaded by living cells, enzymes, flavour oils, pharmaceuticals, vitamins, adhesives, agrochemicals or catalysts and offer considerable advantages in use. Liquids can be handled as solids, odour or taste can be effectively masked in a food product, sensitive substances can be protected from deleterious effects of the surrounding environment, toxic materials can be safely handled, and drug delivery can be controlled and targeted.[2]

In the forementioned laboratories, we started with a new strategy based on phase separation in order to prepare natural particles. Simple or complex coacervation methods involving proteins or protein and polysaccharide mixtures[3] were used to create new matrices dedicated to controlled release applications. The colloidal carriers produced were in the micrometre or nanometre size range depending on the substrates or the methods used. Wheat proteins, gliadins, were implicated in simple coacervation to produce nanospheres. Controlled release experiments with model compounds were conducted in order to evaluate the performance of such matrices. In the case of complex coacervation, the β-lactoglobulin/Arabic gum couple was tested. The mechanism of formation and the structural properties of coacervates were first highlighted in order to better control the stability of these systems.

An alternative and promising research field deals with particles obtained from hydrogel systems. The hydrogels may be sensitive to environmental stimuli such as pH, ionic strength, electric/magnetic fields, light and temperature depending on the substrate used. Totally transparent solid matrices resulting from the dehydration of new protein gels, revealed variable swelling capacities that depend on the solvent used and physical chemical conditions. A dispersion/gelation method was developed to produce micro-beads having potential applications in the encapsulation field.

2 Materials and Methods

2.1 Gliadins Nanoparticles

Nanoparticles were obtained by desolvation of the protein using physiological salt solution as non solvent. Synperonic PE F68 was used to stabilize the nanoparticle suspension.[4] Vitamin E-loaded nanoparticles (1 mg) were digested in 10 ml of an ethanol/water mixture (62/38 v/v) at room temperature in the dark. The vitamin E concentration encapsulated in nanoparticles (C_1) and the residual concentration in the supernatant (C_2) were then assayed by HPLC at 290 nm. Empty nanoparticles were treated in the same way and used as references for these determinations. The drug loading (rate) and the entrapment efficiency were calculated according to the equations:

$$\text{Rate}\,(\%) = \frac{C_1}{m_{\text{gliadins}}} \qquad \text{Efficiency}\,(\%) = \frac{C_1}{C_1 + C_2}$$

In vitro drug release kinetic was also performed in non-sink conditions using decane as solvent (to prevent drug loss from nanoparticles dissolution) and a laboratory designed release cell. About 10 mg of gliadins nanoparticles (containing 824 μg/g gliadins) were resuspended in 10 ml of decane. Aliquots were collected at successive time intervals and replaced by the same quantity of solvent in order to get a constant volume in the release cell. The samples were analysed by HPLC as described above for encapsulation experiments.

2.2 β-Lactoglobulin/Arabic Gum Coacervates

Coacervation process was established at pH 4.2 in water where the two biopolymers interact electrostatically.[5] The structure of the coacervates was explored using both phase contrast and confocal scanning laser microscopies.

2.3 β-Lactoglobulin Hydrogels

The protein hydrogels were formed at pH 8 in 50% ethanol solutions.[6] Swelling

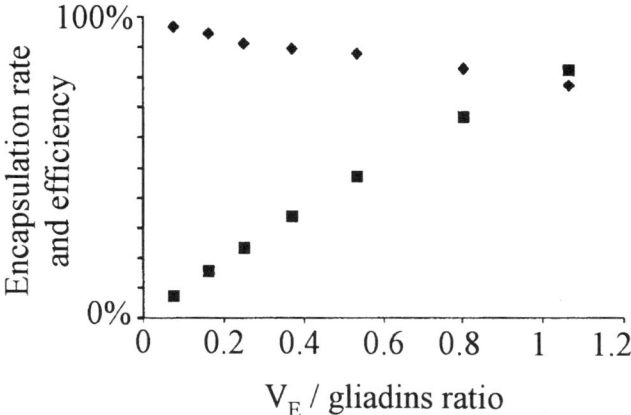

Figure 1 *Vitamin E (V_E) encapsulation rate (■) and efficiency (◆) by gliadin nanospheres as a function of V_E/gliadins ratio*

kinetics of cylindrical gels in both hydrated ($m_0 \sim 4.5$ mg) and dehydrated ($m_0 \sim 1.5$ mg) forms were followed in different solvent conditions.

A new method was also developed based on the dispersion of a β-lactoglobulin pre-gel in an apolar phase in order to produce gelled droplets. These droplets were then washed and dehydrated under vacuum in order to produce particles of 500 μm mean diameter.

3 Results and Discussion

Gliadin nanospheres typical size was around 900 nm.[4] Particles size and polydispersity increased with the increase of the solvent/non solvent ratio and with the aggregated state of the proteins. Encapsulation of vitamin E (V_E) into gliadin nanospheres revealed that the rate and efficiency decreased with the decrease of the V_E/gliadins ratio (Figure 1). For a V_E/gliadins ratio of one, an encapsulation rate of 824 μg/mg gliadins and an efficiency of 77% were obtained. The V_E release kinetic from loaded particles is displayed in Figure 2. The experimental points were fitted with the following equations:

$$\frac{V_E}{\text{initial}V_E} = 6\sqrt{\frac{\tau}{\pi}} \text{ (short time)} \qquad \frac{V_E}{\text{initial}V_E} = 6\sqrt{\frac{\tau}{\pi}} - 3\tau \text{(intermediate time)}$$

with $\tau = \dfrac{Dt}{r^2}$, D being the diffusion coefficient of the entrapped vitamin E and r, the radius of the nanoparticles.

The kinetic profile was thus interpreted by a burst effect coupled with a drug diffusion process through the particle modelled as a homogeneous sphere. The drug diffusion coefficient was considerably reduced when entrapped in the carrier: $D_{VE} = 1.1 \; 10^{-20} \text{ cm}^2 \text{ s}^{-1}$ compare to $D_{VE} = 10^{-9} \text{ m}^2 \text{ s}^{-1}$ in solution. The encapsulation of different drugs into nanospheres showed that carriers had

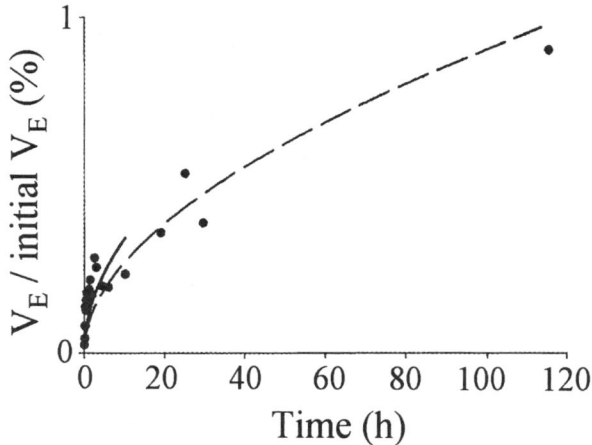

Figure 2 *Vitamin E (V_E) release kinetic by gliadin nanospheres as a function of time (h). See text for equations used in the fitting procedure: (●) released V_E (%); (——) fit (short time); (----) fit (intermediate time)*

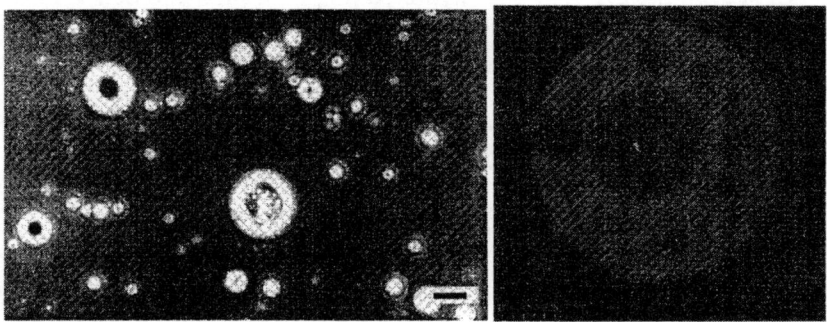

Figure 3 (Left) *Phase contrast optical micrograph of 1wt% β-lactoglobulin/Arabic gum mixtures at pH 4.2 ratio 1:1. Scale bar represents 20 μm.* (Right) *Confocal scanning laser micrograph of a coacervate obtained with a 1wt% β-lactoglobulin/Arabic gum mixture at pH 4.2 ratio 1:1. Scale bar represents 5 μm*

more affinity for hydrophobic drugs and that the ζ-potential of the particles was directly related to the nature of the drug.[7]

β-lactoglobulin/Arabic gum spherical vesicular coacervates revealed by microscopy were the hallmark of these dispersions (Figure 3). Large 'foam-like' coacervates induced by partial coalescence of single coacervates were visible especially at the 2:1 protein to polysaccharide ratio.[8] Increasing dispersions stability was reached by increasing protein to polysaccharide ratio or by decreasing total biopolymer concentration. Another alternative to increase the stability is to produce composite dispersions containing both protein aggregates embedded in protein–polysaccharide coacervates and free coacervates.[5] These systems could thus be used as multifunctional reservoirs for applications in microencapsulation.

The swelling kinetics of β-lactoglobulin hydrogels showed that the increase of mass was the highest in water for both hydrated and dehydrated gels (Figure 4). The capacity of swelling (or not) depended on the solvent conditions and would allow a controlled release of both hydrophilic and hydrophobic drugs. A new method was developed based on the dispersion of a β-lactoglobulin pre-gel in an apolar phase in order to produce gelled droplets. These droplets were then washed and dehydrated under vacuum in order to produce particles of 500 μm mean diameter (Figure 5). Such protein matrices were totally transparent and could be used to encapsulate large molecules or microorganisms.

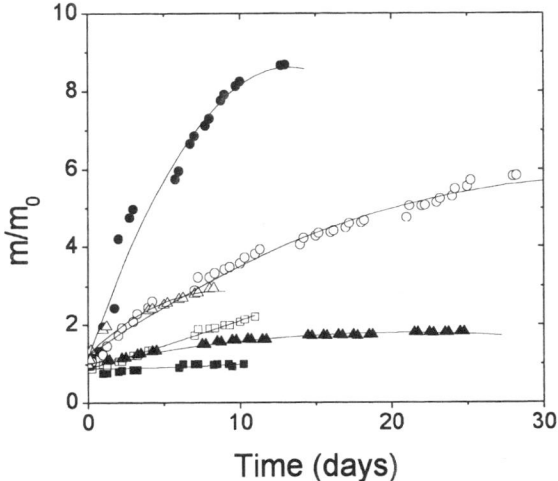

Figure 4 *Swelling kinetics of β-lactoglobulin hydrogels both in dehydrated and hydrated forms for different solvent conditions: (■) ethanol; (□) water/ethanol 50/50 (v/v); (●) water; (○) 0.1M NaCl; (▲) water/ethanol 50/50 (v/v) (hydrated gel); (△) water (hydrated gel)*

Figure 5 *Optical micrograph of β-lactoglobulin beads obtained by a gelation/emulsification method. Scale bar represents 500 μm*

4 Conclusion

Biopolymers represent an interesting alternative to synthetic polymers in order to be used in the route of structured carriers for controlled release and encapsulation applications. In particular, the ability of these carriers to entrap both hydrophilic and hydrophobic drugs may be very promising for many applications. In addition, the absence of chemical compounds and organic solvents used to produce biopolymeric matrices could be very interesting for some industrial applications.

References

1. C. Thies, *Microencapsulation: Methods and Industrial Applications*, S. Benita (ed.), Marcel Dekker, New York, 1996, Vol. 73, Chapter 1, p. 1.
2. J. R. Robinson, *Controlled Drug Delivery: Challenges and Strategies*, K. Park (ed.), American Chemical Society, Washington, 1997, Chapter 1, p. 1.
3. C. Schmitt, C. Sanchez, S. Desobry-Banon and J. Hardy, *Crit. Rev. Food Sci. Nut.*, 1998, **38**(8), 689.
4. C. Duclairoir, E. Nakache, H. Marchais and A.-M. Orecchioni, *Colloid Polym. Sci.*, 1998, **276**, 321.
5. C. Schmitt, C. Sanchez, S. Despond, D. Renard, F. Thomas and J. Hardy, *Food Hydrocolloids*, 2000, **14**, 403.
6. D. Renard, P. Robert, C. Garnier, E. Dufour and J. Lefebvre, *J. Biotechnol.*, 2000, **79**, 231.
7. C. Duclairoir, PhD Thesis, Université de Caen, 2000.
8. C. Schmitt, C. Sanchez, A. Lamprecht, D. Renard, C.-M. Lehr, C.G. de Kruif and J. Hardy, *Colloids Surf. B*, 2001, **20**, 267.

Interfaces, Interphases

Polyelectrolyte–Surfactant Complexes at the Air–Water Interface: Influence of the Polymer Backbone Rigidity

D. Langevin

LABORATOIRE DE PHYSIQUE DES SOLIDES, UNIVERSITÉ PARIS SUD, BÂTIMENT 510, 91405 ORSAY, FRANCE

Abstract

Dilute mixed solutions of non-surface active anionic polymers (polyacrylamide and polystyrene sulfonate, xanthan) and various surfactants have been studied with several methods: surface tension, ellipsometry, X-ray reflectivity, surface rheology and thin film balance. A strong synergistic lowering of the surface tension is found with cationic surfactants in the concentration range where no appreciable complexation of surfactant and polymer occurs in the bulk solution. Despite appreciable differences between surface tension behaviour, the adsorbed layers are very similar for all the polymers: their thickness is small and the polymer chains are stretched along the surface. The behaviour of the linear surface elasticity is also similar, although the non-linear behaviour appears different. When the polymers are confined in thin films, the forces between surfaces are similar and independent of surfactant nature: oscillatory forces are measured, which reflect the existence of a polymer network with a well defined mesh size. The connection of foam stability with surface and bulk complexation is not easy to establish and demands a proper account of the long adsorption kinetics found with these solutions.

1 Introduction

Interactions between surfactants and polymers is now an important field of interest in colloid science.[1] Many aqueous solutions used in industrial applications contain mixtures of surfactants and polymers, in particular polyelectrolytes, which are widely used in water based formulations such as paints, drilling muds, *etc.* An interesting class of natural and biodegradable polymers are the polysaccharides found in plants. Among them, polyelectrolytes with rigid

127

backbone such as xanthan are frequently used, as their solutions in water possess interesting properties like large swelling ability and pronounced shear thinning effects. Polyelectrolyte solutions are less well understood than neutral polymer solutions, although recent work has enhanced the current knowledge.[2] These polymers form more extended structures, with effective persistence lengths much larger than those of neutral polymers.

When two polyelectrolytes of opposite charge are mixed, the two polyions associate, thus releasing the counterions in the solution and increasing the entropy of the solution.[3] It was observed that the behaviour of polyelectrolyte–surfactant solutions is similar to the behaviour of polyelectrolyte–polymer solutions: no association when the surfactant or the polymer are nonionic or when the two species have the same charge and strong association for opposite charges. The case of surfactant–polymer mixtures is however less simple, because the size and the shape of the surfactant aggregates can vary. As for neutral polymers, complexation in bulk occurs at a critical aggregation concentration (CAC) much smaller than the critical micellar concentration (CMC) of the pure surfactant. In the complexes made with intrinsically flexible polymers, surfactant micelles are bound to the polymer chains; their aggregation number is sometimes different from that of the pure micelles, depending on the nature of the polyelectrolyte.[4] In the case of rigid polyelectrolytes, the structure of the complexes appears different.[5]

Polymers and surfactants also form complexes at surfaces, either solid–fluid or fluid–fluid surfaces. Complexation in bulk or at a surface are generally related.[6] Surface complexation is also important for practical applications such as colloidal stabilization, wettability, adhesion, *etc.* Fewer studies have focused on surfactant–polyelectrolyte complexation at a surface. Several recent ones address the problem of polyelectrolyte complexation with insoluble monolayers.[7–10] In this review, we present studies of polymer–surfactant complexation at the free surface of an aqueous solution, the surfactants being soluble. This case is frequent in practice, but more difficult to analyse: indeed, contrary to insoluble monolayers, the amount of surfactant adsorbed at the surface is not known and has to be either measured or inferred from thermodynamic models. The field was pioneered by D. Goddard with studies of cationic polymers and anionic surfactants.[1] We have used in our group several model anionic polymers: polacrylamide sulfonate (PAMPS), polystyrene sulfonate (PSS) and xanthan (Figure 1). PAMPS and PSS are rather flexible polymers, whereas xanthan is more rigid: the intrinsic persistence lengths are about 1 nm for PAMPS and PSS,[2] and between 5 and 140 nm for xanthan, according to its configuration in the solution (coil, simple or double helix).[11] PAMPS is water soluble regardless the degree of sulfonation f (defined as the molar ratio of sulfonated monomers to neutral monomers), whereas PSS is water soluble only if $f > 0.3$.[12] This is because the backbone is more hydrophobic, and as we will see later, it leads to significant differences in surface behavior.

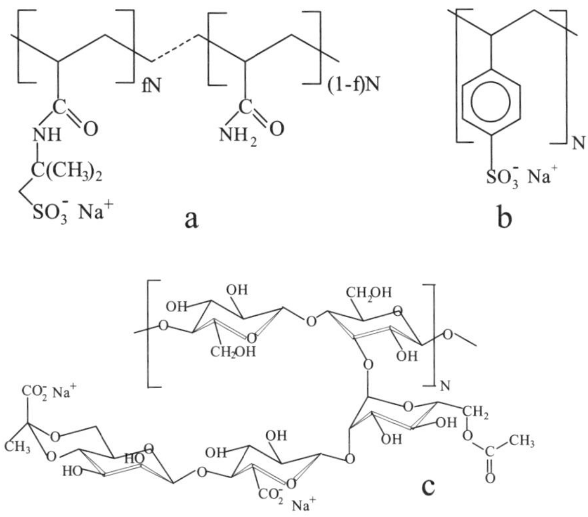

Figure 1 *Molecular structures of (a) PAMPS (b) PSS and (c) xanthan*

2 Results and Discussion

We have worked with polymer/surfactant solutions of fixed polymer concentration and variable surfactant concentrations. We have thus studied the changes induced by the presence of fixed amounts of polymer in the surfactant solutions.

2.1 Surface Tension Studies

2.1.1 Nonionic and Anionic Surfactants. We have used alkyl pentaethylene glycol ethers ($C_{10}E_5$ and $C_{12}E_5$) and octylglucoside (C_8G_1). The surface tension of mixed solutions of $C_{10}E_5$ and C_8G_1 does not differ from the surface tension of the polymer-free surfactant solutions for most polymers.[13,14] One exception is the partially sulfonated PSS, where a large surface tension increase is observed with $C_{12}E_5$.[14] This is similar to the behavior of mixed solutions with a similar surfactant, $C_{12}E_6$, and a water soluble protein, bovine serum albumin.[15] The origin of the surface tension increase is probably due to the fact that the polymer does not bind to the surfactant layer and furthermore removes surfactant from the surface to form bulk complexes in which the polymer hydrophobic groups are less exposed to water. The other exception is xanthan with C_8G_1, where a small surface tension decrease is observed, possibly associated to complexation of the sugar groups of the polymer with that of the surfactant.[16]

The surface tension of mixed solutions of polymers and surfactants of the same charge is slightly smaller than the surface tension of the polymer-free surfactant solution. In this case again, there is no binding of the surfactant to the polymer, but an ionic polymer shifts the critical micellar concentration (CMC) of the solutions to lower values, exactly as a common salt.[13]

2.1.2 Cationic Surfactants and PAMPS. We have studied mainly dodecyl trimethyl ammonium bromide (DTAB). Mixed solutions of DTAB and PAMPS show a synergistic lowering of surface tension at very low surfactant concentrations. As seen in Figure 2, the surface tension curve exhibits two break points: the first one, known as the critical aggregation concentration (CAC), corresponds to the beginning of the formation of a significant number of polymer/surfactant complexes in the bulk. The second one corresponds to the critical micellar concentration (CMC). The surface tension does not change if the polymer concentration is decreased down to very small values (less than 100 ppm, see Figure 3). This remarkable result was also noted by Buckingham *et al.* on polylysine–sodium dodecyl sulfate solutions and explained with a simple model that we will briefly recall below.[17] In our system we have four kinds of ions, the polymer ions P, the surfactant ions S, the sodium and bromide counterions Na^+ and BR^-. The Gibbs equation relating the surface tension variations $d\gamma$ to the chemical potential variations $d\mu_i$ of the species i present in the solution writes:

$$d\gamma = -\Gamma_{Na^+} d\mu_{Na^+} - \Gamma_{Br^-} d\mu_{Br^-} - \Gamma_P d\mu_P - \Gamma_S d\mu_S \qquad (1)$$

where Γ_i is the surface excess concentration of the species i. If we now assume that the complex formation at the surface is a cooperative phenomenon and that the surfactant counterions are replaced by the polymer charged units, the surfactant and polymer counterions are released in the bulk solution. It then follows that: $\Gamma_{Na^+} = \Gamma_{Br^-} = 0$. Because of electrical neutrality at the surface, we also have: $\Gamma_S = fN \Gamma_P$, N being the number of monomers in the polymer chain. Finally, since we deal here with dilute solutions, $d\mu_i \sim kT \ln c_i$, k being the

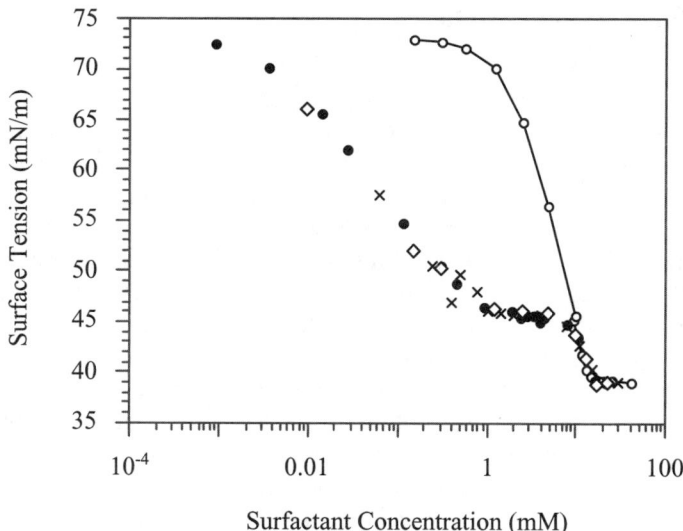

Figure 2 *Effect of PAMPS (f = 25%, $M_w = 2.2 \times 10^6$) on DTAB surface tension: (○)*
DTAB; (◇) DTAB + 75 ppm PAMPS; (×) DTAB + 350 ppm PAMPS; (●)
DTAB + 750 ppm PAMPS. Data from ref. 13

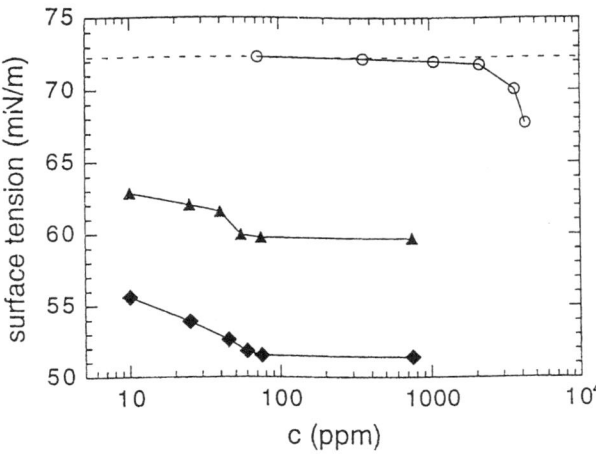

Figure 3 *Surface tension of mixed DTAB–PAMPS solutions* versus *polymer concentration:* (○) *pure PAMPS;* (▲) *PAMPS + 0.5 × 10⁻⁴ M DTAB;* (◆) *PAMPS + 2 × 10⁻⁴ M DTAB. Data from ref. 28*

Boltzmann constant, T the absolute temperature and c_i the bulk concentration of species i, we get:

$$d\gamma = - kT \, \Gamma_s \, d \ln (c_s \, c_p^{1/fN}) \tag{2}$$

Because Nf, the number of charged monomers in the polymer chain is a large number:

$$d\gamma \sim - kT \, \Gamma_s \, d \ln c_s \tag{3}$$

i.e. the surface tension is independent of the polymer concentration. If we use this equation for the data of Figure 2, we find that around the CAC, the area per surfactant molecule A_s at the surface is:

$$A_s = 1/\Gamma_s \sim 78 \pm 5 \, \text{Å}^2$$

The surfactant density is much larger than in the absence of polymer: at CAC, $A_s > 200 \, \text{Å}^2$ for pure surfactant solutions, as calculated with the Gibbs equation for charged monolayers: $d\gamma = - 2kT \, \Gamma_s d \ln c_s$. However, the surfactant chains are not closely packed: for a DTAB solution with a concentration close to CMC, the area per molecule is smaller: $A_s \sim 48 \, \text{Å}^2$. The distance between charges along the polymer chain is about 10 Å, leading to an area of about 100 Å² if one assumes that the chains lie flat at the surface. The value measured is somewhat smaller, suggesting the existence of loops and pendant chains with uncharged monomers.

We have found that the polymer molecular weight does not have any measurable influence on the surface tension, provided that the degree of charge f is the same.[18] This is consistent with the adsorption model recalled above, each surfactant being complexed by a charged monomer. When the degree of charge is increased, the surface tension decreases. Figure 4 shows the surface tension for

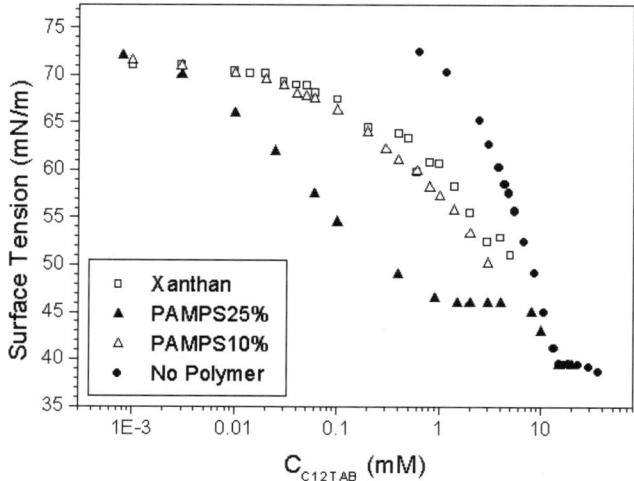

Figure 4 *Surface tension as a function of DTAB concentrations: (●) pure DTAB solution; (□) mixed Xanthan–DTAB solution; (▲) mixed PAMPS 25%–DTAB solution; (△) mixed PAMPS 10%–DTAB solution. Data from ref. 20*

DTAB and PAMPS for which x = 10 and 25%. For the smallest degree of charge, the CAC is no longer well defined. One rather observes an inflection point around 1 mM of DTAB, after which the solutions become turbid. If we apply Equation 3 to evaluate the area per surfactant molecule for x = 10% we find close to CAC:

$$A_s = 1/\Gamma_s \sim 100 \pm 5 \text{ Å}^2$$

This area is larger than for f = 25% (78 Å^2). This is probably associated to the fact that the polymer at the surface acts as a spacer for the surfactant molecules. Indeed, each charged polymer site being complexed with a surfactant molecule, the distance between the molecules should increase if the distance between the sites increases. However, the variation in A_s is smaller than expected for a completely flat adsorption of the polymer (the area A_s should then increase by a factor 2.5, ratio of the two f values). The polymer configuration near the surface is therefore certainly different and the thickness of the polymer layer larger with the smaller degree of charge, with more loops and pendant chains with uncharged monomers. This was confirmed by ellipsometry measurements[13,18] (see Section 2.2).

2.1.3 DTAB and PSS. The behavior of mixed DTAB–PSS solutions is very different from that of DTAB–PAMPS. We will recall here results obtained for f = 100% and M_w = 75 000.[19] First, for given surfactant and polymer concentrations, the decrease of surface tension is much smaller for PSS (Figure 3). Second, there is no evidence of surface tension saturation above a certain polymer concentration and the CAC depends on polymer concentration.[19] For a given surfactant concentration, the surface tension decreases upon addition of polymer and increases again after a minimum. This indicates desorption of species present in the monolayer above this concentration. Let us point out that a similar trend was observed in the experiments of Buckingham *et al.*[17] Obviously, the simple adsorption model presented above fails and more work is needed to clarify this complex surface behavior. It should be noted that contrary to the PAMPS polymers studied, the fully sulfonated PSS is not completely ionized in aqueous solutions, because of Manning condensation of part of the counterions (spontaneous condensation allowing to reduce the electrostatic potential along the polymer chain, so that the chemical potential of free and associated counterions could stay equal).

When the bulk behavior is checked with fluorescence techniques, much smaller CAC values are observed, an order of magnitude smaller than for a more hydrophilic polyelectrolyte, sodium polyacrylate.[4] Bulk association could explain why surface tension changes are only appreciable at high polymer concentration. In this case, the simple adsorption model valid for PAMPS would evidently not work, because it does not account for bulk complexes.

2.1.4 DTAB and Xanthan. Surface tension data for xanthan are plotted on Figure 4 together with those of PAMPS.[20] Although the degree of charge here is equal to one, the surface tension curve is similar to that of PAMPS with x = 10%. Similar results were obtained with DNA, another rigid polyelectrolyte.[16] For these polymers, the distance between effectively charged sites along the chain is about 7 Å (in the case of DNA, this is associated with a large Manning condensation, the distance between charged groups being 1.7 Å). This leads to an area of about 50 Å^2. The measured one is the same as for PAMPS 10%, *i.e.* twice as large, suggesting that this is due perhaps to the helical configuration and to the fact that only half of the charges can directly face the surfactant monomers, the other charges being on the opposite side of the helix. This result is consistent with the observations of Sastry *et al.* of charge reversal in complex surfactant–DNA mixed surface layers.[21]

2.2 Ellipsometry and X-Ray Reflectivity Studies

Ellipsometry studies performed on mixed DTAB–PAMPS solutions confirmed that the adsorbed polymer lies essentially flat. The thickness of the polymer layer is however somewhat larger with the smaller degree of charge, probably with more loops and pendant chains of uncharged monomers.[13,18] The turbid solutions above CAC have non homogeneous interfaces. With ellipsometry, one observes two types of domains with macroscopic extensions: some are similar to the adsorbed layers before the CAC, the others are thicker and denser (larger refractive index and thicknesses around 1000 Å). These last domains are possibly microgels, precursors of the precipitation in the bulk. These microgels have also been observed in the foam films made from these solutions[22] (see Section 2.4.3).

In the case of PSS and xanthan, ellipsometry measurements did not give results significantly different from those of pure water.

X-ray reflectivity measurements were found insensitive to the surfactant layer: because the Br⁻ ions are expelled from the surface, the contrast conditions for X-rays are poor.[23] We checked from neutron reflectivity measurements on deuterated surfactant-protonated polymer, that the thickness of the surfactant layer was reasonably constant and equal to $d_{surf} = 17 \pm 3$Å in the studied concentration range.[24] Because the X-ray wavelength is small, it is not sensitive to the dilute polymer regions far from the interface, observed by using ellipsometry. X-rays probe the polymer layer region where the polymer chains are stretched in order to complex their charged monomers with the surfactant. We have found that the thickness of this region is roughly the same for all the polymers (20–30 Å), as well as the polymer volume fraction (15–30%).[23] Neutron reflectivity experiments done with PSS shows an interesting specificity: below CAC, there is some penetration of the polymer in the surfactant monolayer, whereas at higher concentration, the surfactant monolayer is unchanged when polymer is added.[24] This is similar to what was observed for other polymers.[25]

2.3 Adsorption Kinetics and Surface Viscoelasticity

The equilibration times for surface tension are very long, especially between CAC and CMC, where they can reach hours.[26] This fact is important for the understanding of foam behaviour (see Section 3).

We have also measured the surface compression elasticity and viscosity of DTAB–PAMPS mixed surface layers. These two coefficients describe the resistance of the layer to a uniaxial compression in the surface plane. They were measured with a device in which surface waves are excited at a frequency of a few hundred hertz. It was found that as expected, the layers start to exhibit a measurable elasticity at surfactant concentrations much less than with pure surfactant solutions (Figure 5). The elasticity (both real and imaginary parts, ε_r and ε_i respectively) exhibits a maximum around CAC and decreases to zero around CMC.

The elasticity of pure surfactant solutions also exhibits a maximum, but at higher concentrations, close to CMC (Figure 6). Below this maximum, the monolayer behaves as if it were insoluble, because the frequency of compression is much larger that the inverse characteristic adsorption/desorption time which is here diffusion controlled. The elasticity is then the 'Gibbs elasticity', given by:

$$\varepsilon_G = -\frac{1}{\Gamma}\frac{\partial \gamma}{\partial \Gamma}$$

At larger concentrations, the compression is followed by diffusion and the effective elasticity is smaller than the Gibbs elasticity. This leads to the maximum observed and to the appearance of an imaginary contribution:

$$\varepsilon = \varepsilon_r\left(\omega\right) + i\varepsilon_i\left(\omega\right) = \varepsilon_r(\omega) + i\omega\kappa(\omega)$$

ω being the frequency and κ the surface compression viscosity (sometimes called

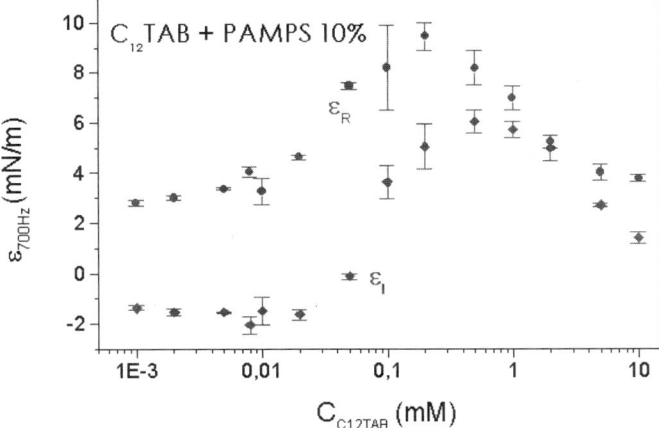

Figure 5 *Surface viscoelasticity as a function of surfactant concentration for DTAB solutions with PAMPS 10%. Data from ref. 20*

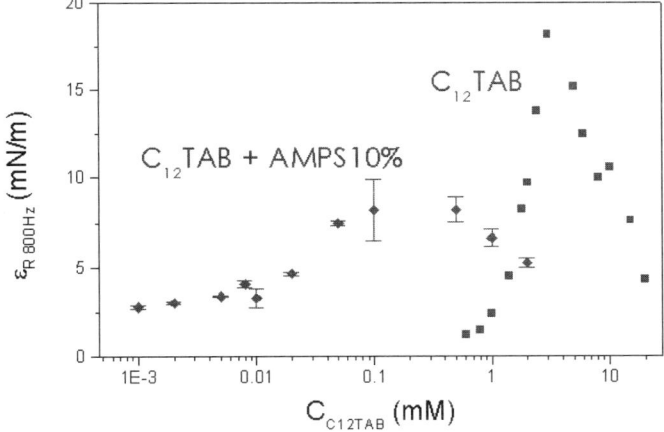

Figure 6 *Surface elasticity as a function of surfactant concentration for pure DTAB solutions and mixed solutions with PAMPS 10%. Data from ref. 20*

dilational). The maximum observed for the mixed laxers occurs in a concentration region where the surface layer behaves as insoluble. Its origin could possibly be associated to a partial desorption of the polymer when the surfactant layer becomes denser as observed in the PSS case (see Section 2.2).

The surface elasticity of xanthan mixed layers is very similar to that of PAMPS mixed layers, with a maximum close to CAC.[20] In these experiments, the relative compression is very small, of the order of 10^{-3}. The behaviour upon large compression is however different. Mixed layers with xanthan can sustain large compressions: when the surface area is decreased by a factor of five, the surface tension can decrease by up to 20 mN/m in a rapid and entirely reversible way. This is accompanied by an increase of the layer thickness (factor up to two), also fully reversible. The PAMPS mixed layers can also be compressed but the surface tension decrease is much more limited, suggesting other exchanges with the bulk are perhaps taking place.[26]

2.4 Thin Liquid Films and Surface Forces

2.4.1 Surface Forces.   Surface forces can be measured with 'thin film balances' where freely suspended horizontal films are submitted to controlled pressures. If the film drainage has been smooth enough, so that no early rupture has occurred, a regime where interactions between the two sides of the film become significant can be attained: this is in a thickness region of about 100 nm and less. Typical interactions are van der Waals forces (attractive) and electrostatic forces (repulsive). Short range forces (steric, hydration) are also present.

Films formed from the mixed aqueous solutions of surfactants and the various polyelectrolytes were studied, at surfactant concentrations well below the CAC. With PSS and PAMPS, the mixed surface layers confer a good stability to the thin film: without the polymer, the DTAB film breaks almost immediately. At the polymer concentrations used, because the polyelectrolyte chains dimensions are large, the solutions are expected to be in the semidilute range. We have checked that this was the case by measuring the viscosity of the polymer solutions. The reduced viscosity $\eta_r = (\eta - \eta_0)/\eta_0 c$ varies as $c^{-1/2}$ for both PAMPS and PSS, as usual for semidilute polyelectrolyte solutions.[27,28] For xanthan, it varies as c^2, as expected for solutions of rigid rods.[29]

A typical disjoining pressure variation with film thickness is shown in Figure 7a. At small thicknesses, a repulsive force is observed which can be fitted with an exponential form exp-(κh), as expected for screened electrostatic repulsion; κ is close to the calculated inverse Debye–Huckel length in the solution:

$$\kappa^2 = 4\pi l_B \sum Z_i^2 \, n_i \quad \text{with} \quad l_B = e^2/(4\pi\varepsilon kT) \quad \text{and} \quad n_i = N \, c_i/M_i \qquad (4)$$

where l_B is the Bjerrum length, n_i the number concentration of the ionic species i, c_i its concentration by weight, M_i its molecular weight, Z_i its valency, N the Avogadro number, e the electron charge, ε the dielectric constant of the solution, kT the thermal energy. For water at room temperature, $l_B \sim 0.7$ nm.

When the polymer concentration is increased, additional branches appear[27–29] (Figure 7b–d). When ΔP increases, h decreases, and when the top of a given branch is reached, the thickness jumps to a smaller value on the next branch. When Δ is decreased, the different branches are returned (although not the jumps). This behavior is similar to that of ionic micellar solutions.[30]

The distance d between the branches of the curves of Figure 7 is the same (when more than two branches are observed) and does not depend on polymer molecular weight. This distance d varies with polymer concentration as d $\sim$ $c^{-1/2}$ (Figure 8). Numerically, d is about four times larger than the Debye length κ^{-1} and is close to the expected mesh size of the polymer network. It does not depend on the type of surfactant used to stabilize the film: for PSS, similar oscillations have been observed with both DTAB and $C_{10}E_5$. Finally, we have checked the effect of added salt (sodium chloride). The effect of increasing salt concentration is similar to the effect of decreasing polymer concentration: first,

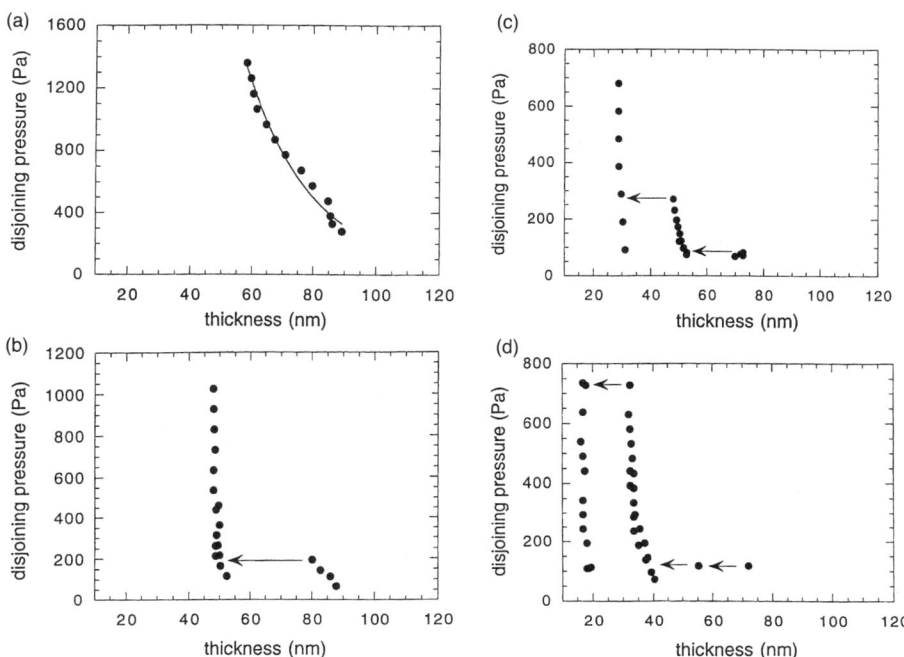

Figure 7 *Disjoining pressure vs film thickness for PSS–DTAB solutions. DTAB concentration is 10^{-4} M, PSS concentration is: (a): 0.02%, (b) 0.2%, (c) 0.4%, (d) 0.8%. The distance d is taken as the distance of the arrows (distance between branches in the limit of vanishing disjoining pressures, after ref 28)*

the outer branches disappear and one is left with a single branch consistent with the electrostatic repulsion $\exp(-\kappa h)$ when the salt concentration becomes comparable to the polymer counterion concentration.

2.4.2 Link with the Structure of Polyelectrolyte Solutions.

This particular behaviour can be explained as follows.[2] If the distance between charges on the chain is smaller than the Bjerrum length l_B, there is a counterion condensation (Manning condensation). This is the case for PSS; for PAMPS, the polymer is weakly charged and no condensation is expected to occur. These polyelectrolyte chains are flexible enough: the intrinsic persistence length (the persistence length when the electric charges are fully screened, for instance by adding excess salt) is comparable to the monomer size. In such a case, the chain starts to coil until the electrostatic repulsion energy becomes larger than kT. The coil size is then ξ_e. If the chain is larger, it starts to form a new coil or 'blob' and the whole chain can be viewed as a rod-like succession of electrostatic blobs of size ξ_e; this picture holds until the rod length reaches a value comparable to the (total) persistence length l_p, after which a coil with larger dimensions (units l_p) is formed. In the semidilute regime, the rods form a mesh of size ξ, the average distance between overlap points of the rods. For rigid polyelectrolytes such as xanthan, the persistence length is large (up to 100 nm) and there are no electrostatic blobs, the chain itself is rod-like.

Small angle X-ray and neutron scattering experiments have evidenced the presence of a peak at a wave vector q^* corresponding to a characteristic distance that varies with polymer concentration as $c^{-1/2}$ and that has been identified with the mesh size.[11,12,31] The peak disappears when salt is added around a concentration comparable to that of the polymer ions. The scattering data, and in particular the structure factor, can be Fourier transformed to obtain the pair correlation function $g(r)$. A peak in the structure factor leads to oscillations in $g(r)$. When the polymer is confined between two surfaces, and when the distance h is comparable to the period of the oscillations of $g(r)$, oscillatory forces between the surfaces are expected. These oscillations have been observed in simple fluids made of spherical molecules in the surface force apparatus with solid mica surfaces.[32] They do not arise because the molecules tend to structure into semi-ordered layers at surfaces, but because of the disruption or change of this ordering during the approach of the second surface. A similar interpretation has been given for the oscillatory forces in micellar solutions.[33] Recently, they have been predicted to exist for polyelectrolyte solutions.[34] The data presented here as well as other recent ones[35,36] confirm that such structural forces do exist in semidilute polyelectrolyte solutions.

We have compared the distance between branches d to the theoretical mesh size ξ:[2]

$$\xi = \frac{\alpha}{c^{1/2}} \qquad (5)$$

with $\alpha = a^{-1/2} \, M^{1/2} \, N^{-1/2} \left(\frac{l_B}{a}\right)^{-1/7} f^{-2/7}$ for a flexible polyelectrolyte without Manning condensation, where a is the monomer size and f the fraction of charged monomers. With Manning condensation, or with rigid chains, $\alpha = A^{-1/2}$, where A is the distance between charges. The agreement between d and ξ is good (Figure 8).

The data are also consistent with the disappearance of the scattering peak by addition of salt, with its sharpening with increasing polymer concentration and with the absence of dependence upon polymer molecular weight.

2.4.3 Behaviour above CAC. Foams films made with DTAB solutions at concentrations comparable to CAC break instantaneously. When small quantities of PSS or PAMPS are added, the foam films become much more stable. At these concentrations, the pure DTAB solutions do not show any measurable viscoelasticity, but the addition of polymer produces a significant increase of surface elasticity (Figure 6). This increase in surface elasticity might stabilize the film against rupture and explain the enhancement of the stability of the foam films. Above CAC, the drainage of the foam films is very different: microgel formation in the film is suggested by the videointerferometry studies[22] (Figure 9). These observations are similar to others on mixed surfactant-protein films.[37,38] In both

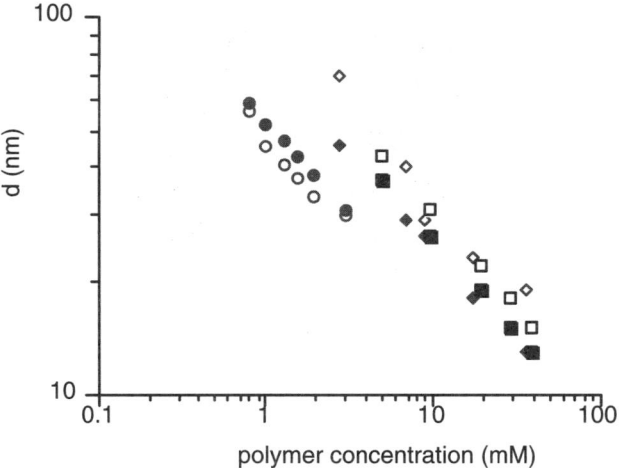

Figure 8 *Comparison between the measured distance between force branches (open symbols) and calculated mesh sizes for the polymer network calculated using the scaling law (Equation 5) (closed symbols); (diamonds) PAMPS; (squares) PSS; (circles) xanthan. Data from ref. 29*

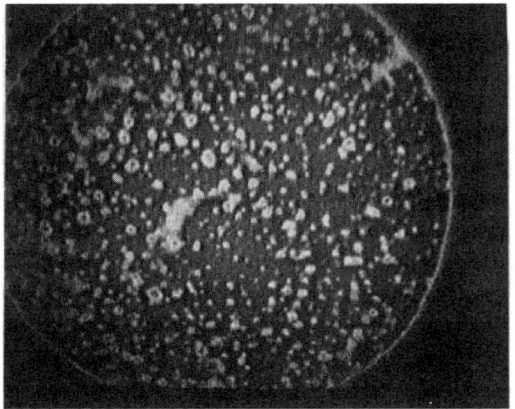

Figure 9 *Image of a horizontal foam film in the concentration range of precipitation; PAMPS/DTAB solutions. The size of the film is about 1 mm*

cases, extremely stable foam films are obtained, visibly associated to an immobilization of the aggregates in the gelified-like film.

The situation is very different for the more rigid polyelectrolyte xanthan. The foam films made with DTAB solutions are very unstable. When using DTAB or CTAB which alone give stable foam films, the incorporation of the polymers in the solutions also leads to unstable films. This behaviour is particular to the cationic surfactants: indeed stable foam films can be made with nonionic surfactants and all the polymers. The data of Figure 8 for xanthan were obtained in this way.[29] We have seen that in the case of cationic surfactants, mixed polymer–surfactant layers are formed. However, their properties seem very similar for all the polymers, except for the behaviour upon large compression. The difference observed in foam film stability might therefore be associated to non-linear surface elasticity: indeed, when a foam film breaks, the surface layer is rapidly and strongly compressed. The mixed surface layers with xanthan, which behave more as insoluble layers might be less resistant to this process than those with PAMPS, potentially able to exchange material with the bulk solutions. Indeed, it is known that solutions of substances that give rise to insoluble surface layers (such as dodecanol for instance) do not foam, even when the surface layers have large compression elasticities (above 80 mN/m for dodecanol).[39]

3 Connection with Foam Stability

The results of standard foaming tests with PAMPS and xanthan are shown in Figure 10: height of foam produced in a glass cylinder by blowing gas through a porous frit disk at a constant flow rate, and time after which the foam height is divided by two if gas flow is stopped.[20] Surprisingly, there is not much difference between the foams made with pure DTAB solutions and the mixed ones.

Foam properties of solutions are generally related to the stability of the films separating the air bubbles. Good foamability demands that the bubble surfaces be rapidly covered by surfactant. In general, therefore, large foamabilities are well correlated to fast adsorption kinetics. Surprisingly here, nothing special happens in the region where the adsorption kinetics is slow, between CAC and CMC. Nothing happens either in the region where the very stable foam films are observed (Figure 9). At larger surfactant concentrations, there is a better correlation between fast adsorption and good foamability. This is probably associated with the fact that at these concentrations, the surface pressure is essentially created by the surfactant, and that even if the polymer–surfactant complexes have not enough time to be formed at the surface, the surfactant monolayer alone confers a good stability to the foam. Indeed good foamability also demands that the foam bubbles do not break too quickly. Foamability and foam stability are therefore closely related. Foam stability is governed by the following processes:

(a) Drainage: removal of liquid from the space between the bubbles and thinning of the films. The drainage rate is slightly less for more viscoelastic layers,[40]
(b) Diffusion of gas from the small bubbles towards larger ones (Ostwald ripening), leading to a bubble size growth. Gas diffusion rate decreases when surface coverage increases,[41]
(c) Coalescence: breaking of the film between adjacent bubbles to form one large bubble. A finite elasticity protects the films from rupture.[42]

All the phenomena depend on the elasticity and viscosity of the surface, and high elasticities and viscosities are thus expected to stabilize the foam. However, since the foam is formed rapidly, the polymer has probably no time to reach its equilibrium adsorption in the CAC region and the foam behaviour should be governed by the surfactant adsorption as observed.

(a)

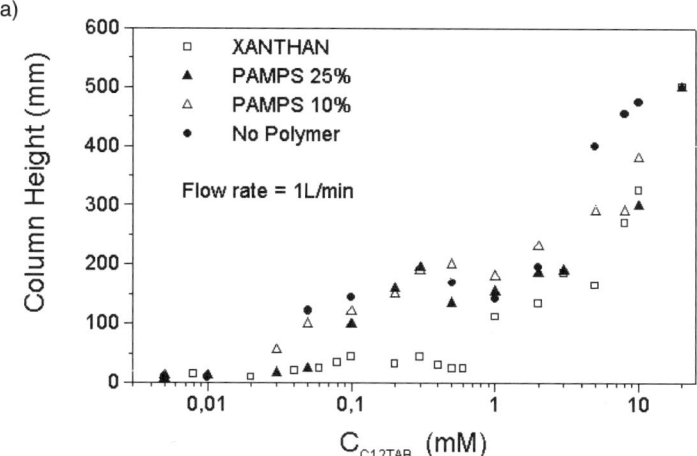

(b)

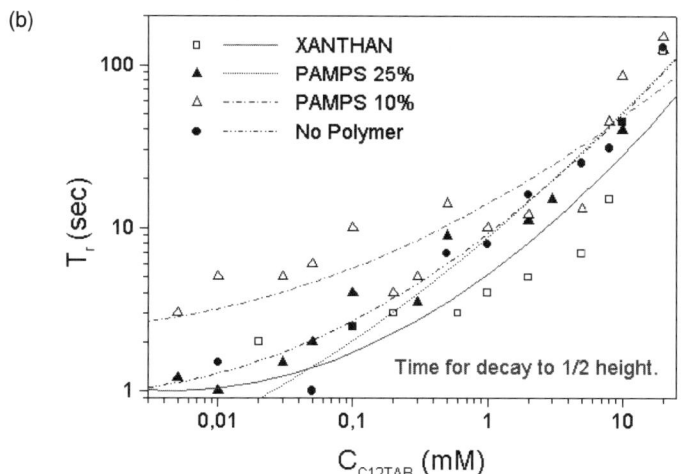

Figure 10 (a) *Foam height as a function of surfactant concentration for the different solutions, and a constant flow rate of* 1 *L/min of nitrogen:* (□) *xanthan,* (▲) *PAMPS 25%,* (△) *PAMPS 10%,* (●) *no polymer.* (b) *Foam lifetimes* T_r *versus surfactant concentration for the different solutions.* T_r *is the time taken by the foam column to drop to half its original height after the gas flow is stopped. The lines are second order polynomial fits to the experimental points and are a guide to the eye;* (□) *xanthan,* (▲) *PAMPS 25%,* (△) *PAMPS 10%,* (●) *no polymer. Data from ref. 20*

Although the surface rheological behaviour of the different polymers is similar, the foaming capacity is much larger with PAMPS than with xanthan. This is probably related to the difference in foam film behaviour. Surprisingly though, the foam lifetime is not significantly different with PAMPS and xanthan.

4 Summary and Conclusion

The addition of polyelectrolytes to dilute surfactant solutions leads to extremely varied effects. Nonionic surfactants can be depleted from the surface by an hydrophobic polyelectrolyte such as PSS. Ionic surfactants of opposite charge lead to a synergistic surface tension lowering due to coadsorption of a non surface-active polyanion and oppositely charged surfactant ions. While a highly surface-active complex forms at the interface, there is no significant binding between the two species in the bulk for hydrophilic polymers such as PAMPS at the very low surfactant concentrations where the synergistic effect first takes place. The surface tension in this case is essentially independent of polymer concentration, in agreement with the ion-exchange model. We have found that the surfactant–polymer complexes are rather thin, suggesting that the polymer chains are stretched out along the surface and that the counterions are expelled in the bulk.

The adsorption kinetics are slow in a surfactant concentration range around CAC and become fast again close to the CMC of the pure surfactant, where the surface layers with and without polymer become similar. Below CAC, they behave as incompressible layers. When subjected to small compression–expansion cycles, the layers exhibit a viscoelastic response, similar with the different polymers. Appreciable differences are seen only when the compression is more important (decrease of the surface area by a factor up to five): the layers with xanthan still behave as insoluble layers (even above CAC), whereas those with PAMPS appear as partially soluble.

Thin liquid films made from the mixed solutions below CAC exhibit a stratification phenomenon, with a stratum thickness corresponding to the mesh size ξ of the polymeric network, *i.e.* the distance between overlap points of two polymer chains. The oscillatory forces are particular to polyelectrolytes and disappear when the electrostatic forces are screened with salt. The study of freely suspended films gives new useful insights into the structure of semidilute polymer solutions which are presently the object of numerous speculations.

The mixed layers of flexible polymers and surfactant of opposite charge enhance the stability of the foam films made from the solutions. The stabilization mechanisms are different below and above CAC: below CAC, the stabilization is probably associated to the increase in surface viscoelasticity, whereas above CAC, the stabilization is visibly associated to the gel-like aspect of the foam films. More rigid polyelectrolytes produce the opposite effect below CAC. This behaviour might be associated to the different behaviour upon large compression and remains to be understood.

As far as the foam properties are concerned, the differences between foams made with and without polymers are surprisingly small. The surface layers and the thin films with and without polymers are indeed strikingly different. This apparent lack of correlation might be due to the very long adsorption kinetics when the polymer is present. Indeed, foams are produced in very short times, and only the surfactant has time to cover the bubble surfaces, explaining why there are little differences between the foams made with and without the polymers.

This behaviour has also been observed with foams made with proteins solutions.[38]

Acknowledgements

This work was supported in part by Institut Français du Pétrole. More details can be found in the original references 13, 14, 16, 18–20, 22–24, 26–29. The work was performed at the Laboratoire de Physique de l'Ecole Normale Superieure, Paris, at the Centre de Recherche Paul Pascal, Bordeaux and at the Laboratoire de Physique des Solides, Orsay with many colleagues: A. Asnacios, J. F. Argillier, V. Bergeron, A. Espert, R. v. Klitzing, A. Colin, F. Monroy, C. Stubenrauch, L. T. Lee, P. A. Albouy, A. Bhattacharrya and H. Ritacco. They are all gratefully acknowledged.

References

1. E. D. Goddard and K. P. Ananthapadmanabhan, *Interaction of Surfactants with Polymers and Proteins*, CRC press, 1993.
2. J. L. Barrat and J. F. Joanny, *Adv. Chem. Phys.*, 1996, **94**, 1.
3. B. Lindman, *Adv. Colloid Interface Sci.*, 1992, **41**, 149.
4. P. Hansson and M. Almgren, *Langmuir*, 1994, **10**, 2115; *J. Phys. Chem.*, 1995, **99**, 16694.
5. J. Merta, V. M. Garamus, A. I. Kuklin, R. Willumeit and P. Stenius, *Langmuir*, 2000, **16**, 10061.
6. B. Cabane and R. Duplessix, *J. Physique*, 1982, **43**, 1529.
7. V. G. Babak, M. A. Anchipolovskii, G. A. Vikhoreva and I. G. Lukina, *Colloid J.*, 1996, **2**, 145 (translated from Russian).
8. K. de Meijere, G. Brezesinski and H. Möhwald, *Macromolecules*, 1997, **30**, 2337.
9. S. Sundaran and K. J. Stebe, *Langmuir*, 1997, **13**, 1729; S. Sundaran, K. J. Ferry, D. Vollhart and K. J. Stebe, *Langmuir*, 1998, **14**, 1208.
10. J. Engelking, D. Ulrich and H. Menzel, *Macromolecules*, 2000, **33**, 9026.
11. M. Milas, M. Rinaudo, R. Duplessix, R. Borsali and P. Lindner, *Macromolecules*, 1995, **28**, 3119; L. Chazeau, M. Milas and M. Rinaudo, *Int. J. Polymer Analysis & Characterization* 1995, **2**, 21.
12. W. Essafi, F. Lafuma and C. E. Williams, *J. Phys. II France*, 1995, **5**, 1269; W. Essafi, PhD Thesis, University Paris, 1996.
13. A. Asnacios, D. Langevin and J. F. Argillier, *Macromolecules*, 1996, **23**, 7412.
14. E. Radlinska, T. Gulik, F. Lafuma, D. Langevin, W. Urbach, C. E. Williams and R. Ober, *Phys. Rev. Lett.*, 1995, **74**, 4237.
15. N. Nishikido, T. Takahara, H. Kobayashi and M. Tanaka, *Bull. Chem. Soc. Jpn.*, 1982, **55**, 3085.
16. C. Stubenrauch, in preparation.
17. J. H. Buckingham, J. Lucassen and F. Hollway, *J. Colloid Interface Sci.*, 1978, **67**, 423.
18. A. Asnacios, D. Langevin and J. F. Argillier, *Eur. Phys. J. B*, 1998, **5**, 905.
19. R. v. Klitzing, A. Asnacios and D. Langevin, *Colloids Surfaces A: Physicochem. Eng. Aspects*, 2000, **167**, 189.
20. A. Bhattacharyya, F. Monroy, J. F. Argillier and D. Langevin, *Langmuir*, 2000, **16**, 8727.

21. M. Sastry, V. Ramakrishnan, M. Pattarkine, A. Gole and K. N. Ganesh, *Langmuir*, 2000, **16**, 9142.
22. V. Bergeron, A. Asnacios and D. Langevin, *Langmuir*, 1996, **12**, 1550.
23. C. Stubenrauch, P. A. Albouy, R. v. Klitzing and D. Langevin, *Langmuir*, 2000, **16**, 3206.
24. C. Stubenrauch, R. v. Klitzing, A. Asnacios, J. F. Argillier and D. Langevin, unpublished data.
25. B. Jean, L. T. Lee and B. Cabane, *Langmuir*, 1999, **15**, 7585.
26. H. Ritacco, P. A. Albouy, A. Bhattacharyya and D. Langevin, *Phys. Chem. Chem. Phys.*, 2000, **2**, 5243.
27. A. Asnacios, A. Espert, A. Colin and D. Langevin, *Phys. Rev. Lett.*, 1997, **78**, 4974.
28. R. v. Klitzing, A. Espert, A. Asnacios, T. Hellweg, A. Colin and D. Langevin, *Colloids Surfaces A*, 1999, **149**, 131.
29. R. v. Klitzing, A. Espert, A. Colin and D. Langevin, *Colloids Surfaces A*, 2001, **176**, 109.
30. V. Bergeron and C. Radke, *Langmuir*, 1992, **8**, 3020.
31. M. Nierlich, C. E. Williams, F. Boué, J. P. Cotton, M. Daoud, B. Farnoux, G. Janninck, C. Picot, M. Moan, C. Wolf, M. Rinaudo and P. G. de Gennes, *J. Physique*, 1978, **40**, 701.
32. J. N. Israelachvili, *Intermolecular and Surface Forces*, Academic Press, London, 1985.
33. M. L. Pollard and C. J. Radke, *J. Chem. Phys.*, 1994, **101**, 6979.
34. X. Châtellier and J. F. Joanny, *J. Phys. II (France)*, 1996, **6**, 1669.
35. O. Théodoly, PhD Thesis, Paris, 1999.
36. B. Kolaric, W. Jaeger and R. v. Klitzing, *J. Phys. Chem. B*, 2000, **104**, 5096.
37. P. J. Wilde, M. R. Rodriguez Nino, D. C. Clark and J. M. Rodriguez Patino, *Langmuir*, 1997, **13**, 7151.
38. J. Senée, R. Robillard and M. Vignes-Adler, *Food Hydrocolloids*, 1999, **13**, 15.
39. H. Fruhner, K. D. Wantke and K. Luckenheimer, *Colloids and Surfaces A*, 2000, **162**, 193.
40. M. Durand, G. Martinoty and D. Langevin, *Phys. Rev. E*, 1999, **60**, R6307.
41. S. Cohen-Addad and D. Quéré, in *Soft Order in Physical Systems*, Plenum, 1994.
42. D. Langevin, *Curr. Opin. Colloid Interface Sci.* 1998, **3**, 600.

Adsorption Layers of β-Casein at the Air/Water Interface: Effect of Guanidine Hydrochloride

Adel Aschi,[1] Abdehafidh Gharbi,[1] Patrick Calmettes,[2] Mohamed Daoud,[3] Véronique Aguié-Béghin[4] and Roger Douillard[4]

[1]LABORATOIRE DE PHYSICO-CHIMIE DE LA MATIÈRE CONDENSÉE, FACULTÉ DES SCIENCES DE TUNIS, CAMPUS UNIVERSITAIRE, 2092 EL MENZAH 9 TUNIS, TUNISIA
[2]LABORATOIRE LÉON BRILLOUIN, CE SACLAY, 91191 GIF SUR YVETTE CEDEX, FRANCE
[3]SERVICE DE PHYSIQUE DE L'ETAT CONDENSÉ, CE SACLAY, 91191 GIF SUR YVETTE CEDEX, FRANCE
[4]ÉQUIPE PAROIS ET MATÉRIAUX FIBREUX, CENTRE DE RECHERCHE AGRONOMIQUE, 2 ESPLANADE R. GARROS, BP 224, 51686 REIMS CEDEX 2, FRANCE

Summary

The adsorption of proteins at interfaces is a key step in the stabilization of numerous food and non-food foams and emulsions. Our goal is to improve our understanding of the relationships between the sequence of proteins and their surface properties. A theoretical approach has been developed to model the structure and properties of protein adsorption layers using the analogy between proteins and multiblock copolymers. This model seems to be particularly well suited to β-casein. However, the exponent relating surface pressure to surface concentration is indicative of a polymer structure intermediate between that of a two-dimensional excluded volume chain and a partially collapsed chain. For the protein structure, this would correspond to attractive interactions between some amino acids (hydrogen bonds, for instance). To test this possibility, guanidine hydrochloride was added to the buffer. A transition in the structure and properties of the layer is noticed for a 1.5 molar concentration of the denaturant. Beyond the transition, the properties of the layer are those of a two-dimensional excluded volume chain, a situation expected when there are no attractive interactions between the amino acids. It may be concluded that in the adsorption layer formed by β-casein at the air/water interface, some attractive interactions occur which are ruptured by guanidine hydrochloride.

1 Introduction

The adsorption of proteins at fluid interfaces is a key step in the stabilization of numerous food and non-food foams and emulsions.[1] Our general goal is to relate the amino acid sequence of proteins to their surface properties, *e. g.* to the equation of state or other structural and thermodynamic properties. To improve this understanding, the effect of guanidine hydrochloride (Gu HCl) on β-casein adsorption is evaluated in the framework of the block-copolymer model for the adsorption of this protein. At first the main features of the model are presented, and then the effect of Gu HCl is interpreted using the previously introduced concepts.

2 Block-copolymer Model of β-Casein Adsorption

The analysis of the hydrophobicity of β-casein along its sequence reveals that it can be viewed as built up of alternating hydrophilic and hydrophobic sequences.[2,3] However, no absolute hydrophobicity scale exists and the actual length of the blocks cannot be determined. Thus, in the theoretical model, an adjustable parameter remains which is α, the ratio of the number of hydrophobic, Z_A, to that of hydrophilic, Z_B, monomers in the diblock model (the simplest one) or in the large regular multiblock model (N diblocks linked sequentially).

The properties of the model have been calculated[4] assuming that each block behaves like a polymer coil and has a fractal structure resulting from the equilibrium between the conformational entropy, the excluded volume effects and the attractions between the monomers[5] (Figure 1). Thus its radius of gyration, R, is $R = Z^{1/D}$ where D is the fractal dimension of the coil ($v = 1/D$ is the Flory exponent characterizing the 'solvent properties' of a 3-D coil in a solvent). The values of $1/D$ corresponding to a random self-avoiding walk of the monomers is 3/4 in 2-D and 3/5 in 3-D. The diblock is built up of a 2-D hydrophobic coil with a pancake conformation and a 3-D hydrophilic coil with a sphere conformation. The junction between the blocks is strictly located in the interface plane. The multiblock model is made of a sequence of N such diblocks. Only the surface pressure, π, and the dilational modulus, ε, are dealt with in this communication, and they have been calculated using scaling law arguments such as:

$$\pi \cong k_B T / \xi^2 \cong k_B T \Gamma^y \tag{1}$$

$$\varepsilon = d\pi / dLn\Gamma = y\pi \tag{2}$$

where the sign $\cong$ means that all numerical coefficients are purposely ignored, k_B is the Boltzmann constant, T is the absolute temperature, ξ is the screening (or correlation) length with the higher dependence on Γ, Γ is the surface concentra-

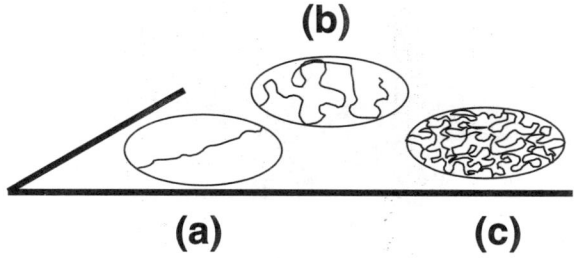

Figure 1 *Examples of fractal dimensions of 2-D coils. The fractal dimension can be calculated as the exponent relating the number of monomers to the radius of gyration of the polymer. In (a), the coil is quasi-linear, the fractal dimension is 1. In (c), the monomers occupy all the surface, the fractal dimension is 2 (the dimension of the space concerned by the coil). In (b), the fractal dimension is intermediate between 1 and 2. A typical coil with an 'excluded volume' geometry has a fractal dimension of 4/3 (in a 2-D space)*

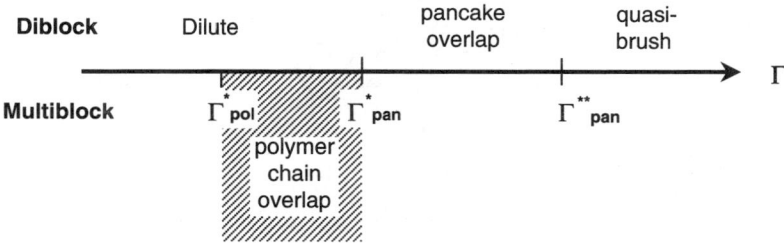

Figure 2 *Effect of surface concentration of monomers, Γ, on the regimes occurring at the interface for medium values of α (the diameter of the 2-D hydrophobic pancakes is slightly larger than that of the 3-D hydrophilic loops). As Γ increases, the regime changes from dilute to semi-dilute when the pancakes overlap in the case of the diblock (Γ^*_{pan}) or when the polymer molecules overlap in the case of a large multiblock (Γ^*_{pol}). At surface concentrations larger than Γ^*_{pan}, the regimes are the same in the diblock and in the multiblock cases. When Γ is larger than Γ^{**}_{pan}, the hydrophobic coils have to grow in the gas in a direction perpendicular to the interface. Then, the properties of the interface are dominated by those of the hydrophilic loops forming quasi-brushes. The hatched region corresponds to the only regime particular to the multiblock case*

tion (number of monomer per unit surface area) and y is an exponent characteristic of the regime and of the fractal dimension of the blocks involved in the definition of ξ. A detailed calculation[4] shows that the equation of state throughout the regimes may fall into three different ranges according to the value of α (small, medium or large). Only the range II (medium values of α) is considered here.

The main features of the model calculated according to Equation (1) and depicted in Figure 2 are that: (i) the diblock and multiblock cases differ only by the semi-dilute regime for the polymer which cannot be distinguished from the semi-dilute regime for the hydrophobic blocks using (1); (ii) the first regime is the dilute one ($\pi \cong \Gamma$, $y = 1$); (iii) the semi-dilute regime for the polymer and semi-dilute regime for the 2-D blocks with constant thickness are characterized

(a)

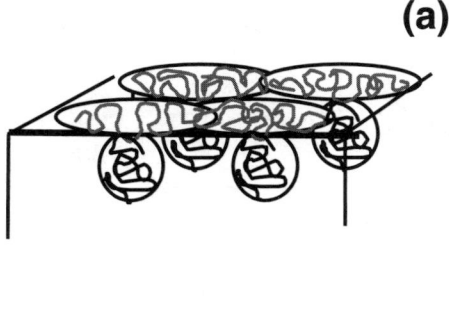

(b)

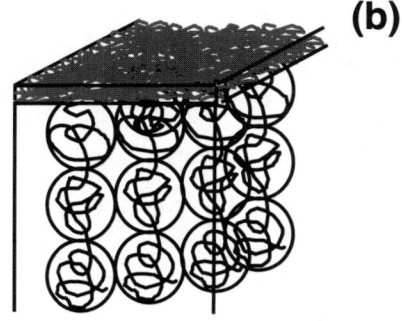

Figure 3 *Examples of regimes found in range II (medium values of α). (a) regime dominated by purely two-dimensional overlap (of the hydrophobic coils here). Within this regime, there is no increase of the thickness of the hydrophobic layer when the surface concentration increases, with y = 3 the coils are in the excluded volume geometry. (b) regime dominated by a partial three-dimensional growth of the hydrophilic coils which form quasi-brush structures. In this regime, the contribution of the hydrophobic coils is independent of the surface concentration because they form a quasi-homogeneous layer, the value of y being 1, whatever the 'quality' of the solvent*

by $y = 3$, corresponding to a 2-D overlap; (iv) the fourth regime, with $y = 1$, corresponds to a quasi-brush structure of the 3-D hydrophilic coils where the hydrophobic blocks undergo a moderate growth in the direction perpendicular to the interface plane and form a kind of homogeneous phase with constant surface properties, except the thickness (Figure 3). However, the consistency of this model is quite difficult to check because it requires the measurement of Γ in (1). Instead, it is possible to use the nearly equivalent relation (2) between ε and π which should allow the evaluation of the y value. This can be done with a bubble tensiometer or any other experimental set-up which allows the simultaneous measurement of the surface pressure and the dilational modulus of the adsorbed layer at various surface concentrations.

3 Experimental

All the experimental conditions have been previously described in detail.[6] Experiments were performed using a 0.1 M phosphate buffer pH 7 containing 0.1 M

NaCl. The β-casein was prepared from the skimmed milk of a single homozygous cow (α_{s1}B, βB, κB).[7] Guanidine hydrochloride (Pierce, Sequanal Grade) was purified from some surface-active compounds by several crystallizations in water.

Surface tension measurements were performed using a bubble tensiometer where the bubble is formed at the tip of a syringe needle (IT Concept, Longessaigne, France). The surface pressure, π, is the difference between the surface tension of the pure solvent, γ_0, and that of the solution under study, γ, *i.e.* $\pi = \gamma_0 - \gamma$. The surface dilational modulus, ε, is defined as the ratio between the variation of surface tension, $d\gamma$, and the relative change of the surface area,[8] $dA/A = dLn(A)$: *i.e.* $\varepsilon = d\gamma/dLn(A)$; it was determined during a sine-wave deformation of the area of the bubble performed by periodically moving the plunger of the syringe.[9]

Neutron reflectivity experiments have been carried out at the reflectometer DESIR in the Orphée reactor (Laboratoire Léon Brillouin, CEA, Saclay). The reflectivity spectra were obtained at a fixed angle of the set-up using a 'white' beam for wavelengths shorter than that of total reflection. A two-layer model was fitted to the spectra to calculate the surface concentration of adsorbed β-casein.[10]

4 Results and Discussion

4.1 Adsorption of β-casein from a Standard Buffer

The comparison of the experimental results with the model according to Equation (2) (Figure 4) shows that the general trends are similar, except that (a) the transition between the regimes is smoother in the experiments than in the model and (b) the slope of the curve in regime two and/or three is significantly larger than three in the experimental data while it is exactly three in the model for an excluded volume coil. The reason for (a) is that the model is said to be asymptotic (or true for very large polymers). The reason for (b) could be that the fractal

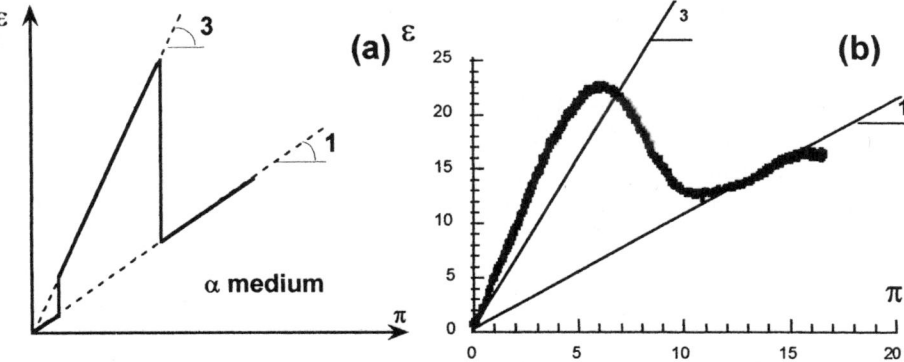

Figure 4 *Variations of the dilational modulus as a function of the surface pressure. Comparison between the model (range II for a medium value of α) (a) and the experimental data (b)*

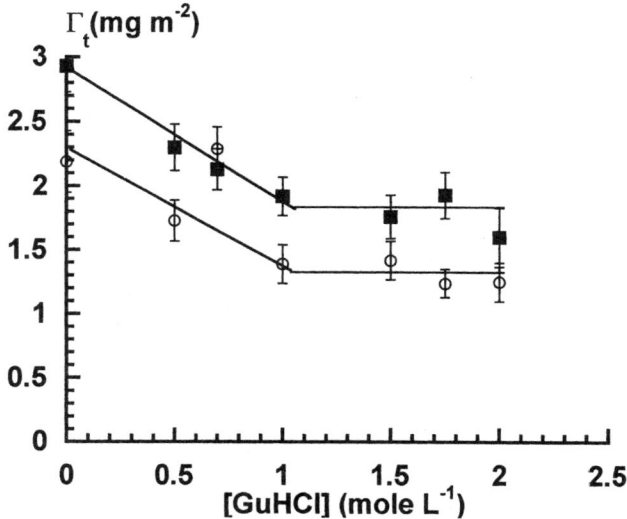

Figure 5 *Effects of temperature and Gu HCl concentration on the surface concentration of β-casein at the air/liquid interface. The surface concentration was calculated by fitting a two-layer model to the neutron reflectivity spectra obtained after 8 hours adsorption from a 100 mg/L β-casein solution. (○), 10°C; (■), 20°C*

dimension of the 2-*D* coil (or polymer) is larger than that predicted for a random self-avoiding walk because of attractions occurring between the monomers (or the coils). This interpretation could be checked by rupturing the attractions between the monomers, and then the value of *y* should be equal to three. Gu HCl is known to break non-covalent bonds between amino acids in solution, and so its effect on the surface properties of β-casein has been checked.

4.2 Effects of Gu HCl on β-casein Surface Properties

The adsorption of β-casein from the solution is significantly lowered when the amount of Gu HCl increases (Figure 5). It can be seen that the total surface concentration decreases but remains significant, and that the adsorption is enhanced by a temperature rise from 10 to 20 °C, indicating that the driving force of adsorption is dominated by hydrophobic effects, whatever the Gu HCl concentration. The effect of the Gu HCl concentration increase is clearly to reduce the *y* exponent relating ε to π in the 2-*D* dominated regime (Figure 6). More precisely, at a 4 M volume concentration of Gu HCl, the *y* value is equal to three as predicted by the model in the case of a 2-*D* random self-avoiding walk. The actual effect of Gu HCl in the interface vicinity is probably not exactly the same as in solution, but, nevertheless, its occurrence in solution seems to disrupt the attractions between the monomers of the 2-*D* adsorbed blocks. Moreover, the effect of temperature between 10 and 30 °C on the *y* exponent at 1.5 M Gu HCl is to slightly lower its value (data not shown). Thus, the attractions responsible for the partial collapse of these coils are not predominantly hydrophobic in nature,

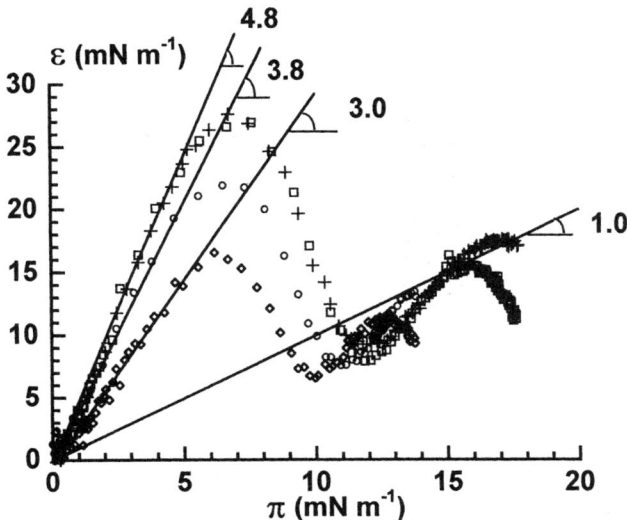

Figure 6 *Effect of the Gu HCl volume concentration on the relation between the dilational modulus and the surface pressure of β-casein adsorption layers formed from a 100 mg/L solution at 10°C. Gu HCl concentration: ($+$), 0 M; ($\square$), 1 M; ($\bigcirc$), 2 M; ($\diamondsuit$), 4 M*

but rather they arise from hydrogen bonds, or van der Waals or electrostatic interactions.

Acknowledgments

Our thanks to B. Monties for his interest in adsorption layer structure and properties and to B. Harzallah and N. Puff who performed some experiments.

References

1. D. Möbius and R. Miller, (eds.), *Proteins at Liquid Interfaces*, Elsevier, Amsterdam, 1998, pp. 498.
2. D. G. Dalgleish and J. Leaver, *J. Colloid Interface Sci.*, 1991, **141**, 288.
3. R. Mackie, J. Mingins and A. N. North, *J. Chem. Soc. Faraday Trans.*, 1991, **87**, 3043.
4. V. Aguié-Béghin, E. Leclerc, M. Daoud and R. Douillard, *J. Colloid Interface Sci.*, 1999, **214**, 143.
5. P. G. De Gennes, *Scaling Concepts in Polymer Physics*, Cornell University Press, Ithaca, 1979.
6. A. Aschi, A. Gharbi, L. Bitri, P. Calmettes, M. Daoud, V. Aguie-Beghin and R. Douillard, *Langmuir*, 2001, **17**, 1896.
7. J. C. Mercier, J. L. Maubois, S. Poznanski and B. Ribadeau-Dumas, *Bull. Soc. Chim. Biol.*, 1968, **50**, 521.
8. E. H. Lucassen-Reynders, in *Anionic Surfactants*, E. H. Lucassen, (ed.), Dekker, New York, p. 1981, p. 173.

9. N. Puff, A. Cagna, V. Aguié-Béghin and R. Douillard, *J. Colloid Interface Sci.*, 1998, **208**, 405.

10. B. Harzallah, V. Aguié-Béghin, R. Douillard and L. Bosio, *Int. J. Biol. Macromol.*, 1998, **23**, 73.

Adsorption and Rheological Behaviour of Biopolymers at Liquid Interfaces

Reinhard Miller,[1]* Valentin Fainerman,[2] Martina O'Neill,[1]
Jürgen Krägel[1] and Alexander Makievski[1,3]

[1]MAX-PLANCK-INSTITUT FÜR KOLLOID- UND
GRENZFLÄCHENFORSCHUNG, AM MÜHLENBERG 1, D–14476
GOLM, GERMANY
[2]INTERNATIONAL MEDICAL PHYSICOCHEMICAL CENTRE,
DONETSK MEDICAL UNIVERSITY, 16 ILYCH AVENUE, DONETSK
83003, UKRAINE
[3]SINTECH SURFACE AND INTERFACE TECHNOLOGIES,
VOLMERSTRAßE 5–7, D–12489 BERLIN, GERMANY

1 Abstract

The kinetics of adsorption and the rheology of surface layers formed by the proteins β-lactoglobulin and β-casein at the water/air interface were investigated by the profile analysis tensiometer PAT1 and the shear rheometer ISR1. The adsorption behaviour can be described by a new thermodynamic model which accounts for a molar surface area of adsorbed protein molecules changing with surface pressure. The dynamics of adsorption is measured *via* the time dependence of interfacial tensions using the PAT1 over a broad time interval. Transient and harmonic relaxation studies performed also with PAT1 show the expected differences for the two proteins which can be understood in the framework of the thermodynamic and adsorption kinetic theory. At low molar protein bulk concentration, the relaxation process is mainly controlled by the reorganisation processes of the adsorbed molecules. The shear rheology of the interfacial layers investigated with the ISR1 providing information about the protein layer structure is in good agreement with the conclusions drawn from the adsorption and dilational rheology results.

2 Introduction

Studies of the equilibrium and dynamic adsorption behaviour of biopolymers at liquid interfaces in combination with thermodynamic and kinetic models allow us to draw conclusions about the structure of the interfacial layer. Additional information can be obtained from rheological investigations, both under shear and dilational deformations.

The state of the art of the thermodynamic description of protein adsorption layers was recently described in detail elsewhere.[1] On its basis also the kinetics of adsorption can be explained much better, *i.e.* taking into account the peculiarities of proteins arriving at the interface.[2] On the other hand, rheological studies are still on a more descriptive level and quantitative models do not exist yet to explain the relaxation processes going on in adsorption layers during perturbations.

Here we report on an adsorption and surface rheology study with β-lactoglobulin (β-LG) and β-casein (β-CS) at the water/air interface using essentially a newly designed profile analysis tensiometer that allows us to investigate adsorption processes at the surface of drops as well as bubbles. The differences in the interfacial behaviour of the two proteins is discussed.

3 Thermodynamic Model for the Adsorption Equilibrium

A generalised model for the adsorption of protein molecules in i different states in the adsorption layer was given in ref. 3. This thermodynamic model assumes that independent of the surface coverage (or surface pressure) adsorbed molecules arrange a respective conformation at the interface. The composition of the surface layer is then given by a certain distribution of molecular conformations such that at small surface coverage molecules occupy a maximum and at high coverage the smallest possible molar area.

To describe the adsorption behaviour of such systems the chemical potentials of the solvent and all configurations of the protein are set equal in bulk and at the surface. For non-ideal adsorption layers (in respect to enthalpy and entropy) a more general equation of state is obtained (*cf.* Equations (22) and (27) in ref. 3):

$$\Pi = -\frac{RT}{\omega_0}\left[\ln\left(1 - \sum\Gamma_i\omega_i\right) + \sum\Gamma_i\omega_i\left(1 - \frac{1}{n_i}\right) + a\left(\sum\Gamma_i\omega_i\right)^2\right] \qquad (1)$$

where R and T are the gas law constant and the absolute temperature, Π is the surface pressure, Γ_i is the partial adsorption of the molecules in state i, a is a coefficient that summarises the interaction between adsorbed molecules and the effect of electric charge, ω_i are the molar areas in state i and ω_0 is the molar area of solvent, $n_i = \omega_i/\omega_0$. Note, that summation in (1) is over the entire number of conformations of the proteins in the interfacial layer. A much simpler equation results when we choose $\omega_0 = \omega$ (dividing surface according to Lucassen-Reynders), where ω is the average molar area for all states. Then, the equations of state and adsorption isotherm, respectively, read:

$$-\frac{\Pi\omega}{RT} = \ln(1 - \Gamma\omega) - a(\Gamma\omega)^2 \tag{2}$$

$$bc = \frac{\Gamma_1\omega}{(1 - \Gamma\omega)^{\omega_1/\omega}} \tag{3}$$

where $\Gamma = \Sigma\Gamma_i$ is the total adsorption, c is the protein bulk concentration, Γ_1 and ω_1 are the partial adsorption and the molar area of the molecules in state 1, b is the equilibrium adsorption constant. The adsorptions of molecules in the i states are given by the relationships

$$\Gamma_i = \Gamma\frac{\exp\left(\dfrac{\omega_i - \omega_1}{\omega}\right)\exp\left[-\dfrac{(\omega_i - \omega_1)\Pi}{RT}\right]}{\sum\limits_{i=1}^{n}\exp\left(\dfrac{\omega_i - \omega_1}{\omega}\right)\exp\left[-\dfrac{(\omega_i - \omega_1)\Pi}{RT}\right]} \tag{4}$$

with the value of ω to be calculated from the relation

$$\omega\Gamma = \sum\limits_{i=1}^{n}\omega_i\Gamma_i \tag{5}$$

Note, that another choice of the dividing surface (for example $\omega_0 \neq \omega$, with a value of ω_0 close to the real one for water molecules, yields $\omega_0 \ll \omega$) does not change the form of the equation of state. One can show that for this case Equation (2) remains the same, however, the constant a then includes also the ratio ω/ω_0.

The equations can be essentially simplified for globular protein molecules, which can exist in the surface layer in two states only. In this case, the adsorption values in the states 1 and 2 (Γ_1 and Γ_2, respectively), characterised by different values of the partial molar area ω_i ($\omega_2 > \omega_1$), are related *via* the adsorption equation

$$\frac{\Gamma_2}{\Gamma_1} = \beta\exp\left[-\frac{\Pi}{RT}(\omega_2 - \omega_1)\right] \text{ with } \beta = \exp\left(\frac{\omega_2 - \omega_1}{\omega}\right). \tag{6}$$

The mean partial molar area (averaged over the two states) can be expressed by

$$\omega = \frac{\omega_1 + \omega_2\cdot\beta\cdot\exp[\Pi(\omega_1 - \omega_2)/RT]}{1 + \beta\cdot\exp[\Pi(\omega_1 - \omega_2)/RT]} \tag{7}$$

The Equations (1)–(7) have been used to describe the equilibrium surface tension isotherms. The fitting was based on a software package that was developed recently and is explained in ref. 4.

4 Dynamics of Protein Adsorption

The equation of state for a surface layer and the respective adsorption isotherm are the basis for calculations of adsorption kinetics and dynamic surface tensions. The equation proposed by Ward and Tordai[5] represents a general relationship between the dynamic adsorption and the subsurface concentration $c(0,t)$. For freshly formed and non-deformed surfaces it reads

$$\Gamma = 2\sqrt{\frac{D}{\pi}}\left[c_0\sqrt{t} - \int_0^{\sqrt{t}} c(0, t-\tau)d\sqrt{\tau}\right].$$ (8)

Here D is the diffusion coefficient, t is the time, τ is a dummy integration variable. Using Equation (8), respective $\Gamma(t)$ dependencies can be obtained, while the Equations (1)–(7) serve as boundary condition for the diffusion model. This set of equations yield a quasi-equilibrium diffusion model which means that at a given surface pressure the composition of the surface layer under dynamic conditions is equal to that in the equilibrium. Another regime of adsorption kinetics, called kinetic model, can also be described by assuming compositions of the adsorption layer that can differ from the equilibrium state. The deviation of the adsorption layer from the equilibrium composition is the result of the finite rate of the transition process between the adsorption states. In case of two adsorption states we have[6]

$$\frac{d\Gamma_1}{dt} = \frac{1}{2}\frac{d\Gamma_\Sigma}{dt} + k\left\{\Gamma_2 - \beta\Gamma_1 \exp\left[-\frac{\Pi}{RT}(\omega_2 - \omega_1)\right]\right\}$$ (9)

$$\frac{d\Gamma_2}{dt} = \frac{1}{2}\frac{d\Gamma_\Sigma}{dt} - k\left\{\Gamma_2 - \beta\Gamma_1 \exp\left[-\frac{\Pi}{RT}(\omega_2 - \omega_1)\right]\right\}$$ (10)

where k is the rate constant for the transition from state 2 to state 1, and $d\Gamma/dt$ is the flux of both 'species' from the bulk phase. Details of the two models have been summarised in ref. 2. The derivation of a new kinetic model is under way, which is more general and also applicable to flexible proteins, while the presented model is certainly restricted to globular proteins under conditions where they do not unfold at the interface but adsorb in two states only, for example in the orientations 'side on' or 'head on'.

5 Material and Methods

The β-casein (β-CS) and β-lactoglobulin (β-LG) from bovine milk were obtained from SIGMA. Molecular weights of 24 000 for β-CS and 18 400 for β-LG were used. The protein samples were used without further purification. The solutions were prepared with phosphate buffer made by mixing appropriate stock solutions of Na_2HPO_4 and NaH_2PO_4. The surface tension of a 10 mM buffer at pH = 7 was 72.7 mN/m. All measurements were performed at room temperature of 22°C.

The methods used in this study are mainly the drop and bubble shape technique and the torsion pendulum rheometry. The drop and bubble shape technique, which is based on fitting the shape of an axisymmetric liquid meniscus to the respective Gauss–Laplace equation, allows to continuously monitor the change in interfacial tension over a very large time interval, starting from a few seconds up to many hours. A proper experiment required accurate control loops to keep the drop or bubble surface area constant over the whole experiment. In addition this method can be used for dilational rheological experiments by performing transient or harmonic relaxations (frequencies of 0.01 to 1Hz). This technique can be applied to both liquid/gas and liquid/liquid interfaces without extensive modification. In the present investigations we used the instrument PAT1 manufactured by SINTECH, Berlin, as shown in Figure 1.

A detailed description of the instrument is given in a recent overview.[7] In brief, the shape of a drop or bubble is determined from a video image and compared with the Gauss–Laplace equation of capillarity. A best fit between the experimental shape coordinates and the theoretical meniscus curves is achieved by tuning the interfacial tension as the only free parameter. The PC controlled dosing system allows the performance of accurate surface tension measurements at constant surface area as well as transient and harmonic relaxation studies. Equipped with the optional piezo drive the PAT1 is able to perform oscillations with bubbles at higher frequencies, up to 1 Hz. Note, however, that experiments above a certain frequency limit results in non-equilibrium drop or bubble shapes which are no longer described by the standard Gauss–Laplace equation.

Dynamic surface tensions have also been measured by the bubble pressure tensiometer MPT2 and the drop volume tensiometer TVT1, both manufactured

Figure 1 *Photo of the Profile Analysis Tensiometer PAT1 from SINTECH, Berlin; 1 – measuring cell with capillary, 2 – dosing system, 3 – objective, 4 – box with CCD camera, 5 – platform with x-y-z stages to adjust the position of the cell*

by LAUDA, Germany. Detailed descriptions of both methods are given in the monograph.[8]

The shear rheology instrument ISR1 (SINTECH, Berlin) is based on free damped oscillations of a ring or disk located in the interfacial layer and suspended on a tungsten torsion wire. The frequency is of the order of 0.1 Hz and can be changed at certain intervals. Details of the instrument have been given elsewhere.[9]

6 Results

The surface tension isotherms γ as a function of protein bulk concentration c are shown in Figure 2. The data are obtained from the time dependencies of the measured $\gamma(t)$ curves shown in Figure 3 and Figure 4 by extrapolation *via* $\gamma(1/\sqrt{t})$.

Fitting the data to the simple 2-states model results in the following values of the isotherm parameters of Equations (1) and (2):

Protein	ω_1 (nm^2/molecule)	ω_2 (nm^2/molecule)	b (l/mol)	a (/)
β-CS	6	110	7×10^6	80
β-LG	7	22	3×10^8	45

The description of the data above the kink point in the isotherm is not possible with this model. In ref. 10 however, it was shown that the further course of the

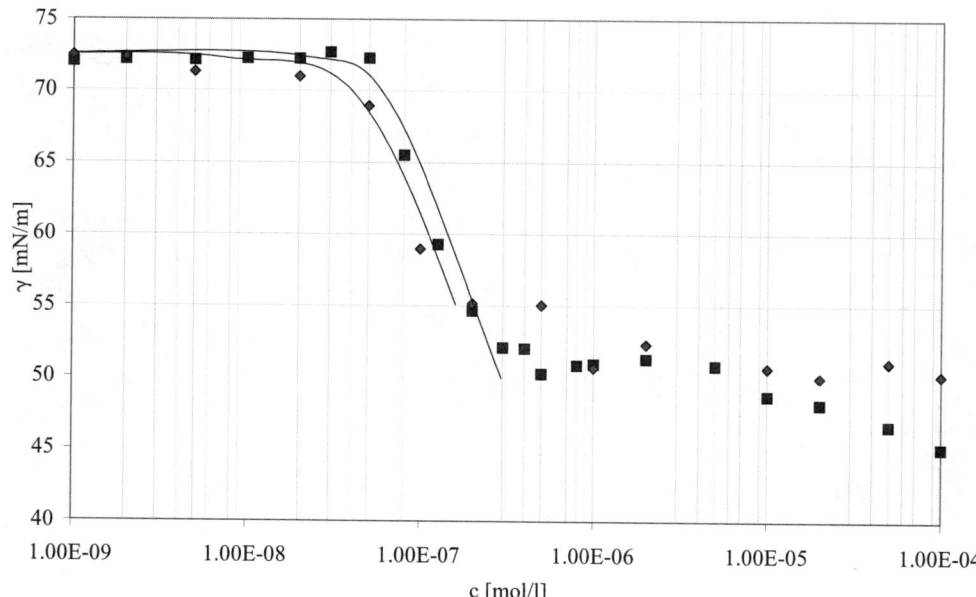

Figure 2 *Isotherms for β-LG ($\bullet$) and β-CS ($\blacksquare$) in buffer solution at 22°C, solid lines calculated according to the thermodynamic model given above*

isotherms can be described by an inset of surface aggregation between the adsorption molecules.

Figure 3 contains dynamic data for β-LG received by three methods: the maximum bubble pressure method in the time range 0.001 s to 100 s, the drop volume method for times in the range 5 s to 500 s, and the profile analysis tensiometer PAT1 in the time range from 10 s up to several hours.

The respective dynamic results for β-CS are given in Figure 4.

As one can see, the three methods complement each other and in the overlapping time interval they give one common curve. For low protein concentrations only the PAT1 yields surface tensions decreasing from the water value at larger adsorption times. For the largest concentrations, especially for the PAT1 experiments, care was taken to start with drops without a significant initial load, as this would lead to initially much lower values than those measured with the MPT2 or TVT1 where the surfaces are continuously freshly formed due to the detachment of the bubble or drop, respectively. In order to get sufficiently fresh surfaces for the PAT1 studies some drops are quickly produced and discarded before the drop for the measurement is formed.

The software of the PAT1 allows the programming of a complete sequence of experiments, such as transient and harmonic relaxations after the equilibrium of adsorption has been established. The oscillations were usually performed at various frequencies, so that sufficient data for the dilational rheology were available. A typical result, as an example, is shown for β-LG in Figure 5, where

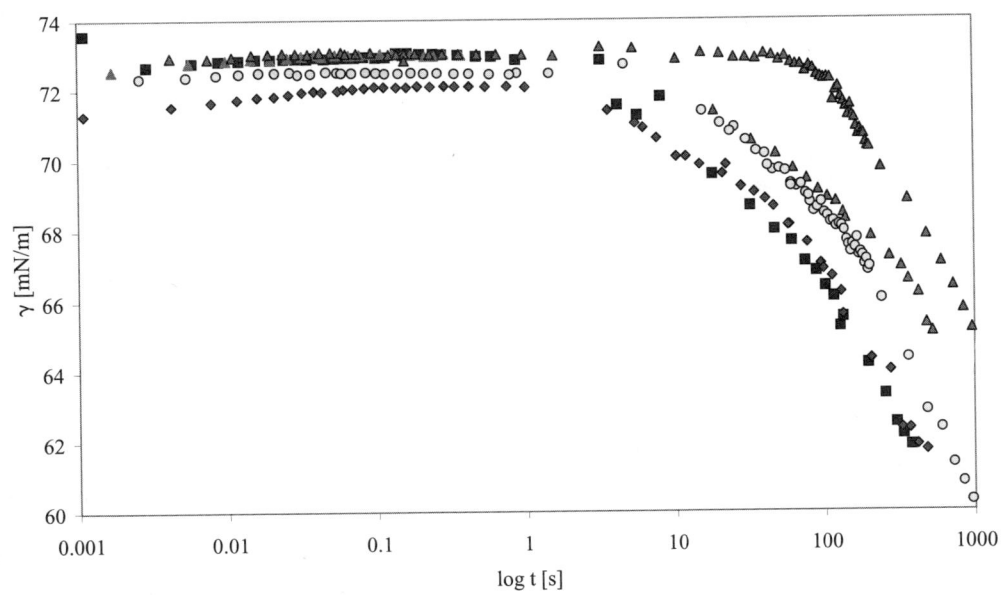

Figure 3 *Adsorption kinetics for different β-LG concentrations in buffer at 22°C, points measured by PAT1, MPT2 and TVT1 in the respective time windows; concentrations from top to bottom and right to left: 5×10^{-7}, 1×10^{-6}, 2×10^{-6}, 5×10^{-6}, 1×10^{-5} mol/l*

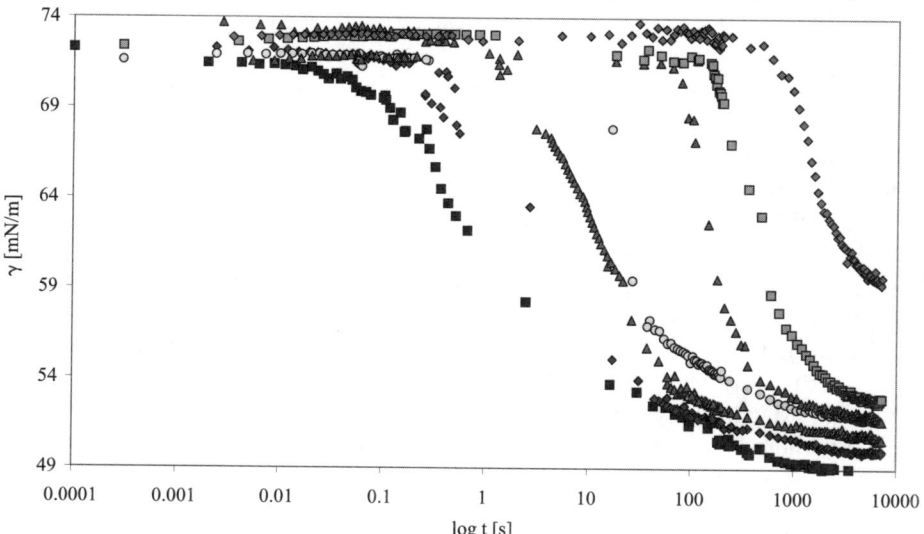

Figure 4 *Adsorption kinetics for different β-CS concentrations in buffer at 22°C, points measured by PAT1, MPT2 and TVT1 in the respective time windows; concentrations from top to bottom and right to left: 2×10^{-7}, 5×10^{-7}, 1×10^{-6}, 2×10^{-6}, 5×10^{-6}, 1×10^{-5}, 2×10^{-5} mol/l*

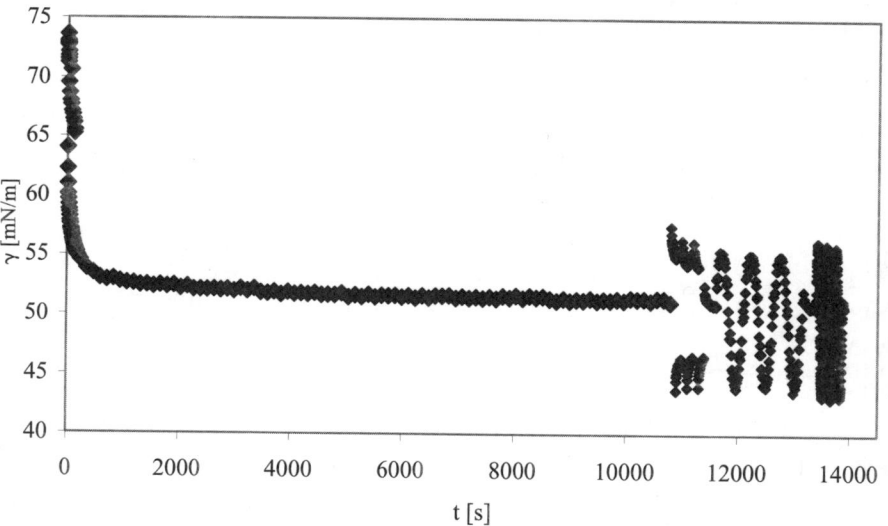

Figure 5 *Typical dynamic experiment consisting of adsorption kinetics, transient and harmonic relaxations for a β-LG concentrations of 1×10^{-5} mol/l in buffer at 22°C*

after three square pulses harmonic oscillations at two different frequencies are performed.

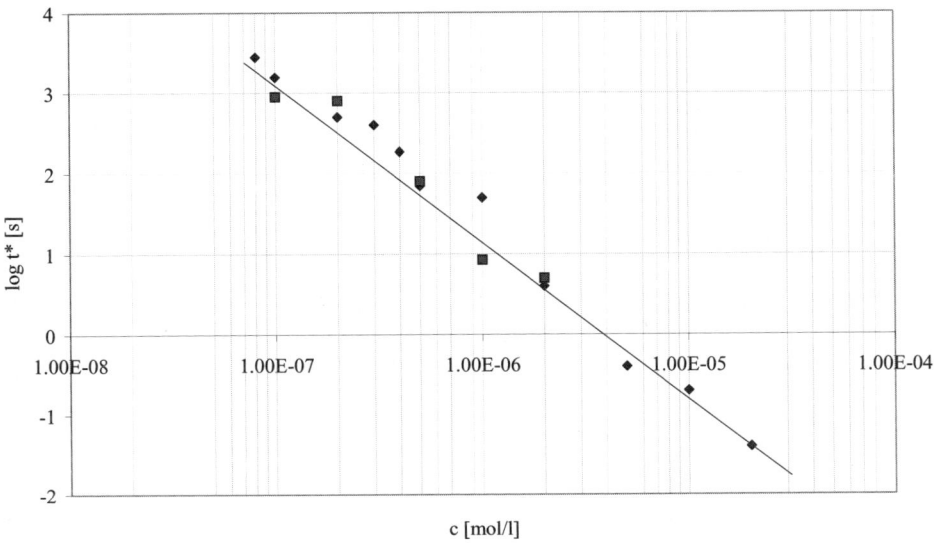

Figure 6 *Induction time as a function of protein concentration as determined from the dynamic surface tensions given in Figures 3 and 4; β-LG (■), β-CS (♦)*

7 Discussion

The dynamic surface tensions for four β-LG concentrations are shown again in Figure 6 together with the curves calculated for the quasi-equilibrium and kinetic regime model. For all concentrations there is better agreement with the experimental points for the kinetic model rather than for the quasi-equilibrium model.

For the highest β-LG concentration of $c = 10^{-5}$ mol/l the deviation from the experimental results is largest. While the kinetic diffusion model is still in acceptable agreement with the experiment, the quasi-equilibrium model predicts a much faster decrease of the surface tension, leading to the equilibrium state after approximately 100 s. To establish agreement between the experimental data and the quasi-equilibrium model a diffusion coefficient two orders of magnitude larger would be required. The agreement between experiment and theory is however not good enough and further work on the kinetics theory is required.

An easy control, if the diffusion mechanism is the one controlling the adsorption process, is to determine the so-called induction time, *i.e.* the time at which the surface tension starts to decrease. According to the diffusion law this time should increase with the square root of the protein bulk concentration. In Figure 7 the data obtained from the dynamic surface tensions are shown in a double logarithmic plot.

We obtain a linear dependence with a slope of about − 2, which is exactly what was expected from the theory. Hence, we can conclude that at least in the short time range, until there is enough material adsorbed so that the surface tension starts to decrease, the controlling adsorption mechanism is diffusion.

Let us now look into the surface dilational rheology of the adsorbed protein

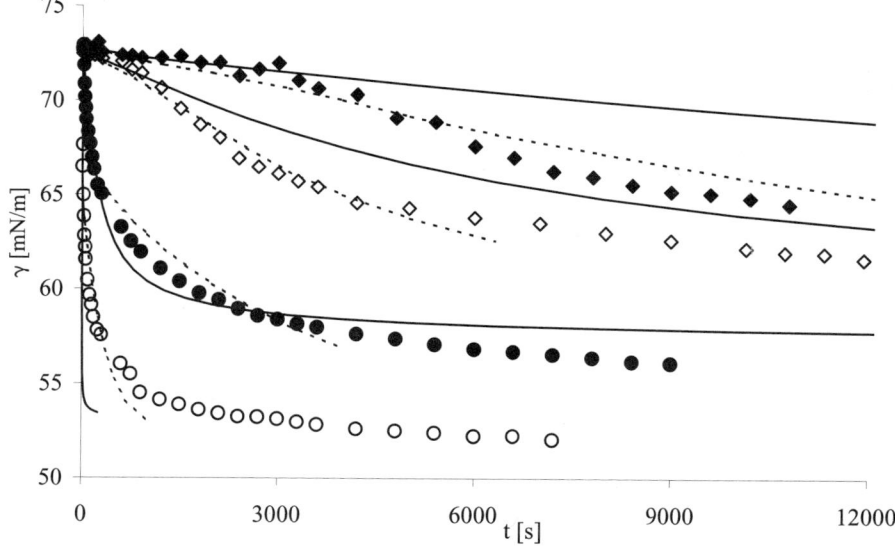

Figure 7 $\gamma(t)$ *for β-LG solutions in buffer at 22°C, solid line – quasi-equilibrium model, dotted lines – kinetic model, $c = 1 \times 10^{-7}$ (♦), 2×10^{-7} (◇), 1×10^{-6} (●) and 1×10^{-5} (○) mol/l*

layers. The data obtained from oscillation experiments can be presented in the form of an ellipse when plotting the surface tension change $\Delta\gamma$ as function of the relative area change $\Delta A/A$. Examples are given in Figure 8. From the tilt of the ellipse and its thickness the elasticity and viscosity, respectively, can be estimated. From (a) over (b) to (c) we see a decrease of the tilt of the ellipse, which is identical with a decrease in the elasticity. The thickness of the ellipses increase relative to the absolute values, hence the viscosity increases.

The elasticities and viscosities calculated from the oscillations as a function of the concentration are given in Figures 9 and 10. The two proteins show a completely different behaviour. The concentration dependencies for β-CS are fully in line with what we would expect for a slightly soluble surfactant. At fixed frequency, the elasticity increases while the viscosity decreases. This behaviour is observed for systems with a characteristic relaxation frequency larger than the applied frequency. Note, there is a peculiarity at the concentration where the isotherm shows a kink in the slope.

For β-LG the behaviour is different The elasticity increases simultaneously with the viscosity, as demonstrated in Figure 10.

Such behaviour is understood only for systems where the relaxation frequency is relatively low as compared with the applied oscillation frequency. Again, at a concentration where the adsorption isotherm shows a decreasing slope, the viscosity as well as the elasticity shows a non-continuous behaviour. Either aggregates in the bulk or in the surface layer come into play and cause these peculiarities.

A full understanding can be gained only from studies over a large frequency interval, which is the aim of studies under way. The instrument PAT1 allows

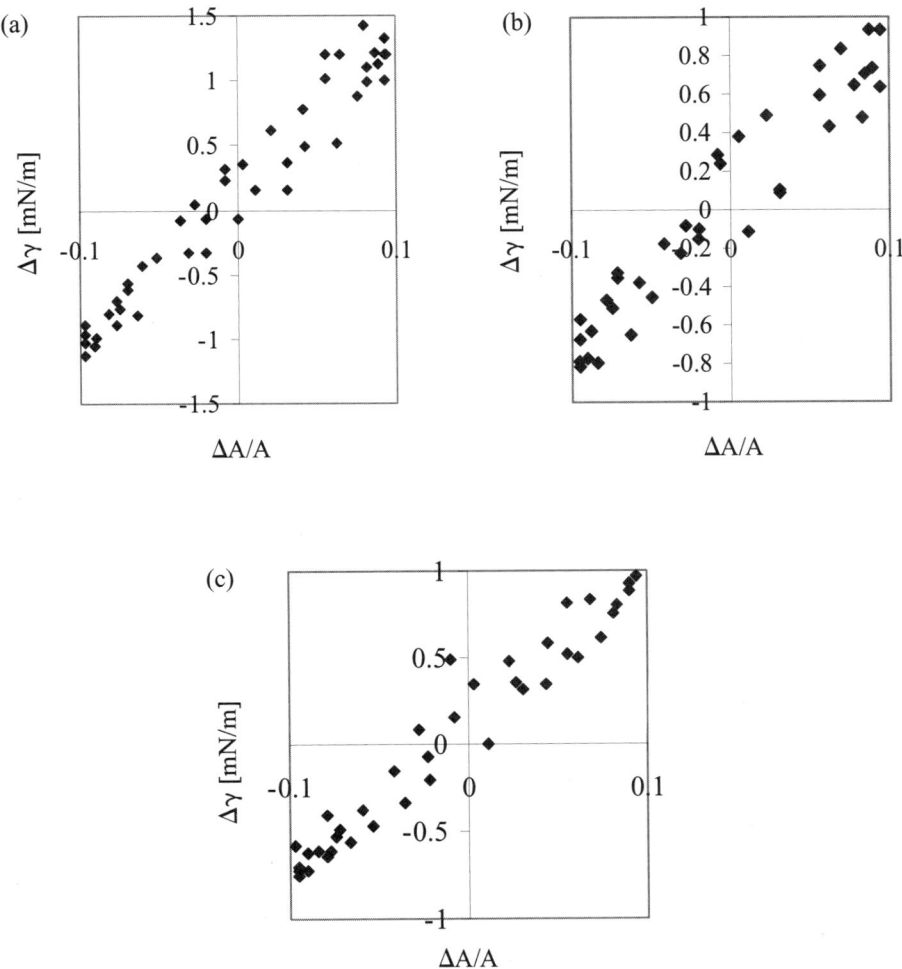

Figure 8 *Presentation of oscillation experiments in form of* $\Delta\gamma$ *as function of* $\Delta A(t)/A$ *for* β*-CS solutions in buffer at 22°C, (a)* $c = 5 \times 10^{-7}$ *mol/l, (b)* $c = 5 \times 10^{-6}$ *mol/l, (c)* $c = 1 \times 10^{-4}$ *mol/l; frequency 0.002 Hz*

studies of frequencies about two orders of magnitude higher than those used here, *i.e.* up to 0.2 Hz, which corresponds to 5 seconds per period. Higher frequencies are not suitable as the resulting drop or bubble shapes are no longer Laplacian and obtained data cannot be interpreted properly.

8 Conclusions

The studied proteins β-CS and β-LG essentially adsorb according to a diffusion controlled model. The isotherm obtained from the kinetics data by extrapolation

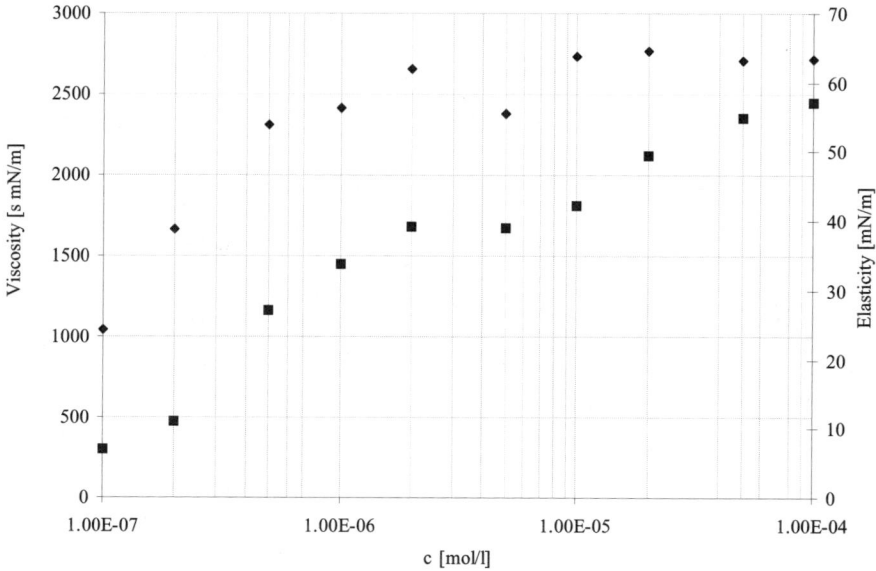

Figure 9 *Dilational elasticity (♦) and viscosity (■) for β-LG in buffer at 22°C at a frequency of 0.002 Hz*

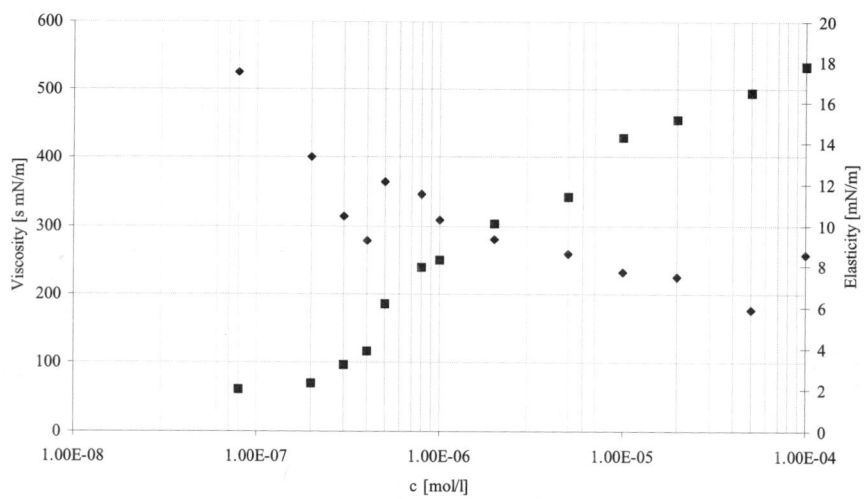

Figure 10 *Dilational elasticity (♦) and viscosity (■) for β-CS in buffer at 22°C at a frequency of 0.002 Hz*

can be described by a thermodynamic model based on a molar surface area changing with surface coverage (surface pressure). The elasticity and viscosity of the adsorption layers show peculiarities at a critical concentration in the adsorption isotherm, which can be called critical aggregation concentration CAC. This could be explained by a second layer or by a strong interaction, *i.e.* aggregation, in the interfacial layer, as it was described in ref. 10. PAT1 from SINTECH allows to measure dynamics of adsorption and gives access to the dilational rheology using transient and harmonic relaxation (0.001 to 0.2 Hz). It also allows for dynamic contact angle studies, which were not shown here.

Acknowledgements

The work was financially supported by a long-term research programme of the Max Planck Society.

References

1. V. B. Fainerman, E. H. Lucassen-Reynders and R. Miller, *Colloids & Surfaces A*, 1998, **143**, 141.
2. R. Miller, E. V. Aksenenko, V. B. Fainerman and U. Pison, *Colloids & Surfaces A*, 2001, **183**, 381.
3. V. B. Fainerman, E. H. Lucassen-Reynders and R. Miller, *Colloids Surfaces A*, 1998, **143**, 141,
4. E. V. Aksenenko in *Surfactants – Chemistry, Interfacial Properties and Application*, D. Möbius and R. Miller (eds.), Elsevier, 2001, Vol. 13, Chapter 7.
5. A. F. H. Ward and L. Tordai, *J. Chem. Phys.*, 1946, **14**, 543.
6. R. Miller, E. V. Aksenenko, L. Liggieri, F. Ravera, M. Ferrari and V. B. Fainerman, *Langmuir*, 1999, **15**, 1328.
7. G. Loglio, P. Pandolfini, R. Miller, A. V. Makievski, F. Ravera, M. Ferrari and L. Liggieri 'Novel Methods to Study Interfacial Layers', in *Studies in Interface Science*, D. Möbius and R. Miller (eds.), Elsevier, Amsterdam, 2001, Vol. 11.
8. D. Möbius and R. Miller (eds.), 'Drops and Bubbles in Interfacial Science', in *Studies of Interface Science*, Elsevier, Amsterdam, 1998, Vol. 6.
9. J. Krägel, S. Siegel, R. Miller, M. Born and K.-H. Schano, *Colloids Surfaces A*, 1994, **91**, 169.
10. V. B. Fainerman and R. Miller, *Langmuir*, 1999, **15**, 1812.

Dynamic Surface Tension and Surface Dilational Properties of an Amphiphilic Polysaccharide

Samuel Guillot,[§] Dominique Guibert, Monique A. V. Axelos*

UNITÉ DE PHYSICO CHIMIE DES MACROMOLÉCULES, INSTITUT NATIONAL DE LA RECHERCHE AGRONOMIQUE, RUE DE LA GÉRAUDIÈRE, BP 71627, 44316 NANTES CEDEX 3, FRANCE
[§]PRESENT ADDRESS: LABORATOIRE DE PHYSIQUE DES SOLIDES, UNIVERSITÉ DE PARIS SUD, BÂTIMENT 510, 91405 ORSAY, FRANCE

1 Abstract

The adsorption of methylcellulose at the air/water interface and the dilational viscoelastic properties of the interface have been studied using a dynamic drop tensiometer. Methylcellulose adsorbs at the air/water interface due to its partially hydrophobic nature. The decrease of the surface tension with time displays a complex non-exponential behaviour. Very long times are necessary to equilibrate the surface tension. The Gibbs adsorption isotherm obtained from the equilibrium surface tension allows the determination of a critical aggregation concentration of 10^{-3} g/l and a surface area of 13 nm^2. The dilational elastic modulus ε is very high and exhibits a maximum at a polymer concentration around 4 g/l. The surface dilational viscoelastic properties have also been measured at one concentration during polymer adsorption. At low surface pressure, a linear increase of the dilational elasticity with the surface pressure π is observed. From the slope of the ε–π curve a value of 0.66 was found for the excluded volume critical exponent. This result suggests that the air/water interface is not a good solvent for the methylcellulose.

2 Introduction

Methylcellulose (MeC) is a neutral polysaccharide which is often used in the food, pharmaceutical and cosmetic industries. As a food additive, under the name E461, it is employed for its emulsifying, thickening and gelling properties.[1] MeC has also been known for a long time for the property of forming gels in water on heating.[2] Methylation of cellulose is generally achieved in a heterogeneous phase (Williamson etherification), leading to highly substituted or hydrophobic zones along the original hydrophilic backbone.[3] Thus, MeC in water can be viewed as a random block copolymer in a selective solvent. The gelation process of MeC is reversible and is associated with a phase separation which is caused by hydrophobic interactions above a Lower Critical Solution Temperature (LCST).[4] Interesting properties of this complex system result from its strong temperature–concentration dependence: below the LCST, the solution is a single phase system; when crossing the coexistence curve by increasing the temperature, a phase separation occurs. Instead of a macroscopic phase separation a turbid gel appears because of the presence of hydrophilic sequences, as shown in Figure 1. At very low concentration aggregation occurs leading to large non self-similar objects.[5]

Dynamic properties of interfaces have attracted attention for many years because they help in understanding the behaviour of polymer, surfactant or mixed adsorption layers.[6] In particular, interfacial rheology (dilational properties) is crucial for many technological processes (emulsions, flotation, foaming, etc.).[7] The present work deals with the adsorption of MeC at the air–water interface. Because of its amphiphilic character MeC is able to adsorb at the liquid interface thus lowering the surface tension. Our aim is to quantify how surface active this polymer is, and to determine the rheological properties of the layer. A qualitative and quantitative evaluation of the adsorption process and the dilational surface properties have been realised by dynamic interface tension measurements using a drop tensiometer and an axisymmetric drop shape analysis.

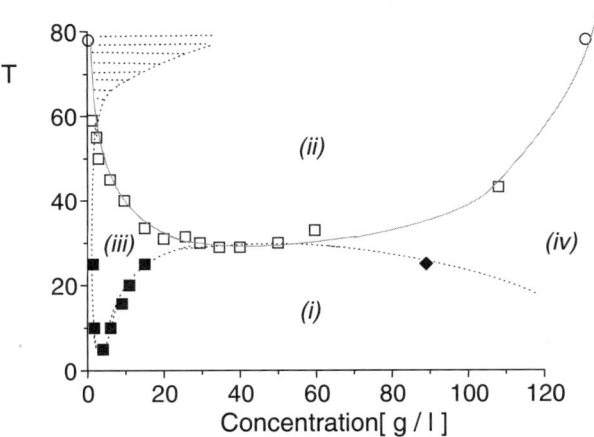

Figure 1 *Temperature/concentration phase diagram of aqueous solutions of methylcellulose A4C,* ■ ◆ *sol–gel transition,* □, ○ *binodal curve. (i) solution, (ii) turbid gel, (iii, iv) clear gel; shaded region: occurrence of the phase separation after one month*

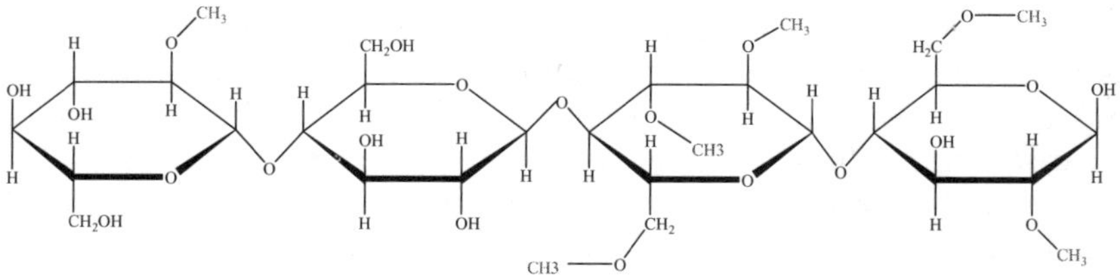

Figure 2 *Schematic illustration of the molecular structure of methylcellulose*

3 Experimental

3.1 Materials

For this study we used a commercial Methocel A4C sample kindly supplied by The Dow Chemical Company. The degree of substitution (DS), which corresponds to the average number of methyl groups per anhydroglucose unit is 1.7. This linear copolymer contains four different monomers with different hydrophobicities (Figure 2). MeC is a rather hydrophobic polymer. The tri- and di-substituted monomers are the most hydrophobic ones; they represent 61% of the backbone chain.[8] Its average molecular mass is $M_w = 129\,000$ g/mol.

After extensive dialysis to remove products due to its synthesis, MeC solutions were prepared by dispersing the polymer, in freeze-dried form, in de-ionised water at 4°C to avoid aggregation.[9] The dispersion is stirred at the same temperature until a clear solution is obtained, and the solution is then carefully filtered.

3.2 Determination of Surface Dilatational Parameters

Dynamic surface tension measurements have been performed using a dynamic drop tensiometer in the rising bubble configuration (IT Concept, Longessaigne, France).[10] The tip of a U-shaped needle attached to an air filled syringe is dipped in the methylcellulose solution. With this experimental set-up it is possible to form an air bubble whose volume can be well controlled. The bubble then is illuminated with a uniform source of white light. The image of the bubble profile is recorded by a CCD camera and is digitised and stored for the analysis of its shape using the Laplace–Young equation.[11]

For the first measurements we followed the evolution of $\gamma(t)$ during the adsorption process. After having reached equilibrium, the complex surface dilatational modulus ε^* is obtained from the response of the surface to a sinusoidal dilatation/compression deformation. As usual, the real part corresponds to the elastic properties and the imaginary part to the dissipative properties:

$$\varepsilon^* = \varepsilon + i\,\omega\,\kappa \qquad (1)$$

where ω is the frequency, ε the dilational elastic modulus and κ the dilational viscosity. The changes in the bubble surface area, A, should not exceed a few

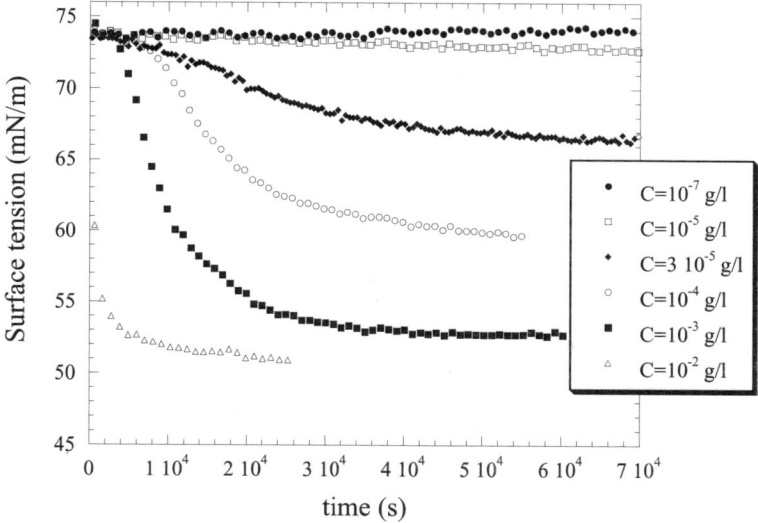

Figure 3 *Dynamic surface tensions of MeC solutions for different polymer concentrations*

percent in order to stay in the linear viscoelastic regime. The dilational elasticity then is obtained from:

$$\varepsilon = - A \, \partial\pi/\partial A \tag{2}$$

with π the surface pressure: $\pi = \gamma_0 - \gamma$, where γ_0 is the surface tension of pure water.

A second more detailed type of investigation is the determination of the dilational elastic modulus at a given frequency during the adsorption process. This allows us to determine the variation of ε with the surface pressure.

4 Results and Discussion

4.1 Dynamic Surface Tension Measurements

The adsorption behaviour is characterised by adsorption isotherms. We have performed dynamic surface tension measurements by determining $\gamma(t)$ for polymer concentrations ranging from 10^{-7} to 20 g/l at 15 °C. Figure 3 shows several adsorption curves as function of time at different concentrations of MeC. The decrease in γ is very slow at low polymer concentration. The curves display a lag time which decreases as the polymer concentration increases, followed by a complex non-exponential behaviour. In all dynamic surface tension measurements, the surface tension decreases and eventually becomes constant indicating a steady state. Due to the large molecular weight of MeC, very long times are necessary to reach this equilibrium. The adsorption time depends, of course, on the MeC concentration, with the time to reach the plateau often being more than 16 hours.

Equilibrium is reached when the dynamic surface tension stays constant. The

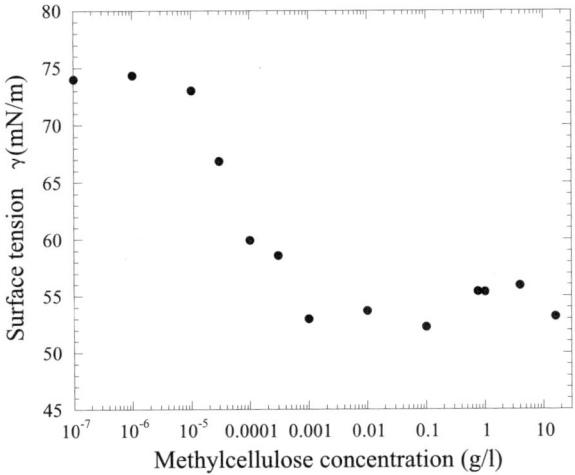

Figure 4 *Adsorption isotherm γ versus logC of aqueous MeC solutions at the air–water interface at 15°C*

equilibrium values of the surface tension are plotted as a function of the polymer concentration in Figure 4. At very low polymer concentration, below $C = 10^{-5}$ g/l, the surface tension is equal to the one of a pure air-water interface. This means that no or very few polymers adsorb at the interface, in any case not enough to have an effect on the surface tension. Above 10^{-5} g/l, the surface tension decreases with increasing polymer concentration until $C = 10^{-3}$ g/l and a value of surface tension of $\gamma = 53$ mN/m. Upon further increasing the polymer concentration, γ remains roughly constant. The Gibbs adsorption isotherm displays the classic behaviour, as found for surfactants, with a critical aggregation concentration of $C = 10^{-3}$ g/l.

The adsorbed amount of polymer Γ as a function of concentration is given by the Gibbs equation:

$$\Gamma = \frac{C}{RT} \cdot \frac{\partial \gamma}{\partial C} \tag{3}$$

where R is the gas constant, T the absolute temperature, γ the equilibrium interfacial tension and C the bulk concentration. For an insoluble monolayer the surface area A occupied per molecule is deduced from the equation:

$$\Gamma \sim 1/A \tag{4}$$

The surface area per molecule using the above equations is equal to $A = 13$ nm^2.

4.2 Surface Rheological Properties

The surface rheological properties of adsorbed layers of MeC at equilibrium have been measured at several polymer concentrations and at different frequencies. As is shown in Figure 5, the real and imaginary parts of the complex dilational modulus are quite constant for all frequencies. Moreover, the real part is much larger than the imaginary part, which indicates that the interfacial layer is almost purely elastic in the concentration range studied. The values of the

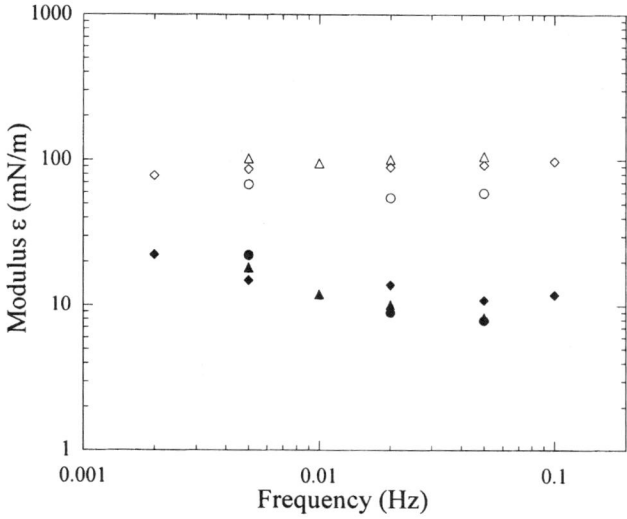

Figure 5 *Real (open symbols) and imaginary (filled symbols) parts of the complex dilational modulus of an adsorbed layer at equilibrium. The circles, diamonds and triangles correspond to the MeC concentration of 3×10^{-4}, 0.1 and 16 g/l, respectively*

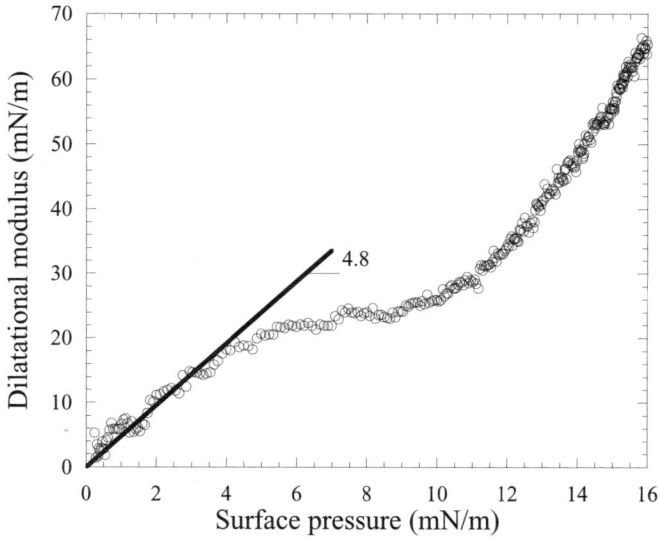

Figure 6 *ε–π curve at $C = 1.7 \times 10^{-2}$ g/l*

dilational elastic modulus displays a maximum for a polymer concentration of 4 g/l.

We have also measured $\gamma(t)$ and $\varepsilon(t)$ during polymer adsorption for a given concentration. In Figure 6, the ε–π curve, the equation state of the layer during the adsorption process, is presented. At low surface pressure, one observes a linear increase of the dilational elastic modulus with the surface pressure π. From the slope of the linear part of the ε–π curve, a value of 0.66 was found for the excluded volume critical exponent. The same value has been measured elsewhere with another technique.[12] This result indicates that, unlike the excluded volume chain behaviour in the bulk, the air–water interface is not a good solvent for MeC. At intermediate surface pressures, the modulus levels off and then increases again until the equilibrium surface pressure is reached.

5 Conclusions

MeC adsorbs at the air–water interface leading to a decrease in the surface tension. An equilibrium state is only reached at very long times, due to the high molecular mass of methylcellulose. The layer which is present at the interface at equilibrium is almost purely elastic with a large dilational elastic modulus. The value of the excluded volume critical exponent extracted from the ε–π curve indicates that the air-water interface is not a good solvent for the polymer.

References

1. D. K. Sarker, M. A. V. Axelos and Y. Popineau, *Coll. Surf. B*, 1999, **12**, 147.
2. E. Heymann, *Trans. Farad. Soc.*, 1935, **31**, 846.
3. P. W. Arisz, H. J. J. Kauw and J. J. Boon, *Carbohydr. Res.*, 1995, **271**, 1.
4. C. Chevillard and M. A. V. Axelos, *Coll. Polym. Sci.*, 1997, **275**, 537.
5. S. Guillot, D. Lairez and M. A. V. Axelos, *J. Appl. Cryst.*, 2000, **33**, 669.
6. A. Bhattacharyya, F. Monroy, D. Langevin and J-F. Argillier, *Langmuir*, 2000, **16**, 8727.
7. R. Miller, R. Wüstneck, J. Krägel and G. Kretzschmar, *Coll. Surf. A*, 1996, **111**, 75.
8. M. Vigouret, M. Rinaudo and J. Desbrieres, *J. Chim. Phys.*, 1996, **93**, 858.
9. W. B. Neely, *J. Polym. Sci.*, 1963, part A, I, 311.
10. S. Labourdenne, N. Gaudry-Rolland, S. Letellier, M. Lin, A. Cagna, G. Esposito, R. Verger and C. Riviere, *Chem. Phys. Lipids*, 1994, **71**, 163.
11. B. Song and J. Springer, *J. Coll. Int. Sci.*, 1996, **184**, 64.
12. Q. Jiang and Y. C. Chiew, *Macromol.*, 1994, **27**, 32.

Polymerization of Coniferyl Alcohol (Monomer of Lignins) at the Air/Water Interface

B. Cathala,* V. Aguié-Béghin, R. Douillard and B. Monties

INRA, UMR FRACTIONNEMENT DES AGRORESSOURCES ET EMBALLAGES INRA/URCA, EQUIPE PAROI VÉGÉTALES ET MATÉRIAUX FIBREUX, CRA, 2 ESP. R. GARROS BP 224, 51686 REIMS CEDEX

Abstract

Polymerization of coniferyl alcohol is attempted at the air/water interface. The polymerization process was monitored by surface tension, ellipsometry and neutron reflectivity. The formation of the interfacial layer was found to proceed according to two steps: formation of a dilute layer and then densification.

1 Introduction

Lignins are natural polymers occurring in plant cell walls and they represent, after cellulose, the most abundant polymers in nature. The biosynthetic reaction mechanism in the formation of lignins is very sensitive to the reaction conditions (pH, temperature, occurrence of polysaccharide, *etc.*), due to the oxidative coupling of monolignols. During that process, delocalised phenoxyl radicals are generated upon phenol dehydrogenation of *p*-hydroxycinnamyl alcohol monomers and intermediate lignin units. These radicals are coupled in a variety of ways to build up the lignin polymer. Each type of inter-unit linkage has a frequency which changes according to several parameters. Among all the parameters it has been hypothesized that the physical environment may change the reactivity of the monolignols by, for instance, a specific orientation,[1] template effect,[2] or changes in the solvation of radicals.[3] This assumption is consistent with the fact that the plant cell wall is a heterogeneous and composite material.

Cellulose and hemicellulose containing a large amount of hydroxyl groups can be considered as hydrophilic substances,[1] whereas lignins, constituted of phenyl propane subunits are rather hydrophobic and these polymers are closely associated within plant cell walls, forming hydrophilic/hydrophobic interfaces.

In order to evaluate the influence of the hydrophilic/hydrophobic environment on the reactivity of a lignin monomer, polymerization of coniferyl alcohol is attempted at the air/water interface. The reaction process is monitored by static tensiometry, ellipsometry and neutron reflectivity.

2 Experimental

Coniferyl alcohol (CA) was synthesised as previously described[4] and was used as a 10 mg/L solution in phosphate buffer (1/30 N, pH 5.5).

Peroxidase solution: 1.5 mg of Peroxidase Type VI (Sigma, 250–330 units/mg) were dissolved in 5 ml phosphate buffer (1/30 N, pH 5.5). Spread surface concentration was 0.0375 mg/m^2 in all experiments.

Hydrogen peroxide solution (30% w/w in water) was deposited at the air/water interface. The total deposited amount was equivalent to 0.9 mg/L of the bulk phase.

Surface pressure was the difference between the surface tension of pure water and that of the experimental system. It was measured at 20°C using a Wilhelmy plate (KSV sigma 70).

Ellipsometric measurements were performed using a spectroscopic phase modulated ellipsometer (UVISEL, Jobin Yvon, Longjumeau, France). The polarizer and the analyser were set to a 45° angle, the photoelastic modulator, activated at the 50 kHz frequency, was set to 0° orientation. The spectroscopic measurements were monitored between 240 and 820 nm and the angle of incidence was 53°4. The two ellipsometric angles Ψ and Δ are linked to the two reflectivity coefficients rp and rs in the direction parallel and perpendicular to the incidence plane respectively by: rp/rs $= \tan\Psi \exp(i\Delta)$. The fixed wavelength chosen for the kinetic measurements corresponds to the Brewster conditions of the substrate (coniferyl alcohol solution) defined by: $\Delta = \pm \pi/2$. The ellipticity coefficient of the adsorption layer measured in the Brewster conditions, $\bar{\rho}_B$, is defined by: $\bar{\rho}_B = \tan\Psi \sin\Delta$.

Neutron reflectivity: reflectivity spectra were determined with a polychromatic beam (3 Å $\leq \lambda \leq$ 18 Å) of neutrons at a fixed incident angle (1.46°) at the EROS reflectometer in the Orphée reactor (Léon Brillouin Laboratory, Saclay). The reflectivity measurements were performed using a teflon trough 6.5 × 13.5 × 0.3 cm in a gas tight cell thermostated at 20°C. The trough was filled up with 11 mL of Coniferyl Alcohol solution. Peroxidase and H$_2$O$_2$ peroxide solutions were spread at the air/water interface. The reflectivity spectra was recorded as two hour scans during twelve hours.

3 Results and Discussion

3.1 Principle of the Polymerization

Enzyme mediated polymerization of various monomers was already achieved on a Langmuir trough and resulted in polymers structurally different from those obtained in a bulk process because of the arranging effect of the interface on the monomers.[5-7] However, these studies concerned surface active monomers which were mostly located at the air/water interface whereas the polymerization agents (peroxidase, $FeCl_3$, *etc.*) were dissolved in the bulk. Consequently the reaction occurred because the polymerization agents diffuse from the bulk to the interface. Such a procedure cannot be applied in our case because of the high solubility of the monomer (coniferyl alcohol) in water and its lack of surface activity. Thus, in contrast to these studies, CA was dissolved in the bulk and the polymerization agent (peroxidase) spread at the air/water interface. Since proteins have a good affinity for the air/water interface, it has been demonstrated that in the case of a low surface deposition, the spreading procedure results in a nearly quantitative adsorption at the interface.[8] In the present study, the peroxidase amount spread is very low (0.0375 mg/m^2) and it can be assumed that all the peroxidase remains at the interface, and that the polymerization reaction occurs only at the interface. This very low amount of spread peroxidase cannot

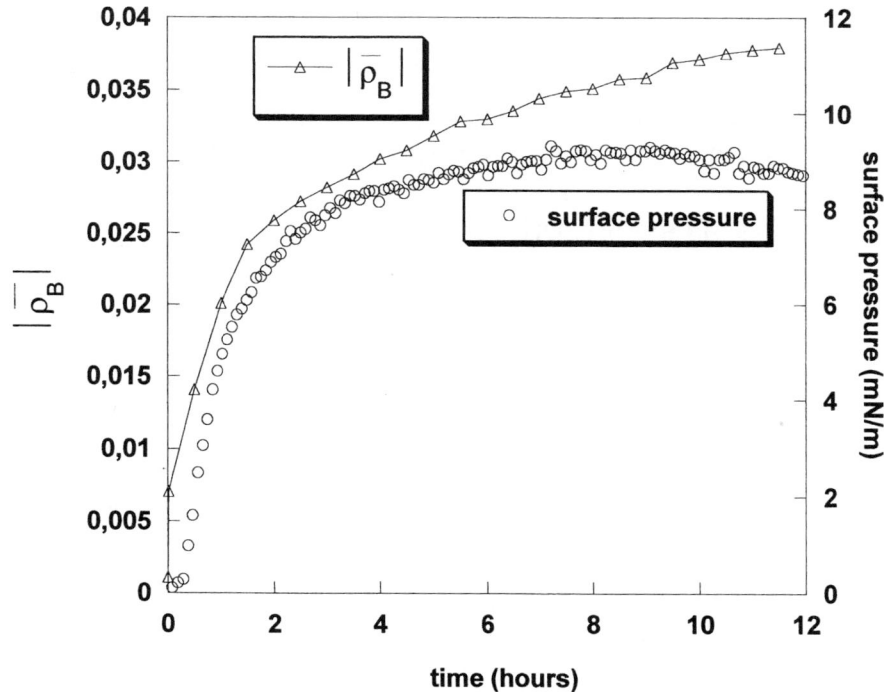

Figure 1 *Surface tension and ellipticity $|\bar{\rho}_B|$ during the polymerization of coniferyl alcohol at the air/water interface*

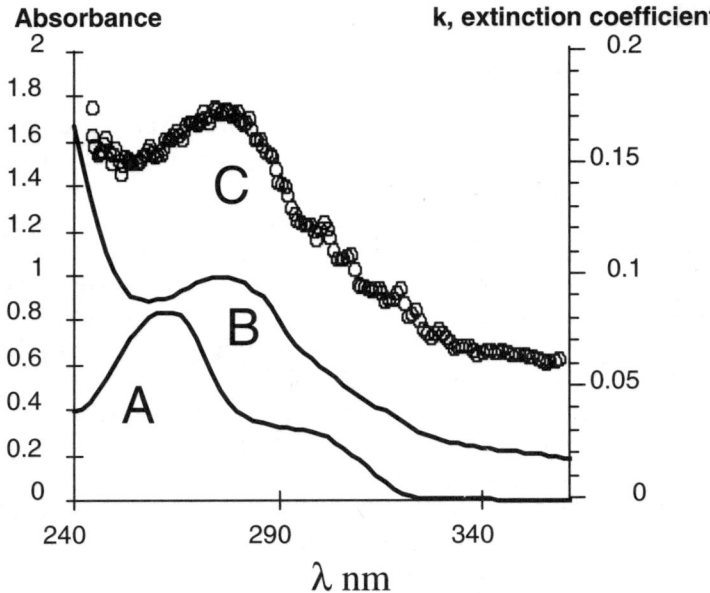

Figure 2 (*A*) *UV spectrum of coniferyl alcohol in solution.* (*B*) *UV spectrum of DHP synthesised in solution.* (*C*) *Extinction coefficient (k) of the adsorption layer after twelve hours of the polymerization process. The quantity k is calculated by point per point inversion and is related to the absorption coefficient of the adsorbed molecules*

be detected by surface tension, ellipsometry, or neutron reflectivity. Only the interface polymerization reveals the occurrence of the peroxidase.

3.2 Kinetics of the Polymerization

The polymerization process was firstly monitored by tensiometry (Figure 1). At the beginning no surface pressure was measured indicating that coniferyl alcohol is not surface active. After peroxidase and hydrogen peroxide spreading, the surface pressure increases indicating the formation of an interfacial layer. After 6–8 hours, the surface pressure levels off around 9 mN/m and then tends to decrease. This behaviour can be explained by a desorption of the molecules from the interface to the bulk or by a change of the interfacial layer structure.

Polymerization was also monitored by ellipsometry at the incidence angle of 53°4. The Brewster wavelength was found to be 387 nm on the coniferyl alcohol solution with or without H_2O_2. The ellipticity of this substrate is slightly positive indicating that the main contribution is the roughness of the interface (no interfacial layer).[9] After peroxidase spreading, the absolute value of the ellipticity increases, indicating the formation of an interfacial layer (Figure 1). After an initial sharp increase, ellipticity steadily and slowly increases in contrast with the surface pressure which stabilizes. This demonstrates that some structural evolution occurs in the layer. The absorption spectrum of the adsorbed molecules in

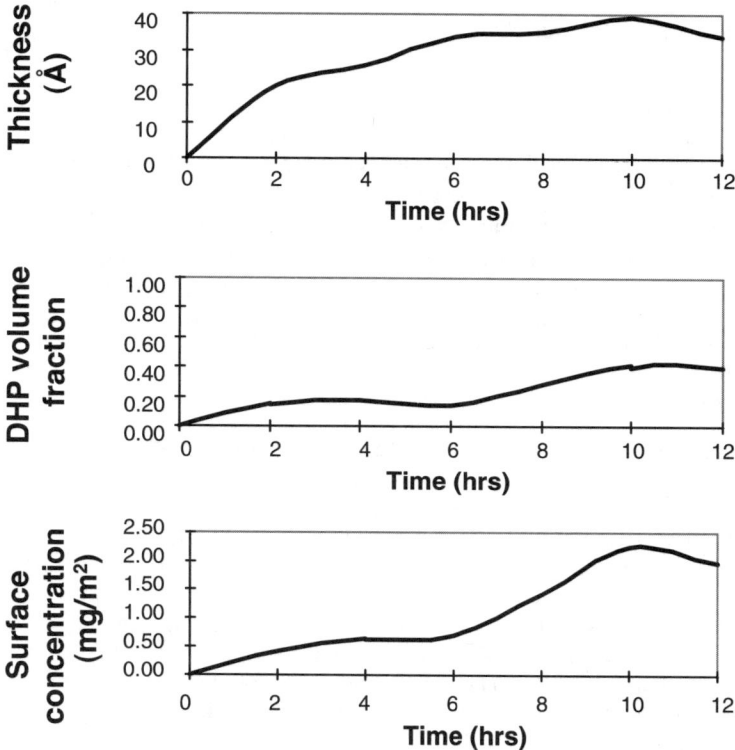

Figure 3 *Variation of thickness (Å), DHP volume fraction (φ) and surface concentration during the polymerization of coniferyl alcohol determined by neutron reflectivity*

the interfacial layer determined by ellipsometry is very different from the spectrum of a coniferyl alcohol solution ($\lambda_{max} = 263$ nm) (Figure 2), showing a maximum of absorbance at roughly 275 nm typical of dehydrogenation polymers (DHPs). This supports the DHPs nature of the interfacial layer resulting from the polymerization of coniferyl alcohol at the air/water interface.

3.3 Interfacial Organisation of the DHP Layers

Neutron reflectivity allows the investigation of the structure of an interfacial layer by determining the thickness and the DHP volume fraction of the layer. Since the energy of the reflected beam is low the polymerization was monitored by averaging spectra obtained during two hours. Thickness and DHP volume fraction increase during the polymerization, resulting in an increase of DHP concentration in the layer. These results are in agreement with those obtained by ellipsometry and indicate the formation of a dilute layer ($\phi_{DHP} = 0.2$) during the first step (roughly 0–4 hours of polymerization) followed by a second step where the layer structure changes towards a more dense phase ($\phi_{DHP} = 0.4$). Accordingly the water concentration decreases to form a 'drier' structure. This behaviour is reminiscent of the removal of water during lignification[10] and also from observations on spread DHP layers.[11]

4 Conclusion

Polymerization of coniferyl alcohol was achieved for the first time at the model air/water interface. The polymerization process examined by ellipsometry and neutron reflectivity, was found to exhibit at least two steps: formation of a dilute layer and then densification. This kinetic pattern will require an in-depth analysis to provide information on the polymerization process of monolignols in concentrated layers similar to those occurring in cell walls.

References

1. C. J. Houtman and R. H. Atalla, *Plant Physiology*, 1995, **107**, 977.
2. S. Y. Guan, J. Mlynar and S. Sarkanen, *Phytochemistry*, 1997, **45**, 911.
3. C. J. Houtman, *Holzforschung*, 1999, **53**, 585.
4. F. H. Ludley and J. Ralph, *J. Agric. Food Chem.*, 1996, **44**, 2942.
5. F. F. Bruno, J. A. Akkara, D. L. Kaplan, P. Sekher, K. A. Marx and S. K. Tripathy, *Industrial & Engineering Chemistry Research*, 1995, **34**, 4009.
6. F. F. Bruno, J. A. Akkara, L. A. Samuelson, D. L. Kaplan, B. K. Mandal, K. A. Marx, J. Kumar and S. K. Tripathy, *Langmuir*, 1995, **11**, 889.
7. M. F. Rubner and K. Hong, *Thin Solid Films*, 1988, **160**, 187.
8. B. Harzallah, V. Aguié-Béghin, R. Douillard and L. Bosio, *International Journal of Biological Macromolecules*, 1998; **23**, 73.
9. J. Meunier, *J. Physique*, 1987, 1819.
10. K. Takabe, M. Fujita, H. Harada and H. Saiki, *Mokuzai Gakkaishi*, 1981, **27**, 813.
11. B. Cathala, L. T. Lee, V. Aguié-Béghin, R. Douillard and B. Monties, *Langmuir*, 2000, **16**, 10444.

Multiphasic Systems

Emulsion-stabilizing Properties of Depolymerized Pectin: Effects of pH, Oil Type and Calcium Ions

Mahmood Akhtar,[1*] Eric Dickinson,[1] Jacques Mazoyer[2] and Virginie Langendorff[2]

[1]PROCTER DEPARTMENT OF FOOD SCIENCE, UNIVERSITY OF LEEDS, LEEDS LS2 9JT, UK
[2]DEGUSSA TEXTURANT SYSTEMS, CENTRE DE RECHERCHE – 50500 BAUPTE, FRANCE

Abstract

The emulsifying properties of depolymerized citrus pectin with varying molecular weight (48–146 kg/mol) and similar degree of esterification (70%) have been investigated in relation to the formulation of oil-in-water (O/W) emulsions containing rapeseed oil (RSO) or D-limonene at pH 4.7 or 7. The time-dependent emulsion stability was followed in terms of average droplet size (d_{43}), creaming behaviour, and microscopic and visual observations. We have discovered especially that, by using a relatively low concentration (4 wt%) of depolymerized pectin of molecular weight 70 kg/mol at pH 4.7, very stable RSO based (20 vol%) O/W emulsions can be formulated. Emulsion stability was reduced (a) with pectin of higher or lower molecular weight, (b) at pH 7, and (c) in the presence of added ionic calcium. The emulsifying properties of depolymerized citrus pectin were also compared with those of sugar beet pectin. The results show that depolymerized citrus pectin is as efficient as acetylated sugar beet pectin in stabilizing vegetable oil-in-water emulsion under acidic conditions.

1 Introduction

Pectin is extracted commercially from citrus peel (lemon, lime and grapefruit) and apple pomace.[1,2] The food industry is the most important field of application of extracted pectin. Pectins from different sources are widely used to stabilize food emulsions and dispersions in products such as fruit drinks, and fruit and tomato pastes.[3] Pectin can form gels under certain circumstances and it is used as a gelling agent in jams, jellies and marmalades. Pectin can also be used to stabilize clouding in beverages, with the effectiveness of the stabilization dependent on the nature and amount of pectin present.[4] For a polysaccharide to function satisfactorily as an emulsifier of flavour oils, both the concentrated flavour oil emulsion and the diluted beverage must remain stable for a period of several months.[5]

Pectin is a mixture of heteropolysaccharides consisting predominantly of partially methoxylated galacturonic acid residues.[6] The polysaccharide structure is based on 1,4-linked α-D-galacturonic acid, interrupted by L-rhamnose residues with side-chains of neutral sugars (mainly D-galactose and L-arabinose). The functional properties are sensitive to the degree of esterification (DE), which is in turn dependent on the type of plant tissue from which the pectin is extracted. High methoxyl pectins ($\geq$ 50% DE) form gels under acidic conditions (pH $\leq$ 3.5) in the presence of a cosolute (typically > 50 wt% sucrose), whereas low methoxyl pectins (< 50% DE) form gels by a different mechanism in the presence of calcium ions.[2,7]

Much research has been carried out on the chemical structural properties of pectin,[8,9] particularly on the composition of the constituent sugars,[10] the solution properties,[11] the mechanism of gelation[12] and the interfacial properties,[13] but so far the emulsion-stabilizing properties of *depolymerized* pectin have received little attention.[14]

In a recent patent,[14] it was demonstrated that depolymerized pectins of molecular weight below 80 kDa derived from citrus fruits and apples can possess good emulsion stabilizing characteristics even though they are low in acetyl groups (< 0.8%). This good emulsifying behaviour was found to be associated with a much higher surface activity of the depolymerized pectin as compared with normal citrus or apple pectin. The purpose of this paper is to further explore the emulsifying properties of depolymerised citrus pectin. We report here on how fine oil-in-water emulsions can be formulated using depolymerized pectin at a relatively low concentration of the biopolymer. The effects of pectin molecular weight, pectin concentration, oil type, and added calcium ions on the emulsion stability are evaluated. For the experiments reported here, a sample of high-molecular-weight pectin (150 kDa) and three samples of depolymerized pectin were prepared from citrus peels by acid hydrolysis. These samples could also be demethoxylated to obtain various degrees of methoxylation, although only one value ($\sim$ 70% DE) is considered in detail here.

2 Materials and Experimental Methods

2.1 Materials

The molecular weight (M_w), degree of esterification (DE), and content of anhydrous galacturonic acid (AG) of the citrus pectin are shown in Table 1. The molecular weight was estimated by intrinsic viscometry (calibrated against known pectin products), and the degree of methoxylation and the galacturonic acid content by titration.

Table 1 *Molecular weight* (M$_w$), *degree of esterification* (DE)[a] *and anhydrous galacturonic acid content* (AG) *of the four depolymerized citrus pectin and sugar beet pectin samples investigated with respect to their emulsification properties*

Sample	M$_w$ (kg mol^{-1})	DE (%)[a]	AG (%)
1	48	68.0	79.0
2	56	72.5	79.6
3	70	70.7	81.0
4	146	69.0	74.8

[a] Based on standard method given in 'Pectins', pp. 87–91, *Compendium of Food Additive Specifications*, FAO FNP No. 52, Add. 1, Rome, 1992.

High-molecular-weight pectin was extracted from dried citrus peels by hydrolysis with nitric acid at pH = 1.6 for 1 hour at 80 °C. Three depolymerized pectin samples were prepared by treating the peels with nitric acid at pH = 2.3 for 3 hours at 85 °C in the presence of various amounts of hydrogen peroxide (4–8 ml of 30% H$_2$O$_2$ per kg of peels). After purifying the slurries by filtration, the slurry syrups were concentrated by ultrafiltration, and the pectin samples were recovered by precipitation into isopropyl alcohol. The products were then dried and ground.

The rapeseed oil (RSO) was a gift from St. Ivel (Swindon, UK) and the flavour oil D-limonene (> 97%, L–2129, Lot 78H3488) was purchased from Sigma Chemicals (St Louis, USA). Sugar beet pectin was provided by Degussa Texturant Systems (Carentan, France).

2.2 Emulsion Preparation

The aqueous buffer was prepared using double-distilled water, citric acid and sodium citrate, with 0.01 wt% sodium azide (Sigma Chemicals) added as an antimicrobial agent. The buffer solution was heated to ∼ 50 °C and pectin powder was added slowly with gentle stirring. The pH of the resulting pectin solution was adjusted to pH 4.7 or pH 7 by adding a few drops of freshly prepared NaOH (1 M) solution.

A laboratory-scale jet homogenizer was used to make O/W emulsions at room temperature.[15] The jet homogenizer block has two chambers, one for the oil phase and the other for the aqueous phase, in different ratios (*e.g.* 11:89, 20:80, 45:55). The water and oil phases were passed through the chamber of the jet homogenizer at a pressure 400 bar to make O/W emulsions. These emulsions were stored at room temperature. In what follows, the pH values refer to the pH of pectin solution before emulsification.

2.3 Droplet Size Determination

Droplet-size distributions and average droplet sizes (volume–surface mean diameter d_{32} and weighted average mean diameter d_{43}) of O/W emulsions were determined immediately after making and as a function of storage time at 25 °C using a Malvern Mastersizer MS2000 (static light-scattering apparatus). The key parameters used in the droplet size analysis were: refractive indices of the disperse phase (*e.g.* RSO; 1.471) and the continuous phase (water; 1.330), general purpose model, absorption (0.001) and the size range (0.020 to 2000 μm). The average droplet size was characterized by two mean diameters, d_{32} and d_{43}, defined by

$$d_{32} = \sum_i n_i d_i^3 / \sum_i n_i d_i^2, \qquad d_{43} = \sum_i n_i d_i^4 / \sum_i n_i d_i^3,$$

where n_i is the number of droplets of diameter d_i. The d_{43} value was used to monitor changes in droplet-size distribution on storage.

A Nikon optical microscope was used to observe states of flocculation of large emulsion droplets. The emulsion samples were regularly viewed with the microscope using the Normarski differential interference contrast technique.[16] Creaming behaviour was examined by visually measuring the height (thickness) of cream and serum layers in emulsions stored at 22 °C at regular time intervals.

3 Results and Discussion

Four pectin samples with a similar high degree of esterification (DE $\approx$ 70%) varying molecular weight (48, 56, 70 and 146 kDa) were tested in relation to the formulation of oil-in-water emulsions (see Table 1). The pectin samples with high degree of esterification were chosen for this study due to their lower calcium ion sensitivity. The emulsions were made with 10 vol% of rapeseed oil (RSO) or D-limonene as dispersed phase and 4 wt% of depolymerized pectin (or sugar beet pectin) as emulsifier at pH 4.7. The long-term stability of the emulsions was assessed over a period of 5–7 weeks on the basis of droplet-size measurements, creaming behaviour, and microscopic and visual observations.

Figure 1 compares the properties of RSO oil-in-water emulsions with D-limonene emulsions stabilized by different molecular weight depolymerized pectin at pH 4.7 over a storage period of 2 months. The initial average droplet size ($d_{43} \sim 16$ μm) and creaming instability profiles for the D-limonene system were substantially higher than for the equivalent RSO emulsions ($d_{43} \sim 2$ μm). The extent of serum separation after 2 months was negligible for the pectin molecular weight 56 kDa and 70 kDa systems, whereas it was significant for the pectin molecular weight 48 kDa and 146 kDa samples. The pectin of molecular weight 70 kDa was a very effective emulsifier of the vegetable oil, with good retention of a low initial droplet size and good creaming stability on extended storage at pH 4.7. There was found to be a good correlation between initial average droplet size and creaming stability of the RSO emulsions. The pectin of lowest molecular weight (48 kDa) and highest molecular weight (146 kDa) were less effective in emulsion stabilizing than the pectin of molecular weight 70 kDa with a similar degree of esterification.

On the basis of the results in Figure 1 (a,b), it is apparent that the depolymerized pectin sample of molecular weight 70 kDa has good emulsifying

(a)

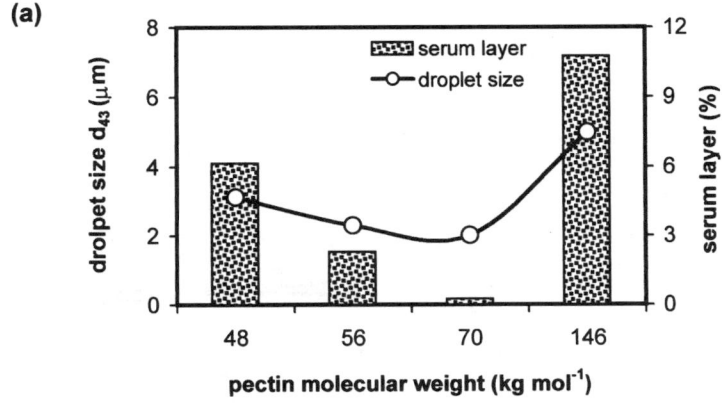

(b)

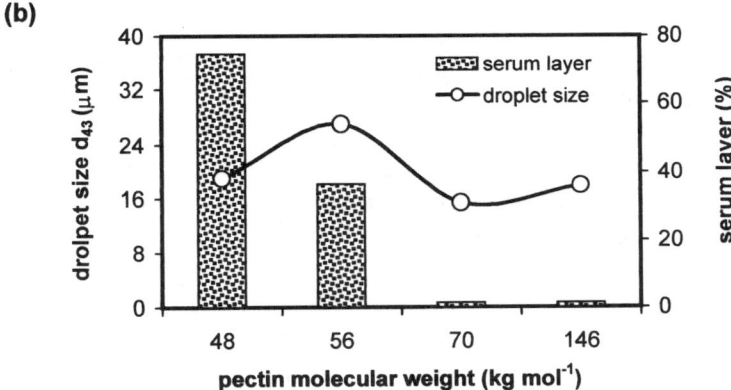

Figure 1 *Average droplet size (d₄₃) and extent of serum separation of oil-in-water emulsions (10 vol% oil, 4 wt% emulsifier, ionic strength 0.2 M, pH 4.7) stabilized by depolymerized pectins (DE 70%) of various molecular weights: (a) RSO; (b) D-limonene*

capabilities for both the vegetable and flavour oil under these acidic conditions. Hence this pectin sample was chosen to investigate the effect of various concentrations of pectin and calcium ions on emulsion stability.

Figure 2 shows average droplet size d_{43} for the pectin-stabilized emulsions made separately with RSO and D-limonene (10 vol%) at various pectin concentrations at pH 4.7. We see that increasing the pectin concentration decreases the average droplet size of the emulsion for both the vegetable oil and flavouring oil. It is clear that the hydrocolloid emulsifier is highly effective at considerably lower emulsifier/oil ratios under acidic conditions. At pH 4.7 the optimum emulsification behaviour is already reached at *ca.* 4 wt% pectin and there is a suggestion of a slight increase in d_{43} at high pectin contents. With D-limonene replacing RSO as the oil phase, the emulsion droplets grow to much larger size after one month of storage. The emulsions made with ≤ 4 wt% pectin become strongly phase separated on storage, whereas those made with ≥ 8 wt% pectin remain homogeneous and stable over the extended storage. Based on these results, the

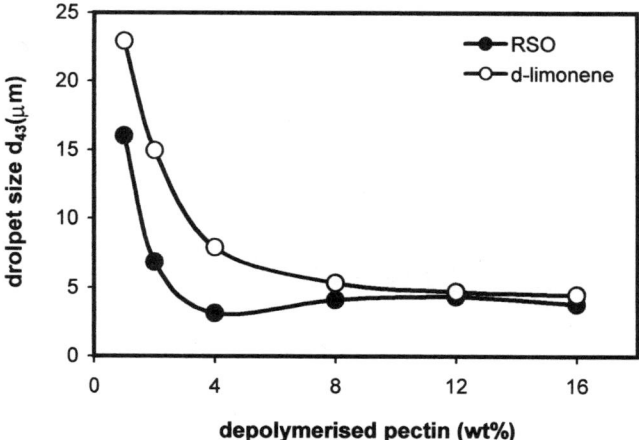

Figure 2 *Effect of pectin concentrations on average droplet size d_{43} and creaming stability of RSO or D-limonene emulsions (10 vol% oil, ionic strength 0.2 M) stabilized by depolymerized pectin of molecular weight 70 kg/mol at pH 4.7 over a storage period of 34 days*

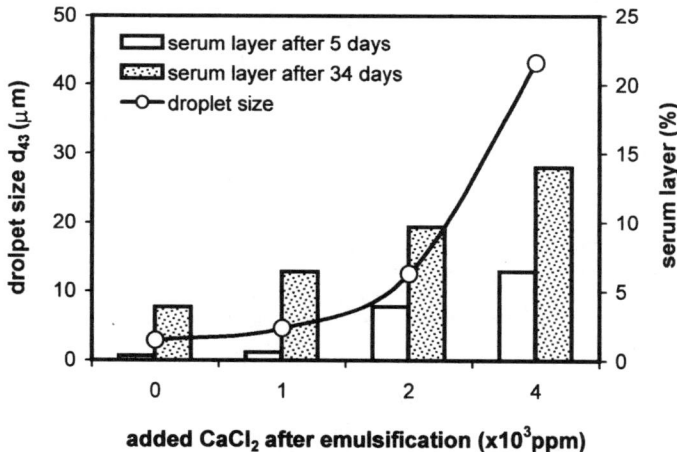

Figure 3 *Effect of ionic calcium addition on average droplet sizes d_{43} and creaming stability of D-limonene emulsions (10 vol% oil, ionic strength 0.2 M, pH 4.7) stabilized by depolymerized pectin molecular weight 70 kg/mol*

minimum pectin concentrations required to formulate a satisfactory stable O/W emulsion with RSO and D-limonene (10 vol%) are roughly 4 wt% and 8 wt%, respectively.

The stability of pectin-based emulsions was found to be sensitive to calcium ion content. The effect of addition of calcium chloride at levels of 10^3 or 4×10^3 ppm (0.1 or 0.4 wt%) to the freshly made emulsions (with continuous gentle stirring for 30 minutes) on the average droplet size and creaming profiles is shown in Figure 3. The results show negligible effect of 10^3 ppm $CaCl_2$ on the time-dependent average droplet size; and neither the calcium-free nor the 10^3

(a)

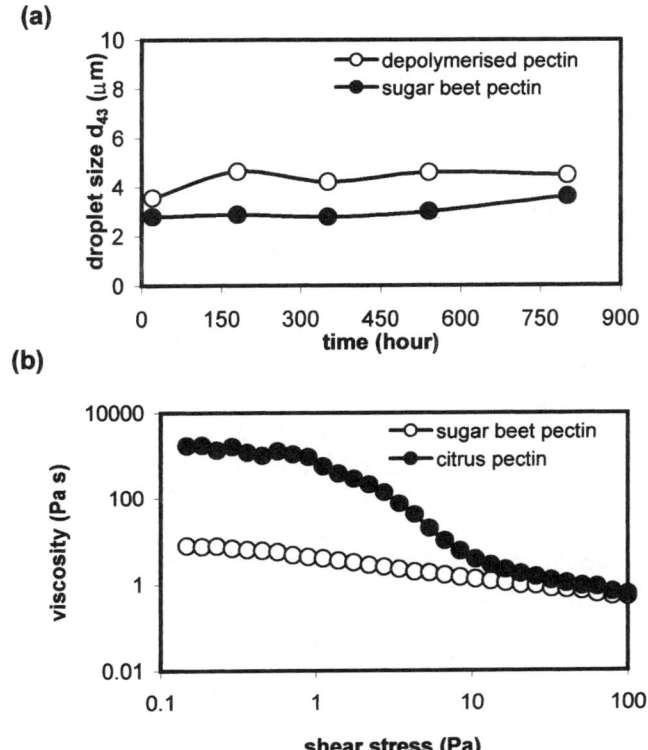

(b)

Figure 4 *Comparison of average droplet size and shear viscosity profiles of RSO (20 vol%) O/W emulsions stabilized by 4 wt% either depolymerized citrus pectin or sugar beet pectin at pH 4.7: (a) average droplet size; (b) shear viscosity*

ppm system exhibits any discernible creaming instability over 5 days, although after 30 days there is slightly more serum separated in the 10^3 ppm system. However, at high levels of calcium ions (4×10^3 ppm) the addition of calcium ions has significant effect on the stability of emulsions as indicated by the high values of d_{43} obtained immediately after calcium salt addition and by the reduced creaming stability on storage. The addition of calcium ions induces the droplet flocculation and hence increases the apparent average droplet size.

Emulsifying properties of depolymerized citrus pectin were compared with those of sugar beet pectin at 4 wt% aqueous phase emulsifier concentration. Emulsions were made with RSO (20 vol%) at pH 4.7. Figure 4 compares the average droplet sizes and shear viscosity profiles of emulsions stabilized by citrus pectin and sugar beet pectin over a storage period of 34 days. The results clearly show that depolymerized citrus pectin is as effective as the highly acetylated sugar beet pectin in formulating stable vegetable oil-in-water emulsions (droplet size $\sim 4\ \mu$m) at 4 wt% of the emulsifier. However, the emulsions stabilized by citrus pectin show relatively higher viscosity values than sugar beet pectin-stabilized emulsions and exhibit more shear-thinning behaviour.

4 Conclusions

It has been demonstrated that depolymerized pectin derived from citrus peel can be used as an effective emulsifying agent for formulating food emulsions under acidic conditions. In particular, rapeseed oil-in-water emulsions made with depolymerized pectin of molecular weight 70 kDa and degree of esterification 70% at relatively low pectin/oil ratios were found to have excellent stability in terms of average droplet size and creaming behaviour over a two-month storage period. Emulsions made at pH 4.7 were found to be more stable than those made under neutral pH conditions.

Based on the combined creaming stability and time-dependent droplet size results presented in Figures 1 and 2, we infer that the depolymerized pectin sample (molecular weight 70 kDa) is an effective stabilizing biopolymer for the vegetable oil-in-water emulsions at pH 4.7. We hypothesize that, following its rapid adsorption during emulsification, the protein moeity of the hydrocolloid anchors the molecule at the oil–water interface, and the polysaccharide chain then provides a sterically stabilizing layer around the droplets – as with gum arabic.[17–19] Presumably, in the much less surface-active native pectin (*i.e.* not depolymerized), the protein moeity is not so available for adsorption during emulsification. Additionally, if pectin chains are too long they may perhaps induce flocculation due to calcium-mediated pectin–pectin bridging, especially if the DE value is low. Conversely, we hypothesize that too much depolymerization is also detrimental to the pectin's emulsifying behaviour, since polysaccharide chains that are too short will not produce adsorbed layers that are thick enough for adequate steric stabilization.[20]

References

1. T. Sakai, T. Sakamoto, J. Hallaert and E. J. Vandamme, *Adv. Appl. Microbiol.*, 1993, **39**, 213.
2. C. D. May, 'Pectins', in *Thickening and Gelling Agents for Food*, A. Imeson (ed.), Blackie, Glasgow, 1992, p. 124.
3. E. Costell, E. Carbonell and L. Duran, *J. Text. Studies*, 1993, **24**, 375.
4. Z. Elshamei and M. Elzoghbi, *Nahrung (Food)*, 1994, **38**, 158.
5. E. Dickinson, D. J. Elverson and B. S. Murray, *Food Hydrocolloids*, 1989, **3**, 101.
6. H. A. Schols, J. M. Ros, P. J. H. Daas, E. J. Bakx and A. G. J. Voragen, 'Structural Features of Native and Commercially Extracted Pectins', in *Gums and Stabilisers for the Food Industry*, P. A. Williams and G. O. Phillips (eds.), Royal Society of Chemistry, Cambridge, UK, 1998, p. 3.
7. C. Rolin, and J. De Vries, 'Pectin', in *Food Gels*, P. Harris (ed.), Elsevier Applied Science, London, 1990, p. 401.
8. Y. A. Antonov, N. P. Lashko, Y. K. Glotova, A. Malovikova and O. Markovich, *Food Hydrocolloids*, 1996, **10**, 1.
9. C. R. F. Grosso and M. A. Rao, *Food Hydrocolloids*, 1998, **12**, 357.
10. A. Kawabata and S. Sawayama, *J. Japanese Soc. of Food Nutrition*, 1975, **28**, 395.
11. A. Kawabata, S. Sawayama and T. Nagoya, *J. Japanese Soc. Food Nutrition*, 1977, **30**, 149.

12. B. R. Thakur, R. K. Singh and A. K. Handa, *Crit. Rev. Food Sci. Nutrition*, 1997, **37**, 47.
13. D. R. Coffin and M. L. Fishman, *J. Applied Polym. Sci.*, 1994, **54**, 1311.
14. J. Mazoyer, J. Leroux and G. Bruneau, 1999, US Patent No. 5,900,268.
15. I. Burgaud, E. Dickinson and P. V. Nelson, *Int. J. Food Sci. Technol.*, 1990, **25**, 39.
16. C. F. A. Cullings, *Modern Microscopy: Elementary Theory and Practice*, Butterworths, London, 1974.
17. R. C. Randall, G. O. Phillips and P. A. William, *Food Hydrocolloids*, 1988, **2**, 131.
18. E. Dickinson, D. J. Elverson and B. S. Murray, *Food Hydrocolloids*, 1989, **3**, 101.
19. E. Dickinson, V. B. Galazka and D. M. W. Anderson, *Carbohydrate Polymers*, 1991, **14**, 373.
20. E. Dickinson, *An Introduction to Food Colloids*, University Press, Oxford, 1992.

Mixed Biopolymer Gels of κ-Carrageenan and Soy Protein Isolate

R. I. Baeza, D. J. Carp, P. Martelli and A. M. R. Pilosof

DEPARTAMENTO DE INDUSTRIAS, FACULTAD DE CIENCIAS EXACTAS Y NATURALES, UNIVERSIDAD DE BUENOS AIRES, CIUDAD UNIVERSITARIA (1428), BUENOS AIRES, ARGENTINA

1 Introduction

κ-Carrageenan is a fraction of the sulfated polysaccharides extracted from red algae which is widely used in food industry as a thickening, gelling and stabilizing agent. In the presence of cations, κ-carrageenan forms thermoreversible gels.

Heating aqueous dispersions of κ-carrageenan at temperatures above 60 °C, the polysaccharide hydrates and adopts a random coil conformation. Gelation occurs on cooling at a critical temperature and has been attributed to a two-stage reaction involving a coil–helix transition followed by aggregation of helices.[1,2] The coil–helix transition is strongly dependent on the concentration and type of cation.[2] Potassium ions are more efficient for gelation than either sodium or calcium. The potassium form of κ-carrageenan is highly dependent on salt concentration and strong gels are formed at high KCl concentrations. In the presence of 50 mM potassium and above, a fine network is first formed where the junction zones are believed to be double helices. Further cooling enhances the association of helices into long stiff superstrands.[3] The balance between the fine and the coarse supramolecular structure varies with KCl concentration and accounts for the increased gel strength.

Amongst the most important applications of κ-carrageenan is its use in milk products. Therefore much work has been done on the gelation of κ-carrageenan in the presence of milk proteins, micellar caseins and real milk systems.[4–10] κ-carrageenan gelation was found to be affected by casein fractions,[8] micellar caseins and whey proteins.[5,10] In the presence of milk proteins, gelation occurs at relatively low carrageenan concentrations and the hardness and viscoelasticity are greatly affected.

Soy proteins are used as functional ingredients in dairy products analogs,[11] due to their excellent ability to impart desirable functionalities. Compared with their dairy counterparts, soy-based analogs are not only lower in cost, but also free of cholesterol and lower in calories and fat. Special formulas based on soy proteins are produced for people who are allergic to cow's milk.

Therefore, the aim of present work is to study the gelling behaviour of κ-carrageenan/soy protein systems at neutral pH, under conditions where the protein does not gel (*i.e.* low protein concentration). To this end a Doehlert uniform shell design for two factors (protein and potassium concentration) was performed.

2 Experimental

A commercial soy protein isolate Supro EX32 was purchased from Protein Technologies Int., USA. The main minerals present in the isolate were sodium (1.2 %wt), potassium (0.1 %wt) and calcium (0.2 %wt). κ-carrageenan was a gift from Sanofi Bioindustrias, Argentina, and was used without further purification.

κ-Carrageenan was dispersed at 1 %wt in 0.005 mol/L NaH_2PO_4/Na_2HPO_4 buffer pH 7.0 under strong stirring and the appropriate KCl amount was added. The systems were then heated at 90°C for 30 min. Soy protein isolate solution was prepared at double the concentration by low stirring at room temperature. The protein/polysaccharide mixed systems were prepared by mixing at a 1:1 ratio the solutions of proteins and κ-carrageenan at 40°C (to avoid protein modification). Final concentrations were 0–4 %wt protein and 0.5 %wt κ-carrageenan. To avoid bacterial growth, 0.02 % NaN_3 was added to all solutions.

2.1 Experimental Design

A Doehlert uniform shell design for two factors was selected.[12] Results were analysed by using the software Statgrafic 5.1. The variables studied were potassium ions (K^+) and protein concentration. The real and coded values are shown in Table 1. The three replicates of the central point allowed the error of the methods to be calculated.

The following full quadratic model containing six coefficients was used to describe the responses, where b_i are the regression coefficients given by the model:

$$Y = b_0 + b_1 P + b_2 (K^+) + b_3 P^2 + b_4 (K^+)^2 + b_5 P (K^+) \qquad (1)$$

The measured responses were the rate of gelation, gelation temperature, melting temperature (T_m), thermal hysteresis ($T_m - T_{gel}$) and gel hardness.

Independent variables that were found significant at $P < 0.05$ in the full model were retained in the reduced models and used to generate contour plots for each response.

Table 1   *Real and coded (in brackets) values of the variables studied*

Variable	Real and coded values
Protein (%) (P)	0 (− 1); 1 (− 0.5); 2 (0); 3 (0.5); 4 (1)
Potassium (M) (K +)	0 (− 0.866); 0.0075 (0); 0.015 (0.866)

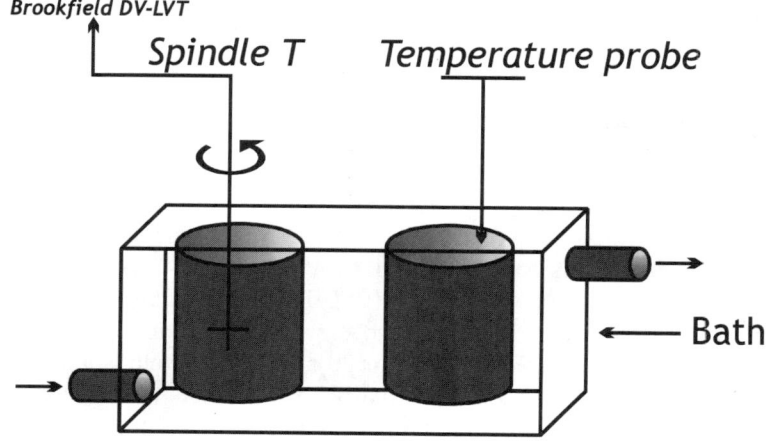

Figure 1 *Experimental set-up for measuring torque and temperature during κ-carrageenan gelation*

2.2 Gel Formation

The experimental set-up with two identical vessels (Figure 1) was used to get simultaneously temperature and torque dataset, in the same heat transfer conditions. Solutions (initially at 40 °C) were cooled in a bath at 19.6 ± 0.2 °C. Apparent viscosity as a function of time was recorded for 1 hour, with a Brookfield DV-LVT viscometer. The T-spindle (type T-C) was rotated at 0.3 rpm and a helipath stand (pathway 0.4 cm) was used to minimize the modification of gel formation. Concomitantly the temperature was recorded. The gelation temperature (T_{gel}) was determined as the temperature where a sharp rise in the apparent viscosity occurs and the rate of gelation was estimated from the initial slope of the apparent viscosity increase as a function of time.

2.3 Gel Texture

Solutions were poured into cylindrical containers, cooled at 5 °C and stored for 7 days. Compression tests were performed on cylindrical specimens of the gels (19 mm diameter × 15 mm high). A texture profile analysis (TPA) was performed in a Texture Analyser (Stable Micro Systems TA.XT2i), using a cylindrical probe (3.6 mm diameter) to compress each sample to 40% of its original height at a compression rate of 0.5 mm/s. The average of two replicates was reported.

2.4 Melting Temperature

A Mettler TA 4000 Thermal Analysis System DSC, was used to determine gel–sol transition temperatures of mixed protein/κ-carrageenan gels. Samples were heated in aluminium pans (40 μl volume) from 5 °C up to 70 °C at 10 °C/min. DSC calibration was performed using ice and indium fusion thermograms and an empty aluminum pan was used as a reference. Thermograms were analyzed using the Mettler TA 72 software to determine the peak melting temperature (T_m). The average of two replicates was reported.

3 Results and Discussion

The dynamics of κ-carrageenan gelation was determined from the continuous evolution of the apparent viscosity and the temperature with time throughout the gelation process. Because the shear rate was kept small and the T-spindle performed an helical pathway, the effect of shear on gel formation was minimized. Typical profiles of the development of viscosity during cooling of κ-carrageenan/soy protein isolate mixtures are shown in Figure 2. In all samples, there was an initial period during which the viscosity was very low and the mixture remained fluid. The onset of κ-carrageenan gelation was signalled by a sharp rise in viscosity (Figure 2a) at the carrageenan coil-to-helix transition temperature (Figure 2b). The presence of soy protein and different concentrations of potassium affected both the gelation temperature and the initial rate of viscosity increase (taken as the gelation rate) (Figure 2a).

Analysis of variance for each dependent variable showed that in almost all cases, R^2 coefficients higher than 0.83 were obtained (Table 2), which means that models were able to explain more than 83% of the observed responses. For the rate of gelation, thermal hysteresis and hardness, the lack of fit test was not significant. For T_{gel} and T_m, the lack of fit was significant, which means that the model may not have included all appropiate function of independent variables. According to Box and Draper,[13] we considered the high coefficients R^2 as evidence of the applicability of the model.

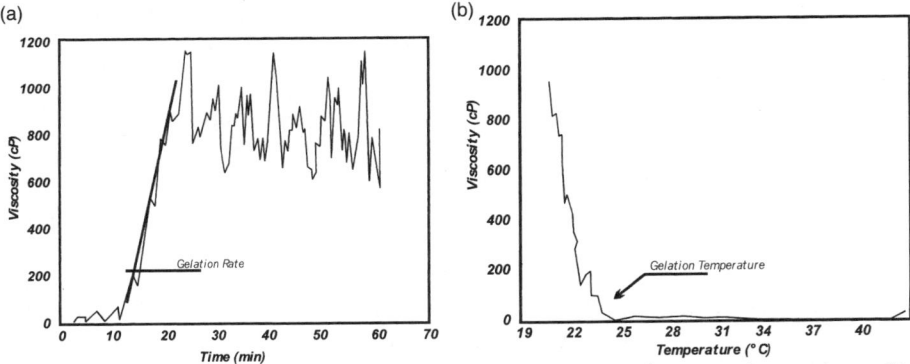

Figure 2 *Apparent viscosity as a function of time* (a) *and as a function of temperature* (b) *during cooling κ-carrageenan/soy protein mixed systems*

Table 2 *Regression coefficients estimated by multiple linear regression analysis*

Coefficients*	Gelation Rate (cp/min)	Gelation Temperature (°C) (T_{gel})	Melting Temperature (°C) (T_m)	T_m-T_{gel} (°C)	Hardness (g)
Constant	263.67	28.87	44	15	48.5
Linear					
P	329.25	1.98	5.83	4.15	21.0
K$^+$	558.17	6.32	3.69	− 2.62	19.9
Quadratic					
P^2	122.33	(− 0.62)[a]	(0.4)[a]	(0.7)[a]	5.97
(K$^+$)^2	307.18	− 1.92	− 4.2	(− 2.00)[a]	(1.2)[a]
Interactions					
(K$^+$) P	245.09	− 1.1	− 0.92	(0.17)[a]	15.5
R^2	0.998	0.976	0.830	0.840	0.998
Lack of fit	NS	**	**	NS	NS

[a]Values in parentheses are not significant.
*, ** Significant (P < 0.05)
NS: not significant

Analysis of variance of the full quadratic model for the rate of κ-carrageenan gelation, showed that all the independent variables were statistically significant. The positive linear regression coefficients for potassium and protein indicate that both variables increase the gelation rate of κ-carrageenan. Nevertheless, the effect of potassium was more important than that of soy protein isolate because of the higher regression coefficient of the linear variable. Interaction between potassium and soy protein was positive, indicating a synergistic effect between them. The contour plots in Figure 3 shows the combined effect of potassium and protein on the gelation rate of κ-carrageenan. Below a potassium concentration of 0.0075 M and protein concentration of 2%, the gelation rate was extremely low. Even in the presence of 2–4% protein, potassium was necessary for the gelation of κ-carrageenan to proceed at a significant rate. The amount of salts (K$^+$, Na$^+$, Ca^{2+}) coming from the soy protein isolate (even at 4%, the maximum protein concentration) was lower than that required for the coil–helix transition of the κ-carrageenan at 20°C.[14] The results are in agreement with the well established role of potassium in the coil-to-helix transition of κ-carrageenan.[2] Soy protein synergistically interacted with potassium, greatly increasing rate of gelation, that was maximum at the highest K$^+$ and protein concentration.

Significant regression coefficients for the gelation temperature in Table 2 showed that it was increased by the presence of both potassium and protein, but potassium was the most important variable as indicated by the higher regression coefficient. Nevertheless no significant interaction between variables was observed. The negative value of the quadratic effect for potassium indicates the existence of a maximum for the gelation temperature as a function of potassium concentration. Contour plots in Figure 4 show that the T_{gel} might be increased up to 12 °C. This increase may be attained in the absence of protein by increasing potassium. The gelation temperature has been shown to raise linearly with the

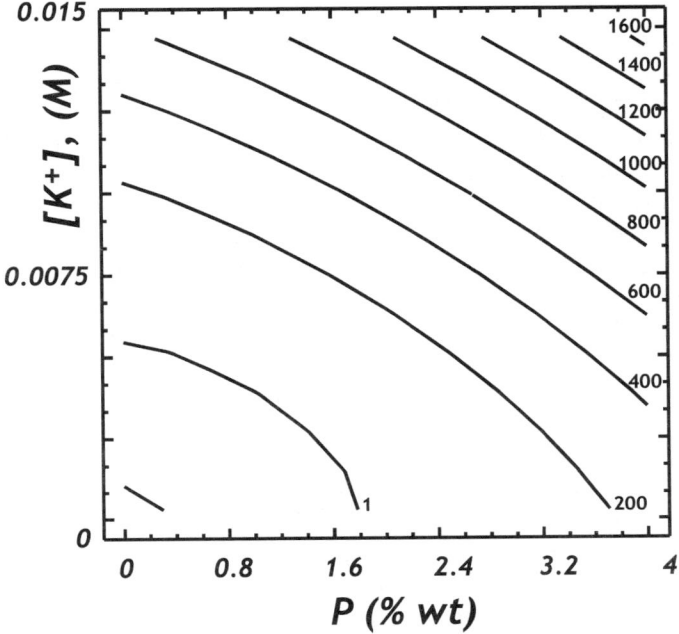

Figure 3 *Contour plots for the rate of gelation of κ-carrageenan/soy protein isolate systems*

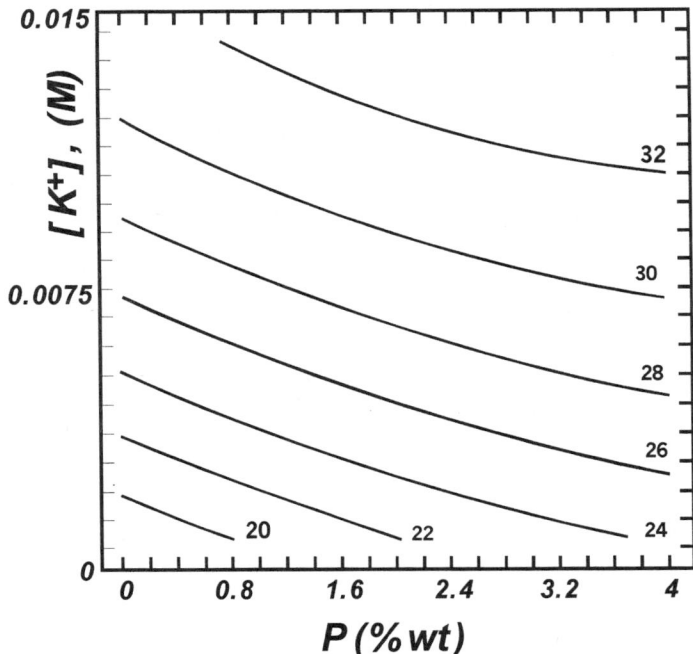

Figure 4 *Contour plots for the gelation temperature of κ-carrageenan/soy protein isolate systems*

logarithm of the concentration of potassium ions.[15] Potassium ions form contact ion pairs with sulfate groups of κ-carrageenan that stabilises the helix by screening of the anionic charge on the polysaccharide.[2] Thus the main effect of protein was to reduce the potassium necessary to reach a target temperature of gelation. Compared to the pure κ-carrageenan system, the gelation temperature in the presence of milk proteins was increased up to 5°C,[8] which is in agreement with the present results.

Similarly to the gelation temperatures, the melting temperatures of κ-carrageenan gels were raised by increasing potassium and protein content. Nevertheless, the effect of protein was more important than that of potassium, as indicated by the higher regression coefficient (Table 2). The contour plots in Figure 5 show that T_m was raised up to 8°C by increasing potassium in the absence of soy protein isolate. However, in the presence of protein (in combination with potassium) T_m increased up to 18°C. Similar effects were observed in the presence of galactomannans, α- and κ-casein and β-lactoglobulin.[8,10,16] The increase in T_m has been related to an increased formation and stability of aggregates of κ-carrageenan helices.[14,17] Electron microscopy studies on the supramolecular structure of κ-carrageenan gels revealed that caseinate or caseine fractions hampers the association of helices in superstrands, probably by association of helices.[3] Soy proteins would perform similarly, promoting mainly the aggregation of carrageenan helices.

Melting temperatures of κ-carrageenan gels are usually higher than gelation temperatures. Experimental evidences suggest that the setting of gels is coincident with helix formation and that a subsequent cation-mediated aggregation of these helices gives rise to hysteresis effects. Thus, an enhanced melting point for the gels is observed because of energy required to break the additional hydrogen bonds involved in the aggregation of helices.[2]

Analysis of variance of the full quadratic model for the thermal hysteresis (T_m-T_{gel}), showed that the only significant variables were the linear terms for soy protein and potassium concentration (Table 2). The protein was the more important variable affecting thermal hysteresis and the positive value of the regression coefficient indicated that it enhanced the thermal effect. This effect is related to the great increase in melting temperatures in the presence of protein. However, the regression coefficient for potassium was negative and indicated that the thermal effect was reduced by increasing this cation. This reflects the small effect of potassium in rising T_m and the concomitantly large effect on increasing T_{gel}. The combined effect of both variables is shown in Figure 6. At low protein concentrations, thermal hysteresis was 8–13°C, but in the presence of soy protein isolate it was increased up to 20°C and shifted to higher temperatures. The largest hysteresis was found to occur in the systems with the highest protein and low potassium concentrations.

Regression coefficients for hardness of gels (Table 2) showed that potassium and protein had positive linear and similar effects. Potassium had also a quadratic effect, that indicates the existence of a minimum for the hardness of gels as a function of potassium concentration. In addition, a large synergistic effect between variables was revealed by the interaction coefficient. Contour plots in

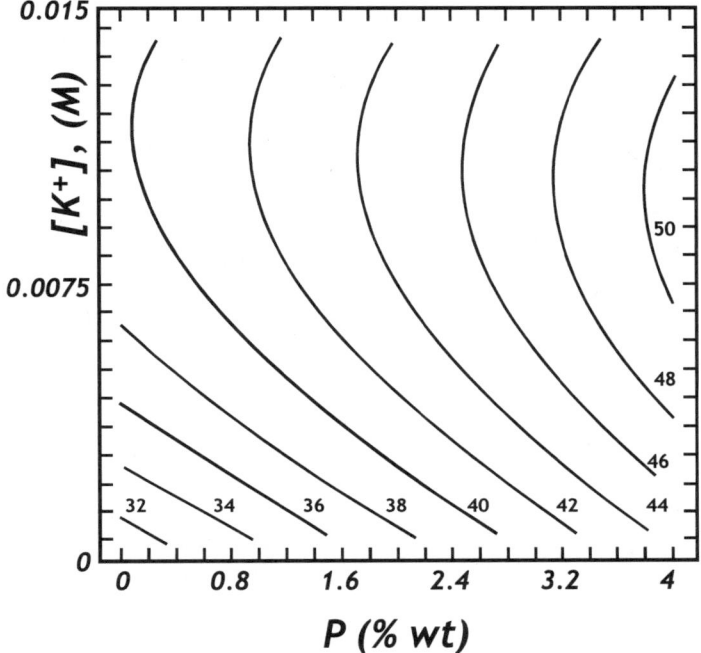

Figure 5 *Contour plots for the melting temperature of κ-carrageenan/soy protein isolate gels*

Figure 7 show that by combining increasing concentrations of soy protein and potassium, very strong gels may be obtained. In addition, those gels showed little syneresis when compared to gels of pure κ-carrageenan (data not shown). Large deformation measurements like hardness have been correlated to parameters describing the structure of strands forming the gels. Stiffer and thicker strands gave rise to stronger gels.[18] Thus the increased hardness of κ-carrageenan gels in the presence of soy protein isolate would reveal an increased protein-mediated aggregation of helices.

A limited biopolymer incompatibility under the conditions of pH (7) and low biopolymers concentration may contribute to increased hardness and thermal properties of gels. Excluded volume effects favour gelation of hydrocolloids. For incompatible biopolymers in mixed solutions, the rate of gelation is higher and the critical concentration for gelation is lower than for each of them individually.[18,19]

Excluded volume effects may increase the effective concentration of κ-carrageenan and the reduction of available water for each macromolecule may lead to increased inter- and intra-helix interactions in the polysaccharide. In the presence of other proteins (*i.e.* milk proteins) κ-carrageenan gel formation involved mainly carrageenan – carrageenan cross-linkages and not polysaccharide – protein interactions.[5,9]

However, the formation of electrostatic complexes between carrageenan and soy protein may not be discarded. Grinberg and Tolstoguzov observed the

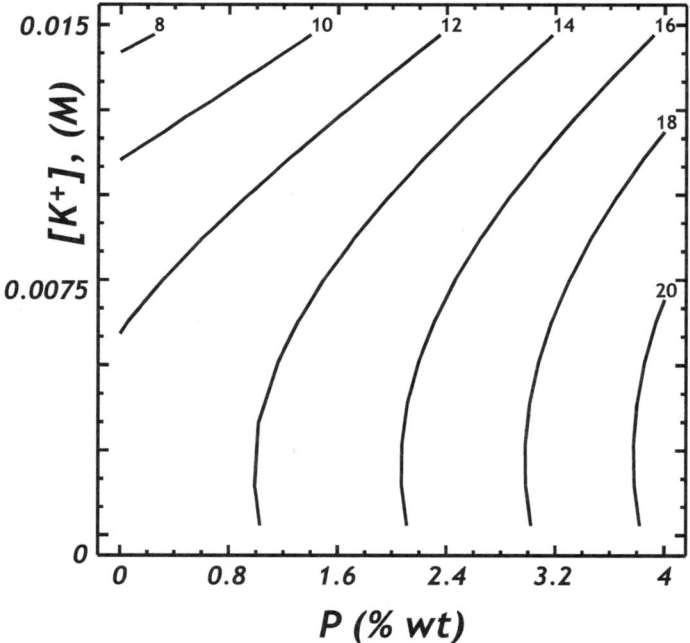

Figure 6 *Contour plots for thermal hysteresis* $(T_m - T_{gel})$ *of κ-carrageenan/soy protein isolate systems*

formation of soluble protein-sulphated polysaccharides complexes at pHs above the isoelectric point of the protein.[20] This was attributed to the formation of ionic pairs between ionised sulphate groups of the polysaccharide and ε-amino groups of the protein.

4 Conclusions

Under conditions of limited thermodynamic incompatibility at neutral pH, soybean protein isolate influences the dynamics of κ-carrageenan gelation and the properties of gels. The presence of soy protein isolate allows the formation of gels at higher temperatures and increases the thermal hysteresis due to a great increase of the melting temperature of gels. It also promotes stronger gels with decreased syneresis.

Soy protein isolate shows an additive or synergistic interaction with potassium that may be successfully exploited to design gel products with desired characteristics. To this end the response surface methodology proved to be an excellent tool.

Acknowledgments

The authors acknowledge the financial support from Universidad de Buenos

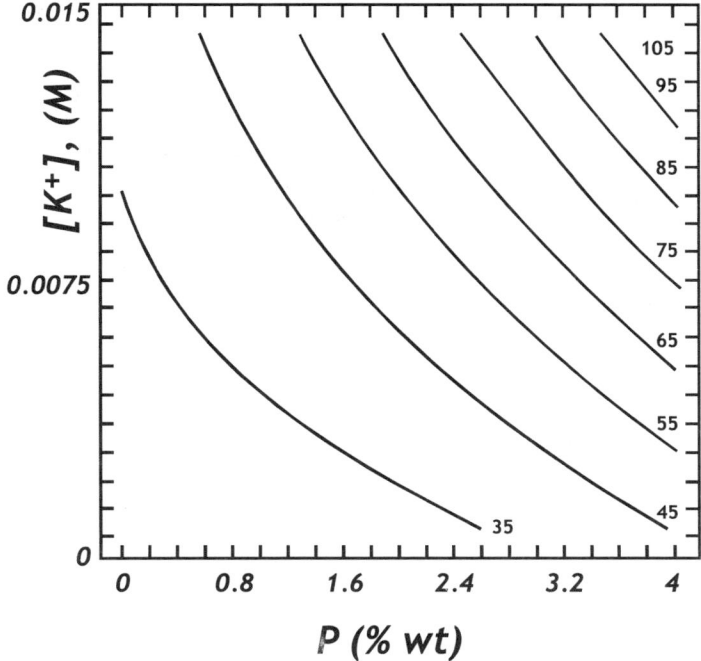

Figure 7 *Contour plots for the hardness of κ-carrageenan/soy protein isolate gels*

Aires, Consejo Nacional de Investigaciones Científicas y Técnicas y Agencia Nacional de Promoción Científica y Tecnológica de la República Argentina.

References

1. E. R. Morris, D. A. Rees and G. Robinson, *J. Mol. Biol.*, 1980, **138**, 349.
2. V. J. Morris, *Functional Properties of Food Macromolecules*, Aspen Publishers, Gaithersburg, Maryland, 1998, Chapter 4, p. 143.
3. A.-M. Hermansson, *Food Macromolecules and Colloids*, Royal Society of Chemistry, Cambridge, UK, 1995, p. 363.
4. C. Schorsch, M. G. Jones and I. T. Norton, *Food Hydrocolloids*, 2000, **14**, 347.
5. D. D. Drohan, A. Tziboula, D. McNulty and D. S. Horne, *Food Hydrocolloids*, 1997, **11**, 101.
6. S. Mleko, E. C. Y. Li-Chan and S. Pikus, *Food Research International*, 1997, **30**, 427.
7. R. Ipsen, *Carbohydrate Polymers*, 1995, **28**, 337.
8. L. Lundin and A.-M. Hermansson, *Food Hydrocolloids*, 1998, **12**, 175.
9. A. Tziboula and D.S. Horne, *Colloids and Surfaces B: Biointerfaces*, 1999, **12**, 299.
10. M. M. Ould Eleya and S. L. Turgeon, *Food Hydrocolloids*, 2000, **14**, 29.
11. K. Liu, *Soybeans: Chemistry, Technology and Utilization*, Chapman and Hall, International Thomson Publishing, USA, 1997.
12. D. H. Doehlert, *Appl. Statistics*, 1970, **19**, 231.
13. G. Box and N. Drapper, *Empirical Model-building and Response Surfaces*, Wiley, New York, 1987.

14. A. M. Hermansson, E. Ericksson and E. Jordansson, *Carbohydrate Polymers*, 1991, **16**, 297.
15. E. E. Braudo, I. G. Plashchina, M. G. Semenova and V. P. Yuryev, *Food Hydrocolloids*, 1998, **12**, 253.
16. P. B. Fernandes, M. P. Gonçalves and J. L. Doublier, *Food Macromolecules and Colloids*, Royal Society of Chemistry, Cambridge, UK, 1995, p. 321.
17. C. Rochas and M. Rinaudo, *Biopolymers*, 1980, **19**, 1675.
18. R. Baeza and A. M. R. Pilosof, *Food Colloids*, Royal Society of Chemistry, Cambridge, UK, 2001, p. 392.
19. V. B. Tolstoguzov, *Food Hydrocolloids*, 1995, **9**, 317.
20. V. Grinberg and V. B. Tolstoguzov, *Food Hydrocolloids*, 1997, **11**, 145.

The Properties of *v/ι*-Carrageenan: Implications for the Gelling Mechanism of *ι*-Carrageenan

Fred van de Velde,[1,2+] Harry S. Rollema,[3] R. Hans Tromp[1,3]

[1]WAGENINGEN CENTRE FOR FOOD SCIENCES, P.O. BOX 557, 6700 AN WAGENINGEN, THE NETHERLANDS
[2]TNO NUTRITION AND FOOD RESEARCH INSTITUTE, CARBOHYDRATE TECHNOLOGY DEPARTMENT, P.O. BOX 360, 3700 AJ ZEIST, THE NETHERLANDS
[3]NIZO FOOD RESEARCH, PRODUCT TECHNOLOGY DEPARTMENT, P.O. BOX 20, 6710 BA EDE, THE NETHERLANDS

Abstract

A series of *ι*-carrageenan samples containing different amounts of *v*-carrageenan (0 to 23% monomer) have been prepared from neutrally extracted carrageenan from *Eucheuma denticulatum*. *v*-Carrageenan is the biochemical precursor of *ι*-carrageenan. The coil-to-helix transition and rheological properties of these samples were studied. The helix-forming capacity of *ι*-carrageenan turns out to decrease with increasing number of *v*-units in the chain. In contrast, the rheological properties of *ι*-carrageenan are enhanced by the presence of a small amount of *v*-units, yielding a two-fold increase in G′ at 4% *v*-units.

It is concluded that the structure building capacity of *ι*-carrageenan containing a small amount of *v*-carrageenan is significantly better than that of pure *ι*-carrageenan. This phenomenon is explained in terms of the balance between the length of helical stretches and the number of 'kinks' in the chain. The former is diminished, the latter is increased by the presence of *v*-units. However, both helical stretches and 'kinks' (giving rise to crosslinks between chains) are needed for gel formation.

1 Introduction

Carrageenan is a generic name for a family of gel-forming polysaccharides, which are commercially obtained by extraction of certain species of red seaweeds

(*Rhodophyceae*). For several hundred years, carrageenan has been used for food in Europe and the Far East. In Europe the use of carrageenan started more than six hundred years ago in Ireland. In Carraghen on the south Irish coast, flans were made by cooking the so-called Irish moss in milk. Nowadays, carrageenan is generally applied as gelling, thickening and stabilising agent in especially food products, such as chocolate, cottage cheese, whipped cream, instant products, jellies, sauces, *etc*. Besides this, carrageenans are used in pharmaceutical formulations, cosmetics and several industrial applications.[1,2]

Carrageenan is a high molecular weight material with a high degree of polydispersity. The molecular weight distribution varies from sample to sample. Commercial (food-grade) carrageenans have a weight average molecular mass (M_w) of 400 to 600kDa. Carrageenans are water soluble, linear, sulfated galactans. They are composed of an alternation of 3-linked β-D-galactopyranose (G-units) and 4-linked α-D-galactopyranose (D-units) or 4-linked 3,6-anhydrogalactose (A-units). The 'ideal' disaccharide repeating unit of carrageenans is shown in Figure 1. The sulfated galactans are classified according to the presence of the 3,6-anhydrogalactose on the 4-linked residue and the position and number of sulfate groups. Typically, for commercial carrageenan the sulfate content falls within the range from 22% to 38% (w/w). Although natural carrageenan is a mixture of nonhomologous polysaccharides, the term disaccharide repeating unit refers to the idealised structure. To describe more complex structures,

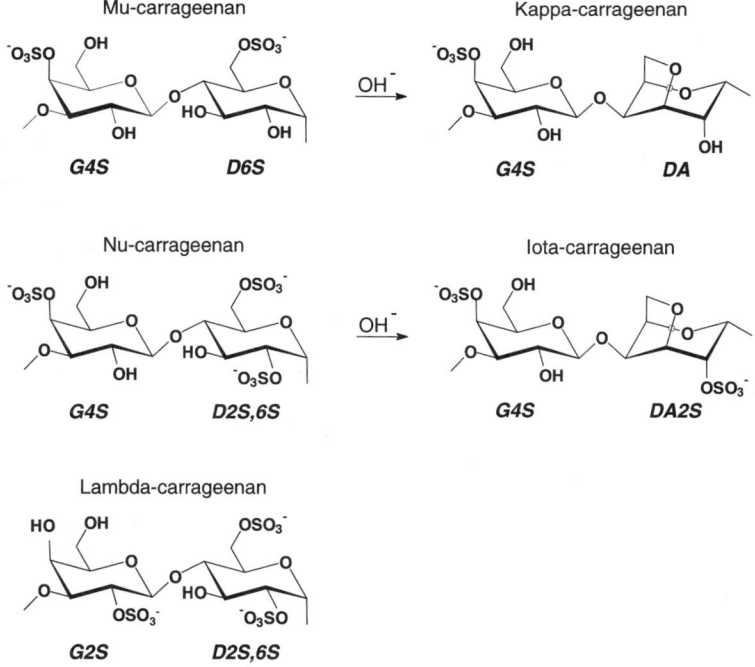

Figure 1 *Schematic representation of the different idealised building blocks of carrageenans. The letter codes refer to the alternative nomenclature, as developed by Knutsen et al.*[3]

Knutsen *et al.*[3] developed a letter code-based nomenclature for red algae galactans.

The most common types of carrageenan are traditionally identified by a Greek prefix. The three main commercial carrageenans are *ι*-, *κ*- and *λ*-carrageenan. *ι*- and *κ*-Carrageenan are gel forming carrageenans, whereas *λ*-carrageenan is a viscosity builder. In general terms, *κ*-carrageenan gels are hard, strong and brittle, whereas *ι*-carrageenan forms soft and weak gels. The gelation of *κ*-carrageenan is promoted by monovalent cations (K^+, Rb^+, Cs^+ and NH_4^+).[4,5] These cations promote the aggregation of *κ*-carrageenan double helices to form so-called aggregated 'domains'. Viebke *et al.*[6] came to the conclusion that the gelation of *κ*-carrageenan occurs at the superhelical level. The association of helical strains explains the hysteresis observed in optical rotation and rheological measurements of *κ*-carrageenan. The gel formation of *ι*-carrageenan appears not to be cation specific. The thermal conformational transition of *ι*-carrageenan shows no hysteresis, which suggests that there is little or no interhelical aggregation. Therefore the gel formation in *ι*-carrageenan gels is assumed to take place at the helical level.[6]

The difference in rheological behaviour between *ι*- and *κ*-carrageenan on the one hand and *λ*-carrageenan on the other hand results from the fact that *ι*- and *κ*-carrageenan contain 3,6-anhydro bridges whereas *λ*-carrageenan does not. The natural precursors of *ι*- and *κ*-carrageenan are called *ν*- and *μ*-carrageenan (letter code G4S-D2S,6S and G4S-D6S, respectively) and are also non-gelling carrageenans lacking the 3,6-anhydro bridge. The 3,6-anhydro bridges are formed by the elimination of the sulfate from the C-6 sulfate ester of the precursors and the concomitant formation of the 3,6-anhydro bridge. *In vivo*, *ι*- and *κ*-carrageenan are formed enzymatically from their precursors,[7] by a sulfohydrolase. In industrial processing, the cyclisation reaction is carried out with OH-as a catalyst. The 3,6-anhydro-D-galactopyranosyl units have the $^1C_4$-conformation, which allows the formation of the helical structure, which is essential for the gel forming properties of *ι*- and *κ*-carrageenan. Occurrence of disaccharide units without the 3,6-anhydro ring and, as a consequence, with a $^4C_1$-conformation causes kinks in the regular helical strands of the gelling carrageenans.[8] An essential condition for gel formation at the helical level is thought to be the occurrence of these kinks, which increase the chance that polymer chains join in helices formed with two or more neighbouring chains. *ν*-Carrageenan is the most important of all kinking units that occur in *ι*-carrageenan.[9] The usual presence of considerable amounts of precursor units in commercial carrageenans shows a strong negative effect on the functional (*e.g.* gelling) properties.[10,11] Therefore, crude carrageenan gets an alkaline treatment prior to use.

However, considering the fact that infinite helical rods are inappropriate network building units we investigated the effect of incomplete removal of kink-inducing *ν*-units in *ι*-carrageenan. To this end, we prepared a range of *ν*-unit contents by applying an alkaline treatment during various times to a neutral extracted *ι*-carrageenan containing 23% *ν*-units. ^1H-NMR spectroscopy was used to determine the molecular ratio of *ι*- and *ν*-repeating units. The helix-forming capacity of the samples was studied by optical rotation measurements and differential scanning calorimetry. Gel strength was studied by small deformation oscillatory rheometry.

2 Experimental

2.1 Materials

Alkaline and neutrally extracted ι-carrageenan samples from *Eucheuma denticulatum* were kindly provided by SKW Biosystems (Boulogne Billancout Cedex, France) and CP Kelco (Copenhagen Pectin A/S, Lille Skensved, Denmark) and listed in Table 1. All samples were dried overnight (40°C, 7mmHg) before use. Solution concentrations of carrageenan are given as millimoles (repeating disaccharide) per L (mM).

2.2 Alkaline Treatment

Carrageenan (1.0 g) was dissolved in water (47.5 ml) and heated to 100°C. A NaOH-solution (2.0 M; 2.5 ml; final concentration 0.1 M) was added and the mixture was heated at 100°C for different periods of time. The reaction was stopped by adding to the reaction mixture cold water (0°C; 100 ml) and a cold HCl-solution (0°C; 0.1 M; 50 ml). The carrageenan samples were dialysed against NaCl-solution (three times 1 L; 0.1 M NaCl with 20 mM Na_2HPO_4) water (three times 1 L) and lyophilised. Phosphate was added to avoid autohydrolysis during dialysis and lyophilisation.[12]

Table 1 *Origin and composition of the samples*

Sample (processing time)	Molar fraction v-units[c] (%)	Mw[d] (kDa)
H1213 [b]	0	539
H1069B [b]	12.3	665
H1069B (8 min)	8.6	640
H1069B (15 min)	4.6	643
H1069B (30 min)	2.6	650
H1069B (45 min)	1.3	631
X6908 [a]	0.4	480
C-192 74-39440-42 [a]	1.4	364
93-35874-77 [a]	3.9	598
93-35851-52 [a]	6.6	585
97-251-14-04 [a]	9.6	495
97-243-10-E [a]	22.8	560
97-243-10-E (5 min)	22.6	472
97-243-10-E (10 min)	20	447
97-243-10-E (15 min)	15	468

a) CP Kelco, Copenhagen Pectin A/S, DK-4623 Lille Skensved, Denmark;
b) SKW Biosystems, 4 Place des Ailes, 92642 Boulogne Billancourt Cedes, France.
c) Determined with ^{1}H-NMR spectroscopy (see VandeVelde *et al.*[10])
d) Determined with SEC-MALLS (see VandeVelde *et al.*[10])

2.3 Optical Rotation Measurements

The coil-to-helix transitions (cooling curves) were monitored by optical rotation at 365 nm on a Perkin Elmer 241 polarimeter in a jacketted cell with a 10 cm path length. The temperature was controlled with a circulating water-bath. Specific optical rotation values are given in degrees $mM^{-1} m^{-1}$ (moles of repeating unit). Samples were prepared by dissolving (at 100 °C) carrageenan (80 mg; 0.2% w/w) in a NaCl-solution (40 ml; 0.8 M).

2.4 Rheological Measurements

The storage modulus (G') and loss modulus (G'') were measured as a function of temperature using a Carri-Med CSL^2 500 (TA Instruments) controlled stress rheometer fitted with a 2° steel cone with a diameter of 6 cm (gap width 55 μm). Carrageenan solutions (1% w/w in a solution containing 35 mM KCl and 20 mM NaCl) for low deformation oscillatory measurements were prepared from dried material, stored overnight at 4 °C and then heated to 80 °C for 10 min. The sample was applied between the cone and plate of the rheometer and covered with paraffin oil to prevent evaporation of the sample. After equilibration for 30 min at 50 °C, measurements were made approximately at 0.2 °C intervals between 50 and 5 °C with a cooling rate of 0.5 °C min^{-1}. Measurements were performed at a frequency of 1 Hz and a strain of 10%.

2.5 DSC Measurements

Calorimetric measurements at a salt concentration of 0.8 M (NaCl) were carried out with a differential scanning microcalorimeter microDSC III (Setaram, Caluire, France). The melting of naphthalene was used to calibrate the apparatus. The sample cell was filled with 850 mg carrageenan solution (0.2% w/w in 0.8M NaCl) and the reference cell with exactly the same amount of NaCl solution. Heating and cooling curves were recorded in the temperature range from 10 to 120 °C at a rate of 1.0 °C min^{-1}.

3 Results

3.1 Structural Analysis

The molecular ratio of the ι- and ν-carrageenan repeating units was determined using ^{1}H-NMR spectroscopy as described.[10] The molecular weight distribution and the weight average molecular weight (M_w) were determined with Size Exclusion Chromatography coupled to Multi Angle Laser Light Scattering (SEC-MALLS).[10] An overview of the ν-content and M_w of the ν/ι-carrageenan samples used in this study is given in Table 1. The molecular mass of the carrageenan samples used in this study falls within the range of 350 to 650 kDa. Within this range of molecular masses, the functional properties of the carrageenans, *e.g.* storage modulus, are independent of their molecular weight.[13]

3.2 Optical Rotation

Optical rotation measurements were performed at a relatively high concentration of salt (0.8 M of NaCl). This high concentration was chosen to obtain as much information about the coil-to-helix transition as possible within the temperature window of our equipment (15 to 95 °C). At the salt concentration chosen the onset of the coil-to-helix transition of ι-carrageenan is observed at 86 °C (Figure 2). The shape of the curve is in good agreement with that observed in the past.[14] The plateau value of the optical rotation for ι-carrageenan is assigned to that of the fully helical conformation. The fraction of helical residues, Θ, in the v-containing samples is calculated using the relation

$$\Theta = \frac{[\alpha] - [\alpha_c]}{[\alpha_h] - [\alpha_c]} \tag{1}$$

where $[\alpha_h]$ ($= 1.14° \, \text{mM}^{-1} \, \text{m}^{-1}$ at 365 nm) and $[\alpha_c]$ ($= 0.68° \, \text{mM}^{-1} \, \text{m}^{-1}$ at 365 nm) are the specific optical rotation values for all-helix and all-coil ι-carrageenan, respectively, obtained from sample H1213 at 15 °C and 95 °C, respectively. $[\alpha]$ is the low temperature plateau value of the specific optical rotation of v-containing carrageenan. The details of the relationship between $[\alpha_c]$ and the v-content is currently under study and will be published elsewhere.

From Figure 2 it is clear that the onset temperature of the coil-to-helix transition is independent of the fraction of v-units present in the sample. For all v-contents probed in this study the onset temperature is found to be 86 °C. The helical fraction in fully transformed material decreases monotonously with increasing v-content. The sharpness of the transition broadens with increasing v-content. The latter is comparable with experimental results and theoretical

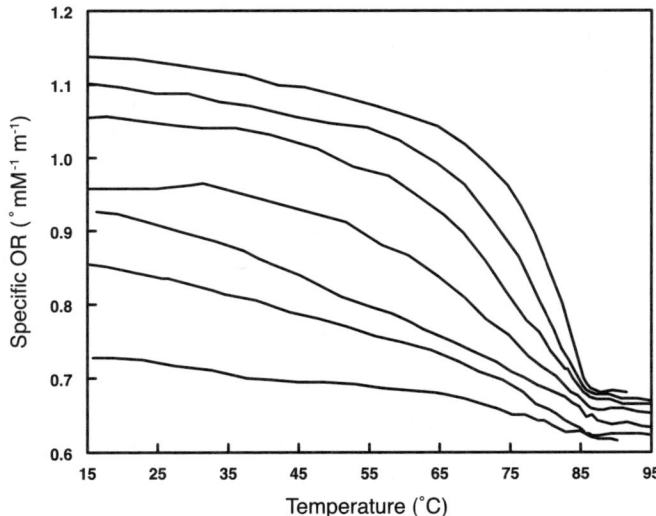

Figure 2 *Optical rotation* versus *temperature curves for* ι-*carrageenan samples (0.2% w/w in 0.8 M NaCl), containing different amounts of v-carrageenan, from top to bottom: 0%, 1.3%, 2.6%, 4.6%, 9.6%, 12.3%, and 22.8% bottom*

calculation on the molecular weight dependency of the coil-to-helix transition of κ-carrageenan. The transition broadened with decreasing molecular weight of the carrageenan sample. In our case, the presence of *v*-units in the *ι*-carrageenan chain effectively reduces the length of the *ι*-stretches.

3.3 Rheological Measurements

In order to calculate the maximum extent of helicity at a certain *v*-content, the polarimetric measurements were done in 0.8 M NaCl, yielding an onset temperature of the coil–helix transition of 86 °C. However, the rheometer used for this study did not allow for temperatures higher than 50 °C. As a consequence, in order to observe a coil–helix transition a lower salt concentration is needed. Therefore, the small deformation oscillatory measurements have been performed in a mixed salt solution containing 35 mM KCl and 20 mM NaCl, in which the coil–helix transition temperature is less than 50 °C. The storage modulus (G′) and loss modulus (G″) were measured as a function of decreasing temperature for the series of *ι*-carrageenan samples containing different amounts of *v*-units. The experimental conditions applied fall within the linear response region as observed by modulus *versus* frequency and strain curves.[10] With decreasing temperature G′ increases. For pure *ι*-carrageenan, a levelling off is observed at low temperatures. The form of the cooling curve of pure *ι*-carrageenan is in good agreement with cooling curves published before.[15,16] The shape of the G′ *versus* temperature curve is similar to that of the curve of the optical rotation *versus* temperature, confirming the relation between the helix formation and gelation. Figure 3 shows that the storage modulus (G′) of the carrageenan samples shows a maximum as a function of the fraction *v*-units. It turns out that the maximum in storage modulus occurs at 4% *v*-units.

3.4 DSC Measurements

Differential scanning calorimetry was used to study the melting behaviour of helical, *v*-unit containing *ι*-carrageenan. Ionic conditions were equal to those of the optical rotation measurements (0.8 M NaCl). The excess heat capacity, C_p, for a pure *ι*-carrageenan and *ι*-carrageenan containing 12.3% *v*-units is plotted against the temperature (Figure 4). Comparable results were obtained with a series of samples measured using a capillary microcalorimeter.[10] The melting temperature is independent of the fraction of precursor units present in the sample. This is in agreement with the conclusion from the optical rotation measurements.

4 Discussion

This study on *ι*-carrageenan is aimed at exploring the gel formation in aqueous systems of *ι*-carrageenan containing small amounts of *v*-units (kinking units).

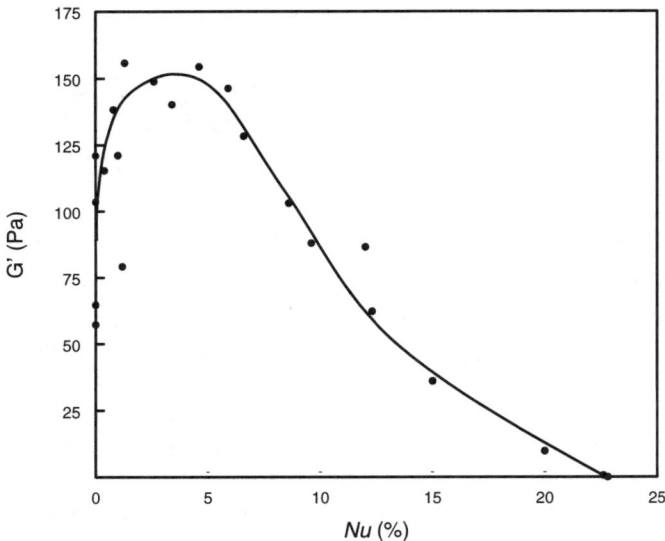

Figure 3 *Storage modulus at 15 °C (G') as a function of the fraction v-units (1% carrageenan in 35 mM KCl and 20 mM NaCl; 1 Hz; 10% strain)*

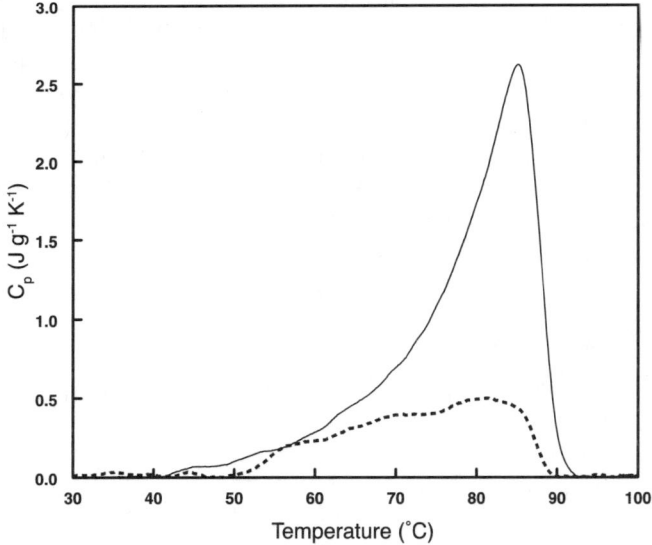

Figure 4 *Excess heat capacities* versus *temperature curves of ı-carrageenan (solid line) and ı-carrageenan containing 12% v-units (dotted line)*

The maximally attainable helical content is strongly dependent on the amount of precursor units present in the chain (Figure 2). Both data from polarimetry as well as from DSC indicate that the temperature of the coil-to-helix transition is independent of the amount of precursor units present (Figures 2 and 4). It is therefore concluded that the presence of v-units does not lower the stability of the helical stretches. On integration of the DSC signal from the temperature of the onset of helix formation downwards, one obtains a curve which is practically identical in shape to that from polarimetry (Figure 5). DSC data and polarimetry data are therefore found to reflect the same process in the sample. This process is the same for pure *ι*-carrageenan and for *ι*-carrageenan containing 12% of *v*-units. It supports a notion of *v*-units only reducing the amount of helix, not the structure of the helix. However, too many *v*-units, *i.e.* more than about 25%, prevent the chain from forming any helix at all. The simplest interpretation of this fact is that a *v*-unit disturbs the helix-forming capacity of the helical conformation of *ι*-units in its direct neighbourhood, roughly across one or two repeating units.

The effect of different helical fractions on the functional properties of *ι*-carrageenan has been studied using small deformation oscillatory rheological measurements. As shown in Figure 3 the storage modulus is dependent on the number of kinking units. For *ι*-carrageenan the change of G' with temperature proceeds parallel with that of the helical content as determined by optical rotation, resulting in a high plateau value for G' at low temperatures. This finding is in agreement with results reported in the past.[15,16] However, as a function of the fraction of precursor units, the storage modulus plateau value shows a maximum at about 4% *v*-units. Such enhanced functional properties were already mentioned by Hansen *et al.*[11] We explain this by the opposing effects of helical content and kinking frequency on gel strength. For building a three-dimensional network, kinks in the chain are necessary. For the gel to have strength, the average length of stretches of double helix should not be too small. On increasing *v*-unit content from zero, kinking increases and helicity decreases, with a maximum at 4% *v*-units as a result.

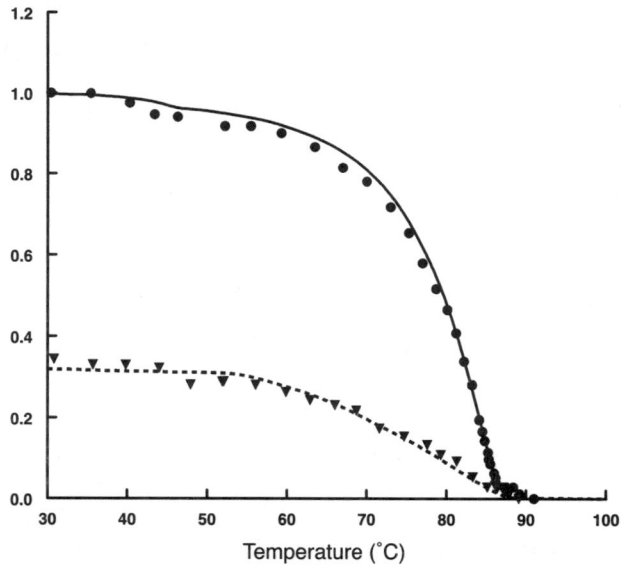

Figure 5 *Normalised enthalpy and optical rotation vs temperature curves of pure ι-carrageenan (solid line: enthalpy; circles: optical rotation) and ι-carrageenan containing 12.3% v-units (dotted line: enthalpy; triangles: optical rotation)*

5 Conclusions

v-Carrageenan is the precursor of ι-carrageenan in the biosynthesis of ι-carrageenan. The presence of v-carrageenan units (kinking units) in ι-carrageenan, lowers the maximally attainable helical content of ι-carrageenan. The coil-to-helix transition temperature and the gel melting temperature are not influenced by the presence of v-units. The structure of helical stretches is independent on v-unit content. The storage modulus G', considered here to be a measure of the gel strength, shows a maximum near a v-unit content of 4%. This maximum appears to reflect the opposing influences of double helix content and kinking frequency. In practical terms, incomplete alkaline processing of crude ι-carrageenan can lead to a considerable increase in functionality.

Acknowledgements

The authors gratefully acknowledge Heleen A. Peppelman (Wageningen Centre for Food Sciences and TNO Nutrition and Food Research Institute) for carrying out the rheological measurements of the carrageenan samples. Dr Sokol Ndoni and Sonja Johansen (CP Kelco, Copenhagen Pectin A/S, Lille Skensved, Denmark) are acknowledged for providing us with a series of ι-carrageenans containing different amounts of v-units and to Prof. V. Y. Grinberg (Russian Academy of Sciences), we are grateful for fruitful collaboration.

References

1. G. H. Therkelsen. 'Carrageenan', in R. L. Whistler and J. N. BeMiller (eds.). *Industrial Gums: Polysaccharides and their Derivatives*, Academic Press Inc., San Diego, 1993, p. 145.
2. F. Vande Velde and G. A. DeRuiter, 'Carrageenan', in S. DeBaets, E. J. VanDamme and S. Steinbüchel (eds.), *Biopolymers. Vol. 8: Polysaccharide II*, Wiley-VCH, Weinheim *to be published.*
3. S. H. Knutsen, D. E. Myslaboski, B. Larsen and A. I. Usov, *Bot. Mar.*, 1994, **37**, 163.
4. E. R. Morris, D. A. Rees and G. Robinson, *J. Mol. Biol.*, 1980, **138**, 349.
5. C. Viebke, J. Borgstrom, I. Carlsson, L. Piculell and P. Williams, *Macromolecules*, 1998, **31**, 1833.
6. C. Viebke, L. Piculell and S. Nilsson, *Macromolecules*, 1994, **27**, 4160.
7. K. F. Wong and J. S. Craigie, *Plant Physiol.*, 1978, **61**, 663.
8. C. J. Lawson and D. A. Rees, *Nature*, 1970, **227**, 392.
9. D. A. Rees, F. B. Williamson, S. A. Frangou and E. R. Morris, *Eur. J. Biochem.*, 1982, **122**, 71.
10. F. VandeVelde, H. S. Rollema, V. Y. Grainberg, N. V. Grinber and R. H. Tromp, to be published.
11. J. H. Hansen, H. Larson and J. Groendal. *Carrageenan Compositions and Methods for their Production*, Patent application US patent applications: US 6,063,915 (2000).
12. R. A. Hoffmann, A. R. Russell and M. J. Gidley. 'Molecular Weight Distribution of Carrageenans: Characterisation of Commercial Stabilisers and Effect of Cation Depletion on Depolymerisation', in G. O. Phillips, P. J. Williams and D. J. Wedlock

(eds.), *Gums & Stabilisers for the Food Industry 8*, Oxford University Press, Oxford, 1996, p. 137.

13. C. Rochas, M. Rinaudo and S. Landry, *Carbohydr. Polym.*, 1990, **12**, 255.
14. S. Nilsson, L. Piculell and B. Jönsson, *Macromolecules*, 1989, **22**, 2367.
15. A. Parker, G. Birgand, C. Miniou, A. Trespoey and P. Vallée, *Carbohydr. Polym.*, 1993, **20**, 253.
16. L. Piculell, S. Nilsson and P. Muhrbeck, *Carbohydr. Polym.*, 1992, **18**, 199.

Films and Foams of Sparkling Wines

M. Vignes-Adler[1] and B. Robillard[2]

[1]LABORATOIRE DE PHYSIQUE DES MATÉRIAUX DIVISÉS ET DES INTERFACES, UNIVERSITÉ DE MARNE-LA-VALLÉE, FRE2395 DU CNRS, BÂTIMENT LAVOISIER, CITÉ DESCARTES, 77454 MARNE-LA-VALLÉE CEDEX 2, FRANCE
[2]LABORATOIRE DE RECHERCHE DE MOËT & CHANDON, 20, AVENUE DE CHAMPAGNE, F-51205 EPERNAY CEDEX, FRANCE

1 Introduction

Every consumer expects a wine to be tasteful, to have a typical aroma and to display a rich colour. Since champagne is a festive drink, it is also expected to fizz lively out of the bottle and *last but not least*, it "must" have pretty foam. Indeed, the champagne foam should be generous just after pouring and it should collapse in a few seconds to form a collar at the wine surface, fed by trains of submillimetric bubbles. As emphasised by McLeod,[1] sight is the sense impression which is the most rapidly analysed by the brain, the other ones, touch, hearing, smell and taste only confirming or contradicting the visual expectations. Since the first quality that a consumer judges is the bubble and foam quality, the produce appearance is a major factor of appreciation. The foam of champagne is transient, even evanescent, and tiny changes in the composition, concentration, or quality of the various compounds can impair, inhibit or enhance the foamability with severe commercial consequences. Reliable informations on the factors and the compounds controlling the foamability and foam stability of champagne are therefore of considerable interest for the winemaker.

As any other foam, the champagne one is a random dispersion of gas bubbles in a liquid. As the foam is just generated, the liquid drains by gravity and thick films appear. Then the films drain into the Plateau border under the action of the capillary pressure and become thinner until they rupture. It is well known that pure liquids do not form stable foams. The presence of a surface-active solute is required, although it does not ensure the persistence of the film against gravity drainage or against stresses that might tend to destroy the film. Surface rheology, surface elasticity, diffusion of gas out and into foam bubbles, and presence of

212

colloidal particles are, to name a few, factors that affect foam properties.

Three lengthscales are generally considered in foam:

(1) The local scale deals with the surface properties of the liquid and with the evolution of a single, isolated film.
(2) The intermediate scale deals with a few bubbles like in the collar at the champagne surface in a glass.
(3) The global scale which is the consumer one deals with the liquid foamability, the foam expansion and stability.

The quantification of the effects of the change of scale between the various levels of investigation of foam properties, *i.e.* from the surface and film scale towards the scales of a few bubbles and a foam column, is still unclear. The foam global behaviour is therefore to some extent unpredictable from the surface and film properties. However, qualitative correlations can be very usefully derived. Against this background, we have investigated the film and foam properties of wines.

2 Sparkling Wine Composition

The average composition of a base wine is given in Table 1. Wines can be viewed as hydroalcoholic solutions containing many organic compounds, some at concentrations as low as a few mg/L, which may show surface activity by themselves (nitrogenous monomers and polymers, *e.g.* aminoacids, proteins and glycoproteins, *etc.*) or by association with other compounds (polysaccharides associated with proteins, for example). It is not possible to decide which constituents are actually adsorbed at the surface and which one is the more foam active.

The positive role of proteins on foam behaviour is attributed to their adsorba-

Table 1 *Average composition of Champagne base wine and composition of the model solvent (MS). Except for ethanol, all concentrations are g/L. pH = 3.2 and the ionic strength is 0.02 M*

Constituent	Base wine	MS
Ethanol	11.3% (V/V)	12% (V/V)
Glycerol	4–7	4.7
Tartaric acid	2.5–4	3.7
Lactic acid	4–6	4.8
Proteins	0.005–0.05	
Polysaccharides	0.2	
Polyphenols	0.1	
Amino acids	0.8–2	
Lipids	0.01	
K^+	0.2–0.45	0.45
Ca^{2+}	0.06–0.11	0.083
Mg^{2+}	0.05–0.08	0.0782
Na^+	0.005–0.015	0.021

bility at the interfaces. Since, in the native protein structure most proteins have hydrophobic and hydrophilic domains, the presence of an air–water interface can induce conformation change that maximises exposure of the hydrophobic domains to the air and the hydrophilic domains to the aqueous medium favouring the protein adsorption at the surface. However, codissolved protein and ethanol interact specifically in bulk aqueous solutions.[2] Ethanol acts as a poor protein solvent, the more so at the protein isoelectric point when the ionic forces are weak. It is also known as a protein denaturing agent, acting primarily by weakening the hydrophobic bonds and exposing the hydrophobic side chains to ethanol in the denatured state. The protein adsorbability is severely modified in the presence of ethanol.[3–5]

Actually, about 70% of the champagne proteins are glycoproteins.[6] Carbohydrates (*e.g.* sugar) tend to protect proteins against denaturation by ethanol and to increase their solubility; the degree of the protection is, to a first approximation, a function of the number of hydroxyl groups. Hence, one can expect that the interactions between glycoproteins and the alcoholic solvent, and therefore their adsorbability, will be intricately dependent on the quality and the sugar content of the glycoproteins, which changes with the vintage.

However, champagnes are too difficult to manipulate because of their natural effervescence which generates uncontrollable bubbles, and experiments are usually done on base wines.[7] Besides, their chemistry is very complex. Better analysis of the chemical and physical influential factors can be obtained when experiments were done on model hydroalcoholic solutions with known composition.

Model solutions of glycoproteins partially mimicking a base wine with same alcohol content, pH and cationic strength as a typical base wine have been used. The solvent does not contain any lipids or volatile substances (Table 1). Two kinds of glycoproteins are used:

(1) Endogenous yeast glycoproteins (YGP) obtained by alcoholic fermentation in a synthetic medium and ultrafiltered on a molecular weight cut-off membrane at 10 kDa.[6] The YGP is essentially a mixture of mannoproteins and glucanes. The proportion of sugar in the YGP mixture is higher than 90% and most of the molecular weight of each structure ranges between 40 kDa to more than 100 kDa. The pI of the YGP ranges between 3 and 4.3. Let us note now that sugars are hydrophilic, and the more sugar the more hydrophilic domains in the glycoprotein.

(2) Exogenous maltosyl bovine serum albumine at two concentrations. MBSA (Sigma Chemical lot 98F8120) is prepared from bovine albumin and maltose coupled *via* reductive amination.[8] One mole of albumin contains 14 moles of disaccharide linked to lysine residues. The sugar content represents 5% of the molecule. The MBSA molecular weight and pI are 71 kDa and 4, respectively. It was used without further purification.

3 Global Properties of Wine Foam

Bulk foams are characterised by global parameters which are often measured by foaming the liquid with a sparging method.[11–12] Typically, a gas is blown at a constant rate Q into a test tube containing a volume V_0 of base wine through a sintered glass plate located at the bottom of the tube. Usually, the foam volume $V(t)$ increases to a maximum, V_{max}, then it decreases towards a plateau V_s, eventually it collapses when the gas flow is stopped (Figure 1). Three global parameters are defined, the foam expansion, E, the foamability, Σ, and the foam stability:

$$E = \frac{V_{max}}{V_{liq}} \qquad \Sigma = \frac{V_s}{Q} \qquad L_f = \frac{1}{V_0} \int_{t_0}^{\infty} V \, dt$$

where V_{liq} is the initial liquid volume, and V_0 and t_0 are, respectively, the foam volume and the time when gas flow was stopped. Σ characterises the average lifetime of a rising bubble in the foam column.

Foam global parameters generally depend on the solution and on the wine vintage. Usually, the decay of the base wine foam occurs in two steps (insert in Figure 1). There is a fast decay of the freshly made foam over a timescale L_{f1} of about 10 s (global scale), then a residual foam, consisting of a 1 to 5 bubbles thick layer, which ages at the wine surface and collapses after a much longer time denoted L_{f2} (intermediate scale).

In the examples given in Table 2, the sparging gas being nitrogen, the values obtained for the 1990 wine are significantly lower than the ones of the 1993 and 1994 wines. In particular, the residual foam of 1990 wine is much less stable. On the contrary, the parameters of the 1993 and 1994 wines foams are almost the same. Malvy *et al.*[13] have shown that the foam lifetime drastically decreases when the wine is ultrafiltered (Figure 2). It is interesting to know that the wine

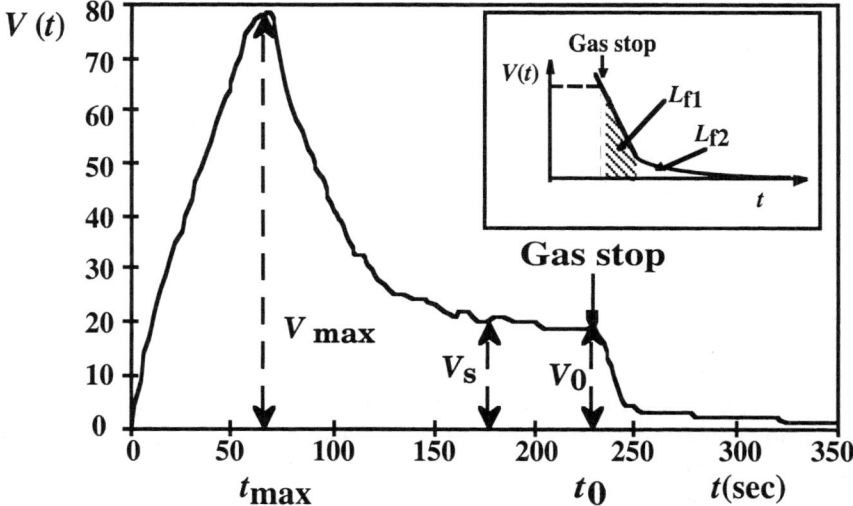

Figure 1 *Typical evolution of foam volume with time obtained by a sparging method. $Q = 5l/h$. Insert shows thee definition of foam decay characteristic times*

Table 2   *Global foam parameters and film equilibrium thickness of base wines and of model solvent and model solutions. The analogy between the 1993 and 1994 base wines on one side, and, except for the existence of a residual foam, between the 1990 base wine and the YGP 1 model solution is remarkable. The reported experiments were done early 1996. Nitrogen was used as sparging gas. (n.m.) = non-measurable with the available test tube*

Liquid	E_{max}	L_{f1} (s)	L_{f2} (s)	Σ (s)	τ (s)
wine 1994	0.55	15	1470	17	∞
wine 1993	0.89	11.5	1090	18	∞
wine 1990	0.17	8	61	10	∞
MS	0.14	8	0	10	5
YGP 1	0.15	10	0	10	
YGP 2	0.5	16.4	0		48
MBSA 1	2.5	38	1980		2
MBSA 2	$\gg 5$	n.m	∞		∞

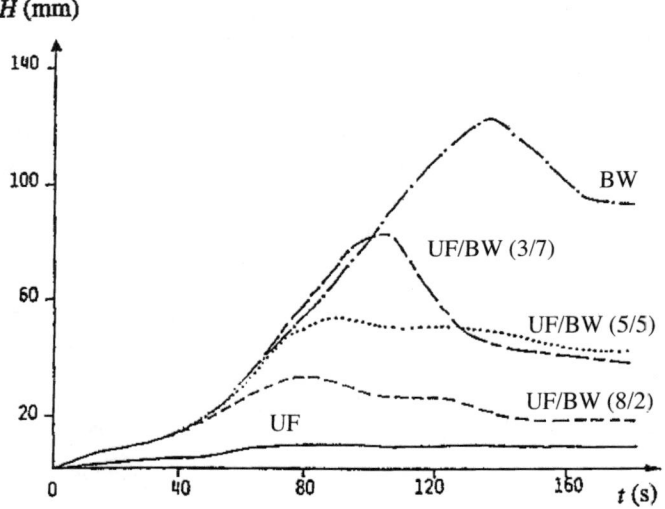

Figure 2 *Time evolution of foam height obtained in sparging experiments with mixtures of ultrafiltered (UF) base wines (MWCO = 10 kDa) and base wines (BW) in various proportions*

foam parameters are also much lower than those of beer are, for example the L_f value can be 10 times, even 100 times larger with beer than with champagne.[14]

The foam parameters of the YGP solutions have the same order of magnitude as the base wine ones although the parameters of the more concentrated solutions are higher. The MBSA solution parameters are drastically higher. They were not measurable with our device for the higher MBSA concentration; besides, some gel-like liquid remained stuck on the wall of the test tube.

However, the knowledge of the global parameter is not sufficient to understand the foaming properties of the wines and local scale measurements are needed to have better insight.

4 Local Properties of Wines

4.1 Wine Surface Tension

The alcohol content, the protein concentration and the surface tension of wines depend on the vintage (Table 3). The wine surface tensions, σ, slightly vary from 46.4 to 47.1 mN/m and the model solution ones from 46 to 48.5 mN/m. The range of variations is small, however, it is significant with respect to the experimental precision ± 0.1 mN/m.

Now, surface tension measurements are done on surfaces at equilibrium, which is rather far from the surfaces of the freshly made foam films. Dynamic experiments on films provide more significant information.

4.2 Wine Films

Let a bubble naturally nucleate in the liquid on the bottom surface of a closed container. It rises in the liquid dragging on its path some surface-active molecules, macromolecules or colloids.[15] When the bubble is approaching the liquid

Table 3 *Surface tensions of model solutions and base wines of three different vintages, 1990, 1993 and 1994 from Moët & Chandon winery. They are blends of various wines (Chardonnay, Pinot Noir and Pinot Meunier). The alcohol content is determined by infrared equipment. The equivalent BSA concentration in wines is measured by a Bradford test;[9] it is expressed in (equivalent BSA concentration). The surface tension σ is measured by means of a De Noüy type tensiometer once the solution surface has reached an equilibrium.[10] The reported experiments were done early 1996*

	Alcohol content (% V/V)	(glyco)-protein (mg/L)	Π (mN/m)	Π_{hyd} (mN/m)
	11.30	13.7		
wine 1994	(0.01)	(0.06)	26.6	24.1
	11.45	13.0		
wine 1993	(0.02)	(0.05)	26.05	24.5
	10.88	14.2		
wine 1990	(0.01)	(0.02)	25.9	23.5
MS	12	0	24.8	
YGP 1	12	3	24.5	
YGP 2	12	10	25.6	
MBSA 1	12	10	26.9	
MBSA 2	12	40	27	

surface, a film is formed, and it drains under the action of gravity and capillary forces (Figure 3).

Microinterferometry images of the films formed from the different base wines and solutions are shown in Figures 4–6.

Whatever the wine and the solution, the film drainage occurs in two steps. When the bubble approaches the liquid free surface, the newly created film presents thickness variations which can be as large as 400 nm and its surfaces appear mobile: this is known as the dimple effect.[16] Then, the dimple rapidly decays; further evolutions drastically differ according to the wine vintage and to the solution.

Wine films do not rupture while films formed from model solvent do in less than 6 s. At $t \sim 50$ s the overall film thickness become uniform and about 100 nm. Afterwards, all the wine films continue draining slowly while small aggregates progressively appear, as small black spots, in the film. With the 1990 base wine, the aggregates become organised in regular patterns inducing the film a granulous aspect. With the 1993 and 1994 base wines, there are two kinds of evolution, which have been observed at random for successive bubbles in the same experiment independently of any external conditions. In some films, the aggregates move to form dendritic patterns typical of diffusion growth of

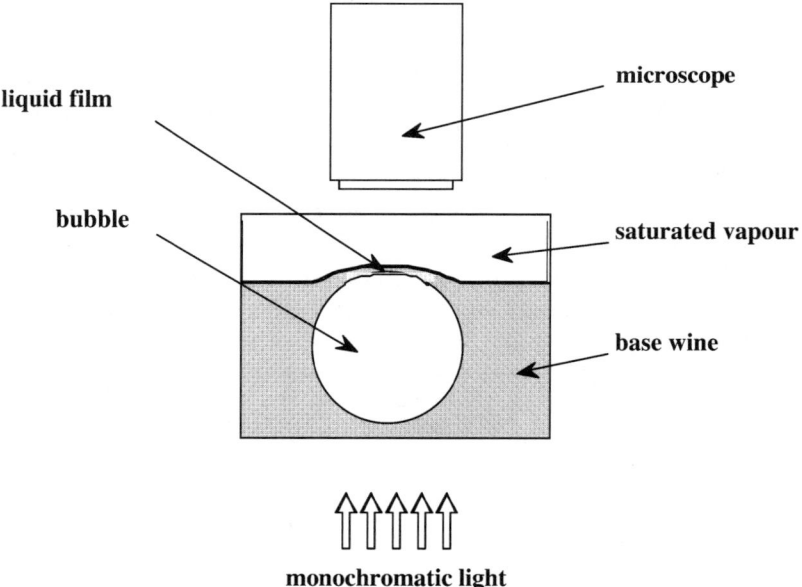

Figure 3 *Principle of a foam film microinterferometry cell. Single, thin liquid films are formed above a bubble attached to an air–water surface in a tightly closed cell placed in the field of a metallographic microscope (×200), and illuminated by reflected heat-filtered monochromatic light. The microscope can be focused either on the film or on the bubble diameter. The interferometry patterns change with the film local thickness. In Figures 4–6, the bubble diameter is 1.21 mm and the film diameter is 0.42 mm*

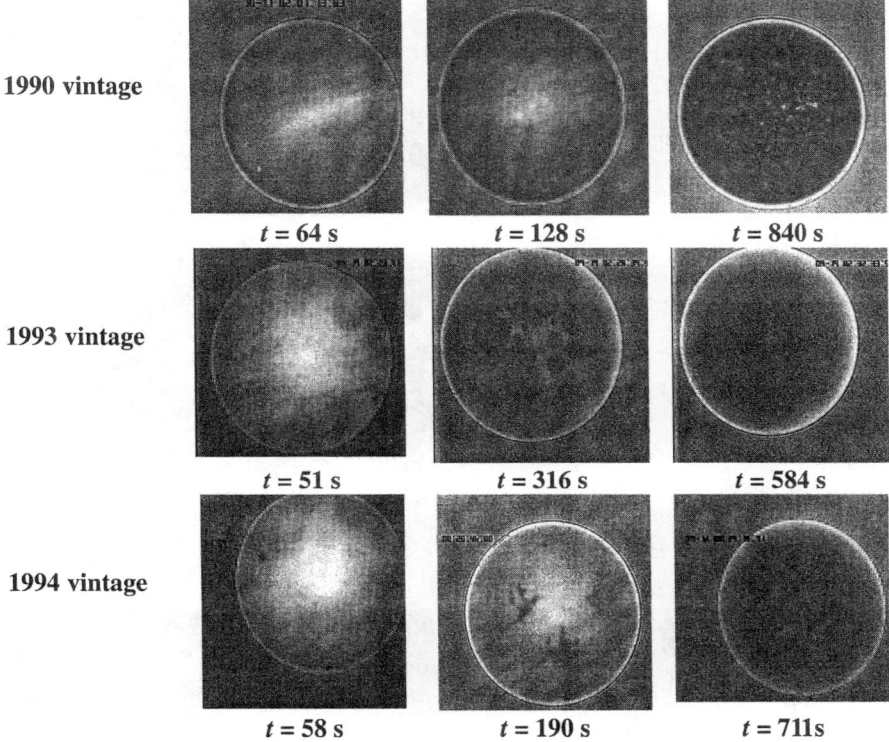

1990 vintage

t = 64 s   *t* = 128 s   *t* = 840 s

1993 vintage

t = 51 s   *t* = 316 s   *t* = 584 s

1994 vintage

t = 58 s   *t* = 190 s   *t* = 711s

Figure 4 *Interference pictures of draining films formed from base wines*

bidimensional aggregates, in others they remain fixed and constitute sites for black film nucleations.

Likewise, the aspects, drainage rates and lifetimes of the films drastically depend on the nature and the concentration of the glycoproteins. Remarkably, the films formed from the dilute model YGP solution ($C = 3$ mg/L) behave like the 1990 wine (Figure 5). They do not rupture and they also become granulous, the grain typical diameter being 6 μm whereas the film thickness is 15 nm. At the higher concentration $C = 10$ mg/L, one can observe formation of dark and irregular clusters which are polydisperse mostly bidimensional aggregates (~ 15 μm). The film ruptures at a finite time equal to 50 s as the film thickness is about 25 nm.

The drainage of the MBSA films is very different from the base wine ones (Figure 6). At moderate MBSA concentration ($C = 10$ mg/L) the dimple is sucked into the film Plateau border leaving behind a grey film with constant thickness (*transition TF* $\rightarrow$ GF). When the grey film is 120 nm thick, several 20 nm thick black films are nucleated therein (*transition GF* $\rightarrow$ BF), which are expanding with a rim of liquid ahead at the expense of the thicker grey film. A second series of black spots are nucleated behind the first series. The rims thicken by steps, their maximum thickness being 500 nm, which means that the black spots are like 480 nm deep craters. These films are very unstable with about 3 s

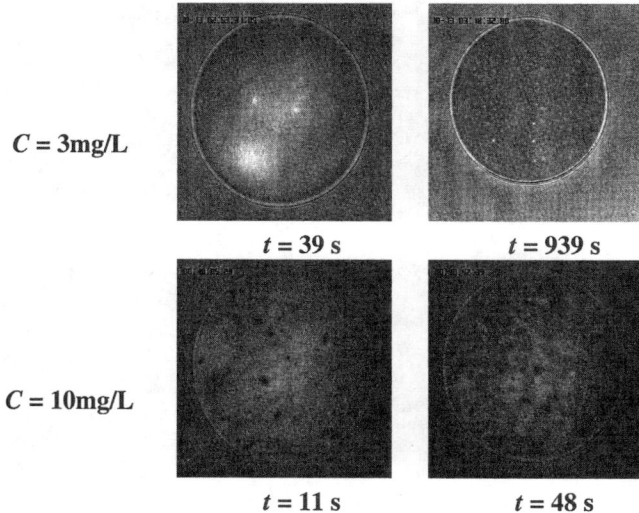

$C = 3mg/L$

$C = 10mg/L$

Figure 5 *Interference pictures of draining films formed from YGP solutions*

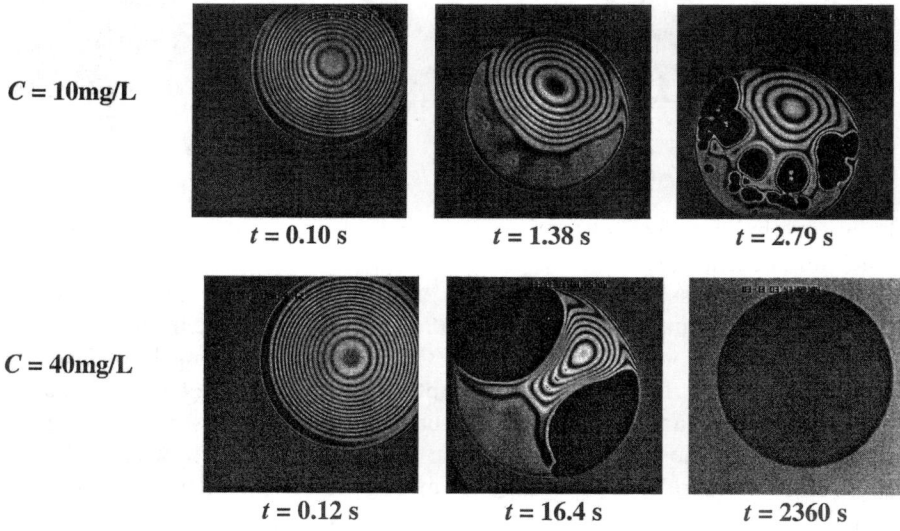

$C = 10mg/L$

$C = 40mg/L$

Figure 6 *Interference pictures of draining films formed from MBSA solutions*

lifetime. At large MBSA concentration ($C = 40$ mg/L) the phenomenon is essentially the same as at $C = 10$ mg/L, but now everything happens very slowly. Only two black films with thick rims are formed and finally coalesce into a single stable black film. The rim can be as thick as three interference fringes between the black spots and the grey film which corresponds to a 350 nm step.

5 Discussion

It is remarkable that the 1993 and 1994 base wines compare so well within the experimental errors, at the local scale with same kinetics of film drainage and same aggregates, and at the global scale with same order of magnitude of the three foam parameters. It is amazing that the foam expansion E, the foam lifetimes L_{f1} and L_{f2}, and the foamability Σ all have smaller values with the 1990 wine than with the two others. Not less remarkable is the comparison between the 1990 and the dilute YGP model solution, the main difference between the wine and the model solution being the existence of residual foams with wines which do not exist in the case of the model solution. And finally the behaviours of the films and foams of the MBSA solutions are completely different from the wine and YGP solutions ones.

These very different results are assigned to the specific ethanol/glycoprotein interactions. Let us remember that the surface tensions of the various wines and solutions do not differ dramatically. Viewing wines as aqueous solutions, the surface activity of their constituents measured by the surface pressure $\Pi = \sigma_w - \sigma$ is about 26 mN/m, whereas the surface pressure, Π_{hyd}, of the simple hydroalcoholic solution with the same alcohol content is about 2 mN/m lower (Table 3). This very small difference is necessary to explain the foamability of a wine, albeit evanescent compared to the lack of foamability of simple hydroalcoholic solutions, but it is not sufficient to explain the observed difference from one wine to another wine (likewise, $\Pi \approx 26 \pm 1$ mN/m for both glycoproteins whatever the concentration). Except for the more dilute YGP solution, the surface pressure is slightly higher than the MS one. Besides, it was found[17] that in the absence of ethanol $\Pi_{eth\ free}^{MBSA} = 13.1$ mN/m and $\Pi_{eth\ free}^{YGP} = 9$ mN/m when the glycoprotein concentration is 10 mg/L.

The interaction forces between the macromolecules and the ethanol molecules at the interface can be appreciated by the following expression deduced from thermodynamic considerations:

$$\Delta\Pi = \Pi - (\Pi_{eth\ free}^{macromol} + \Pi_{eth}^{macromol\ free})$$

With the above values, $\Delta\Pi^{MBSA} = -11$ mN/m and $\Delta\Pi^{YGP} = -8.2$ mN/m. $\Delta\Pi$ is non vanishing and it is negative for both macromolecules. It means either that the ethanol molecules and the glycoproteins interact attractively at the surface, or that there is a competition between them in the adsorption process, the alcohol hindering, albeit not preventing, the glycoprotein adsorption. Other *in-situ* measurements would be necessary to conclude which model is valid in the present cases.

Analysis of the films data helps to qualitatively understand which model is valid. Film thicknesses are plotted *versus* time in Figure 7. All films drain fast, but the ones formed from the 1990 wine do it much more rapidly than the others toward an 'equilibrium' thickness $h^f \cong 16$ nm. Films formed from the 1993 and 1994 wines have the same drainage rate towards a minimum thickness $h \cong 25$ nm but it is slower than the 1990 wine film, and they thicken back. It is again remarkable that the drainage kinetics of the YGP solution is the same as the

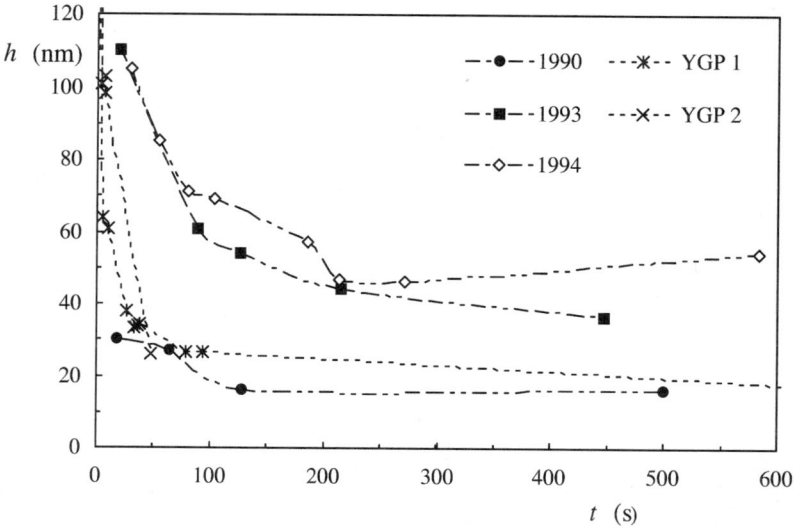

Figure 7 *Drainage kinetics of films formed from base wines and YGP model solutions*

1990 wine one.

Drainage kinetics laws are usually dependent upon the interfacial viscous stresses, which exist in the film and at the film interfaces,[16] and also of the dynamic viscosity of the liquid.

Let us analyse first the film drainage kinetics of the more concentrated MBSA solution (Figure 8). They drain much more slowly than any other ones. Moreover, the drainage rate is very close to the Reynolds value that is usually obtained when the film surfaces are rigid.[21] It is therefore consistent with the hypothesis that the MBSA and the ethanol molecules interact attractively to form a rigid surface with very high interfacial viscosity.[22] Another conclusion is that MBSA is definitely not an appropriate model glycoprotein for champagne wines.

The YGP films drain very rapidly and it is likely that the viscous friction at the interfaces is very weak and that ethanol prevents glycoprotein adsorption at the film surfaces. The role of the aggregates should be examined. Obviously, observed aggregates stabilise the films and increase the dynamic viscosity of the liquid in the film. The origin of the aggregates is rather easy to identify in the YGP solutions. Aggregates are visible when the film thickness h has decreased to a value ranging between 50 nm and 100 nm. This value is comparable or smaller than the hydrodynamic diameter 120 nm of the YGP macromolecule in the solution.[7] The film is a confined system, which hinders Brownian motion in its reduced dimension, and which squeezes the polymeric sugar chains during its thinning. Solvent depletion and increase of the YGP volume concentration result. The processes occur as if the macromolecular concentration increases to a limiting value corresponding to a supersaturation level where spontaneous precipitation occurs and flat-shaped aggregates, with a $\sim 6\mu m$ diameter and ~ 15 nm thickness, are formed.

In wines, the macromolecules are initially below their solubility limit since

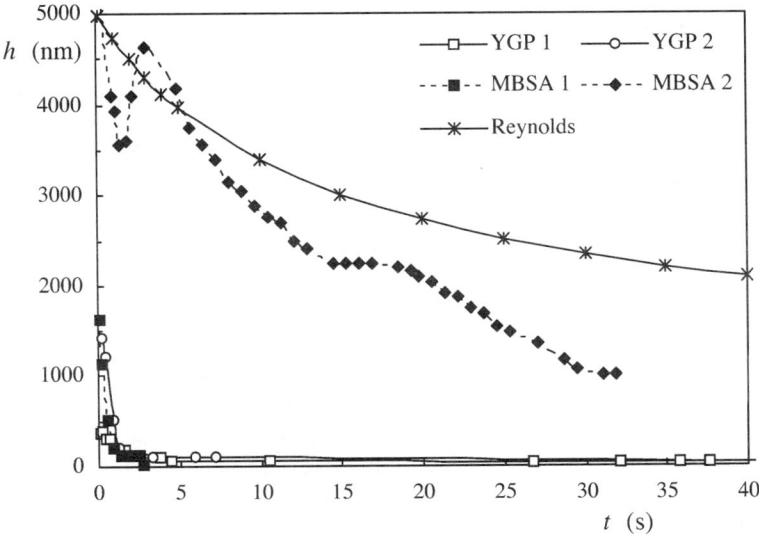

Figure 8 *Drainage kinetics of films formed from model solutions. For MBSA solutions, the thicknesses are measured at the top of the dimple. The Reynolds drainage kinetics is also reported*

otherwise they would have already precipitated inside the vat before racking. However, oenologists have observed that, as the alcohol concentration slightly increases during the second fermentation, proteins are less and less soluble and, either they precipitate or they adsorb on the various endogenous particles in the wines.[18–19] This means that the wine macromolecules (essentially proteins) are not far from their solubility limit, and if for any reason the volume concentration of the macromolecules increases, everything else being equal, precipitation may occur and aggregates can be formed. Actually, in wines there are YGP, which are produced by the yeasts during the alcoholic fermentation, and also polyphenols, which already existed in the grape juice. Vernhet *et al.*[20] have shown that polyphenols give a complex with YGP-like macromolecules in wines, which leads to their precipitation. When the film thins, the volume concentrations of all the macromolecules increase, as long as they are not expelled towards the meniscus. Therefore, a displacement occurs in the phase diagram towards the insolubility limit, which generates small aggregates.

The occurrence of one of the observed behaviours of the aggregates in the wine films depends upon the quality and concentration of the macromolecules in the wines, of their vicinity relatively to a critical line of solubility in the phase diagram, and of the aggregate–aggregate, aggregate–film surface interaction forces. This is drastically vintage dependent. Regular patterns are obtained when net repulsive forces exist between aggregates while dendrites and clusters occur when the net force between the particles is attractive. Thin black films can develop inside the thicker film when layer of aggregates has been expelled.

6 Conclusion

The ultimate purpose of studies on champagne foams and films is to identify whether a given compound is more foam-active, and how it operates. This is a very difficult task, mainly because wines are natural products whose resulting fine flavour and foam aspect depend upon a lot of uncontrollable factors, sunlight, precipitations, *etc.* during vine blossoming and ripening. The oenologists used to hide all the year subtleties behind the vintage concept, and they only make taste assessments to appreciate wine quality. Laboratory experiments have shown that the films and foams properties of wines also are dependent on the vintage. Investigations on model solutions, which mimic the wine foam at local and global scales, have already greatly enhanced the understanding of these differences although more studies are required to confirm the findings. All the base wines of champagne have essentially the same values of alcohol concentration, pH and ionic strength. The differences between the various vintages are mostly related to compounds existing in microscopic amounts, *e.g.* organic volatile compounds, proteins and glycoproteins, polyphenols, *etc.* Their quality and concentration can significantly vary with the vintages. Now, small variations of surface active compounds can considerably modify the surface related properties of a solution whether it is due to the concentration, the quality of the surface active compounds or to the solvent composition. This explains why foams can significantly change with the vintage, sometimes with the bottle and even with the glass.

References

1. McLeod, *Arômes, Additifs & Ingrédients*, 2000, **29**, 13.
2. F. Franks, in *Characterization of Proteins*, ed. F. Franks, Humana Press, New Jersey, 1988.
3. M. Ahmed and E. Dickinson, *Colloids and Surfaces*, 1990, **47**, 353.
4. A. Dussaud, G. B. Han, L. Ter-Minassian-Saraga and M. Vignes-Adler, *J. Colloid Interface Sci.*, 1994, **167**, 247.
5. N. Puff, R. Douillard, A. Cagna, V. Aguié-Béghin and R. Douillard, *J. Colloid Interface Sci.*, 1998, **208**, 405.
6. R. Marchal, S. Bouquelet and A. Maujean, *J. Agric. Food Chem.*, 1996, **44**, 1716.
7. J. Senée, B. Robillard and M. Vignes-Adler, *Food Hydrocolloids*, 1999, **13**, 15.
8. B. A. Schwartz and G. R. Gray, *Arch. Biochemistry Biophysics*, 1977, **181**, 542.
9. M. M. Bradford, *Anal. Biochem.*, 1996, **72**, 248.
10. J. Guastalla, A. Lize and N. Davion, *J. Chim. Phys.*, 1971, **68**, 5, 822.
11. A. Rudin, *J. Inst. Brew.*, 1957, **63**, 506.
12. B. Robillard, E. Delpuech, L. Viaux, J. Malvy, M. Vignes-Adler and B. Duteurtre, *Am. J. Enol. Vitic.* 1993, **44**, 4, 1–6.
13. J. Malvy, B. Robillard and B. Duteurtre, *Sci. Aliments*,1994, **14**, 87.
14. B. Robillard, unpublished data.
15. G. Liger-Belair, R. Marchal, B. Robillard, T. Dambrouck, A. Maujean, M. Vignes-Adler and P. Jeandet, *Langmuir*, 2000, **16**, 4, 1889.
16. I. B. Ivanov and D. S. Dimitrov, in *Thin Liquid Films*, ed. I. B. Ivanov, Marcel Dekker,

New York, 1988.

17. J. Senée, B. Robillard and M. Vignes-Adler, in *Food Emulsions and Foams* eds. E. Dickinson and J. M. Rodriguez Patino, Royal Society of Chemistry, Cambridge, England, 1999.

18. I. Correa-Gorospe, M. C. Polo, R. Rodriguez-Badiola and R. Rodriguez-Clemente, *Food Chem.*, 1991, **41**, 69.

19. J. Senée, L. Viaux, B. Robillard, B. Duteurtre and M. Vignes-Adler, *Food Hydrocolloids*, 1998, **12**, 217.

20. A. Vernhet, P. Pellerin, C. Prieur, J. Osmianski and M. Moutounet, *Am. J. Enol. Vitic.*, 1996, **47**, 1, 25.

21. A. Sheludko, *Adv. Colloid Interf. Sci.*, 1967, **1**, 391.

22. K. J. Mysels, K. Shinoda and S. Frankel, *Soap Films*, Pergamon, New York, 1957.

Structure–Texture Relationships of Starch in Bread

S. Hug-Iten, F. Escher and B. Conde-Petit

INSTITUTE OF FOOD SCIENCE, SWISS FEDERAL INSTITUTE OF
TECHNOLOGY (ETH) ZURICH, CH–8092 ZURICH, SWITZERLAND

1 Introduction

The texture of bread is determined by two well defined bulk phases, the crust and the crumb. At the macroscopic level, bread crumb can be described as a porous material with flexible elastic cell walls. The force required to compress fresh bread crumb is rather low since the structuring agents, starch and protein are present in the rubbery state due to high levels of plasticizer (water). A salient feature of bread crumb is firming during aging, also known as bread staling, which leads to decreasing acceptance of bread by the consumers. The mechanism of bread staling has intrigued researchers for decades, and the literature on this topic is abundant.[1-3]

On cooling and aging of bread crumb, rearrangements in the starch fraction lead to a series of changes including gelation and crystallization also known as starch retrogradation. The kinetics of starch reorganization are different for the two starch polymers amylose and amylopectin. Pure amylose solutions gel within hours whereas amylopectin solutions require several days to gel.[4] Firming of bread also occurs over a time period of several days, and based on a recent staling model changes in the amylopectin fraction are primarily responsible for bread firming.[3] Regarding the amylose fraction, little attention has been paid to understand its role in bread and during staling. The addition of starch degrading enzymes is a common strategy to retard firming of bread.[5]

The main goal of the present investigation was to expand the knowledge on the role of starch as texturogen in bread. A second focus was the elucidation of the antifirming mechanism of different α-amylases to obtain a better understanding of bread staling. Bread and bread models were considered in the study and the characterization of the starch fraction in the systems spanned the range from macro- to nanoscale.

2 Experimental

2.1 Materials

Commercial low-extraction wheat flour was received from Swissmill (Zurich, CH). Native wheat starch was supplied by Blattmann Cerestar (Wädenswil, CH). A thermostable α-amylase (BAN, specific activity 256 KNU/g) and a maltogenic α-amylase (Novamyl, 1500 MANU/g) were received from Novozymes (Bagsward, DK). The two enzymes have a temperature optimum of 60 to 80 °C and 45 to 75 °C, respectively. Hereafter, the enzymes are referred to as α-amylase and maltogenic α-amylase, respectively.

2.2 Preparation of Bread and Model Systems

Pan bread based on low-extraction wheat flour was prepared as described in ref. 6. Enzymes were added at concentrations of 750 MANU/kg flour and 3 KNU/kg flour, respectively. Wheat starch gels (40%) were prepared in cylindrical moulds (ϕ 30 mm, height 20 mm) as described in ref. 7. The thermal treatment involved heating at 96 ± 1 °C for 60 min. Wheat flour gels were prepared by filling dough, without yeast, into the moulds and heating them as described for the starch gels.

2.3 Uniaxial Compression

The texture of bread was evaluated based on an instrumental Texture Profile Analysis (TPA). Bread slices (height 30 mm) were compressed with a universal testing machine (Zwick 1445, Ulm, G) to 15 mm with two repeating cycles. The crumb firmness and the elastic recovery were evaluated as described in ref. 8. The modulus of deformability of starch and flour gels was evaluated as described in ref. 7.

2.4 Microscopy

Samples of fresh and aged bread crumb were frozen, cryosectioned and stained with Lugol's solution and Light Green and prepared for light microscopy as described by Hug-Iten et al.[6] The samples were observed in the bright field and polarized light mode.

2.5 X-Ray Diffraction

Bread crumb samples were frozen, freeze-dried and ground to powder. The freeze-dried material was equilibrated over saturated barium chloride solution for at least 12 h before compressing it into disks of 1–2 mm thickness and a diameter of 13 mm. The samples were measured in the transmission mode at ambient temperature on a powder diffractometer (Siemens Kristolloflex D500, Karlsruhe, G) as described in ref. 9.

3 Results and Discussion

The textural changes of bread crumb during aging and the influence of starch degrading enzymes were studied by instrumental TPA and the results are presented in Figure 1 (a) and (b). The firmness of control bread increased during aging (Figure 1 (a)), and at the same time the elastic recovery decreased (Figure 1 (b)). The elastic recovery is a measure for the reversibility of the imposed bread crumb deformation which is equivalent to the springiness. In contrast, bread baked with α-amylase presented low firming rates, the α-amylase being more effective than the maltogenic α-amylase. Large differences were found between the two enzymes regarding the elastic recovery. Bread with α-amylase showed a reduced elastic recovery after baking which further decreased during aging, while

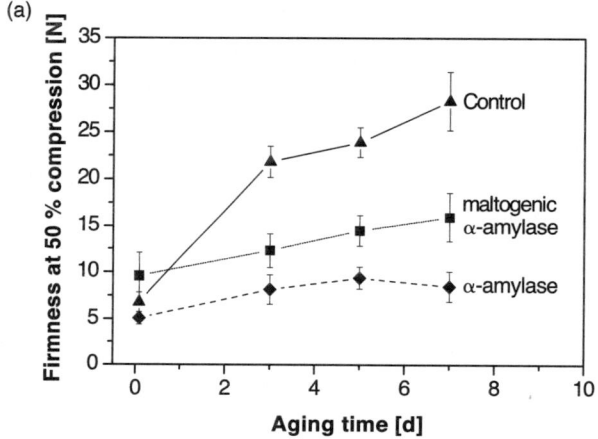

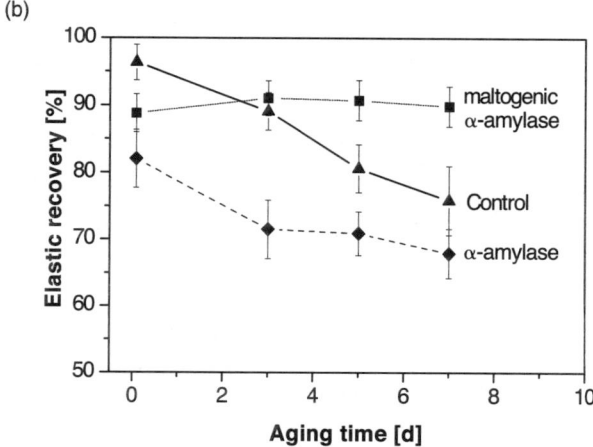

Figure 1 *Influence of aging time on the firmness (a) and elastic recovery (b) of bread without enzyme addition (control) and with addition of α-amylase and maltogenic α-amylase*

the maltogenic α-amylase only slightly reduced the elastic recovery of fresh bread without further reducing it during aging.

Low firmness values and low elastic recoveries, as found for bread with α-amylase, are known to correlate with a sticky and chewy sensory perception of bread crumb during mastication. The addition of maltogenic α-amylase results in a more favorable texture since firmness increase during aging is prevented without weakening the structure to the extent that structure collapse occurs upon deformation. The different textures obtained with the two enzymes reflect different starch degrading mechanisms. The α-amylase cleaves starch statistically in an endo-way producing dextrins of intermediate length. In contrast, the maltogenic α-amylase cuts off oligosaccharides with a degree of polymerization between 7 and 9 which are further degraded to maltose in the α-configuration.[10,11]

Since firmness readings are always influenced by the pore size distribution, experiments were also carried out with flour and starch gels, which were considered as bread models without pores. Figure 2 (a) and (b) show the firmness of flour and starch gels with and without enzyme as revealed by the modulus of deformability at small deformation. Both α-amylases reduced the firming rate of flour gels. The addition of maltogenic α-amylase to starch gels enhanced the initial rigidity of the gels but the firming rate was strongly reduced compared to the control. After an aging period of 7 d, the starch gels with maltogenic α-amylase were less firm than the enzyme-free control. No results for starch gels with α-amylase are presented since the latter enzyme resulted in gels with inhomogeneous moisture distribution.

The above results confirm that the mechanical properties of bread are determined by starch to a large extent, and that starch degrading enzymes may contribute to weaken the structure and act as antifirming agents. Although both α-amylases reduced the firmness of bread and bread models, the increased initial firmness of systems with maltogenic α-amylase suggests a particular antifirming mechanism. An increased initial firmness was found in starch gels, and to a lesser extent also in flour gels and in bread. It is an indication that the gelation of starch was accelerated by the action of the enzyme. A similar phenomenon, *i.e.* an increased initial firmness and a reduced firming rate, is found when emulsifiers that are able to form amylose–lipid complexes are added to high or low concentration starch systems.[12,13]

The microstructure of starch in bread was investigated with light microscopy. Figure 3 (a) shows a cryosection of bread crumb where the starch granules are swollen and partly fused with neighboring granules. The starch granules are elongated and oriented as a consequence of elongational deformation during oven rise of dough. The dark areas in the center of starch granules correspond to amylose-rich phases as they stain blue with iodine. The less intensively stained surrounding phases are violet and correspond to amylopectin-rich phases. Protein forms a separate phase with a sharp interface between protein and starch while the amylose/amylopectin interface is rather diffuse. During aging, both amylose and amylopectin-rich phases become birefringent as detected by polarized microscopy (Figure 3 (b)).

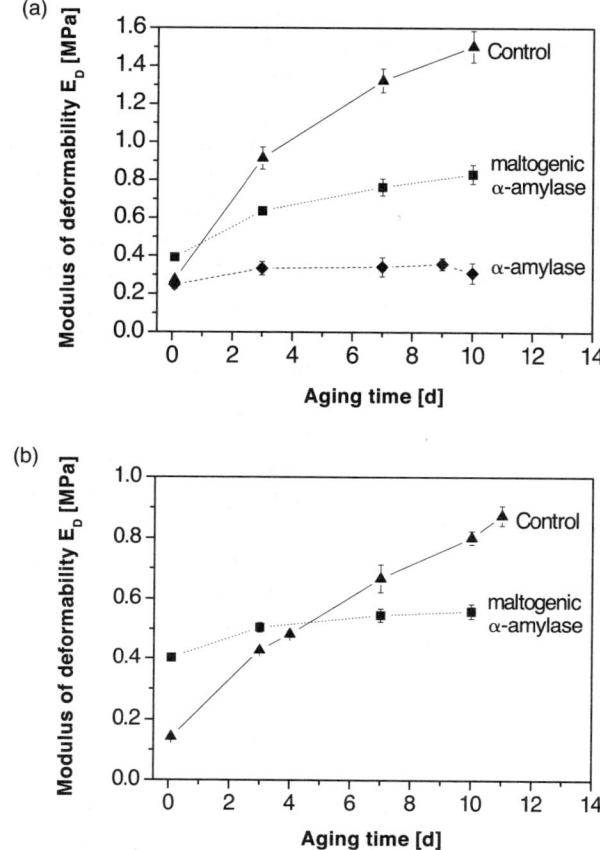

Figure 2 *Influence of aging time on the modulus of deformability E_D of flour gels (a) and of starch gels (b) without enzyme addition (control) and with addition of α-amylase and maltogenic α-amylase*

The solid phase of bread crumb can be viewed as a composite material where amylose, amylopectin and protein form separated phases due to poor thermodynamic miscibility of the different polymers. Composites are characterized by exhibiting mechanical properties that cannot be achieved with the individual constituents alone, but are dependent on the interface between the components. A sharp interface as found between starch and protein provides strong evidence of little polymer interdiffusion and weak interfacial adhesion.[14] The present results suggest that starch forms a continuous phase in bread which has also been confirmed with confocal scanning laser microscopy.[15] The presence of a protein phase reduces the continuity of the starch phase and, thus, reduces the cohesion of the material as revealed by a comparison of the breaking stresses of aged flour and starch gels (data not shown).[16]

The amorphous starch phase in bread crumb is far from equilibrium and the formation of ordered structures reduces the free energy of the system. Starch has a strong tendency to associate since the interaction between the starch polymers

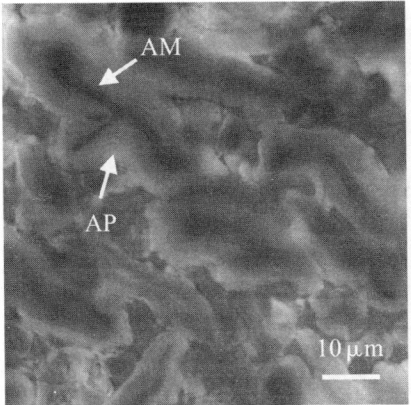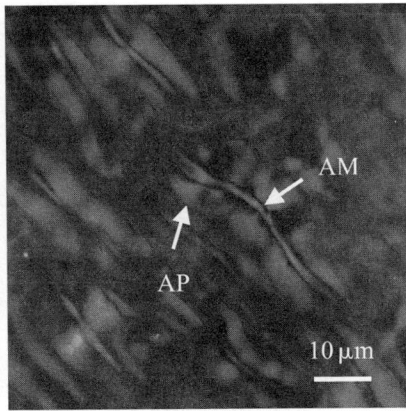

Figure 3 *Micrographs of fresh control bread crumb in the bright field mode (a) and after an aging period of 7 d in the polarized light mode (b). Amylose-rich regions (AM) and amylopectin-rich regions (AP) are pointed out*

is more favorable than the interaction between starch and water. The long-range ordered structures that are formed during aging are birefringent and can thus be detected by microscopy. Figure 4 reveals that the maltogenic α-amylase influences the extent and the kinetics of starch reorganization. As expected, fresh control bread presented no birefringence and during aging amylose and amylopectin became birefingent. In contrast, in fresh bread with maltogenic α-amylase the amylose-rich phase was birefringent. During aging of the latter system almost no changes occurred, and the amylopectin-rich phase did not develop birefrigency. Similar results were found for maltogenic α-amylase in flour and starch gels (results not shown).[16] Polarized light microscopy of bread crumb was complemented by X-ray diffraction. Diffractograms of fresh and aged (7 d) bread crumb with and without maltogenic α-amylase are presented in Figure 5. Fresh control bread presented a pronounced reflection at $2\theta = 20°$ and small reflection at 7.5 and 13° which are characteristic for the V-type pattern[17] caused by the formation of amylose complexes with endogenous wheat lipids. In aged control bread, reflections were detected at $2\theta = 5.6$, 15 (weak) and 17, and between 22 and 24° (broad peak). This corresponds essentially to a B-type pattern.[18] In addition, the reflection at 20° is still visible. Thus, the crystalline organization of aged starch in control bread can be described as a mixture of V- and B-type structure. Bread with maltogenic α-amylase exhibited a slight mixed V and B pattern already in the fresh state and during aging the crystallinity only slightly increased. Similar results were found for α-amylase where bread with enzyme presented the highest level of crystallinity after an aging period of 7 d. It should be added, that the latter enzyme also induced the formation of birefringent amylose and amylopectin structures upon aging (data not shown).[16]

X-ray diffraction does not allow us to identify which starch fraction contributes to the formation of crystalline structures since both amylose and amylopectin crystallize as B-type upon aging. On the other hand, the detection of birefringence means that long-range ordered structures are formed which are not

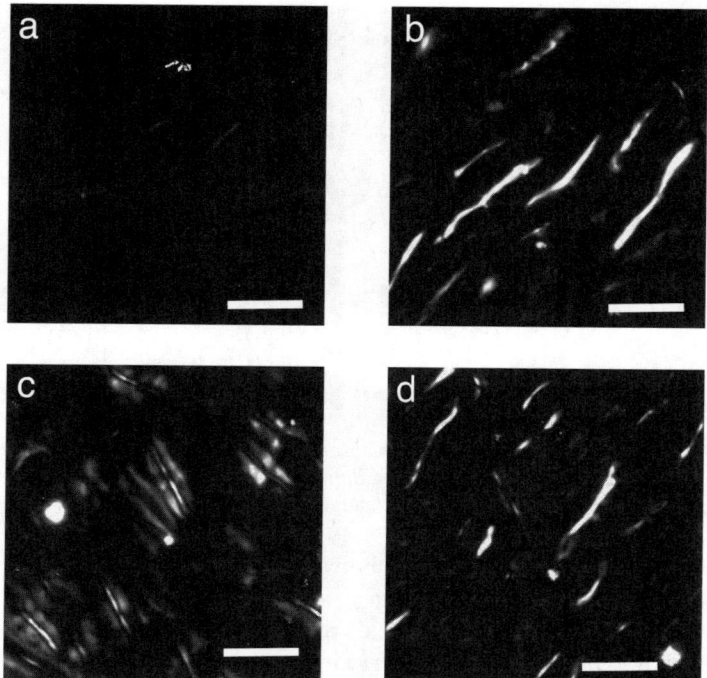

Figure 4 *Polarized light micrographs of bread crumb. (a) Control bread crumb, fresh; (b) bread with maltogenic α-amylase, fresh; (c) control bread crumb, aged for 7 d: (d) bread with maltogenic α-amylase, aged for 7 d. Scale bar corresponds to 25 μm*

necessarily crystalline. The combination of both methods leads to the conclusion that maltogenic α-amylase induces the formation of a partly crystalline amylose network and hinders the recrystallization of amylopectin. This conclusion is supported by DSC measurements, where no retrogradation of amylopectin was found for bread with maltogenic α-amylase.[19] The changes in the molecular order of starch as induced by α-amylase are linked to the particular starch degrading mechanism of the enzyme. Most probably the enzyme reduces the average molecular weight of amylose which, in turn, increases the mobility of the molecule and promotes its aggregation. Regarding the amylopectin fraction, it is likely that the degradation of the side chains hinders its reorganization (Figure 6).

4 Conclusions

The mechanical properties of bread crumb result from the combination of protein and starch that form separate phases. Based on microscopy the hypothesis is that starch forms a continuous phase in bread. Firming of bread during aging is primarily due to an increase of order of the starch polymers involving changes in both the amylose and the amylopectin fraction as shown in Figure 5.

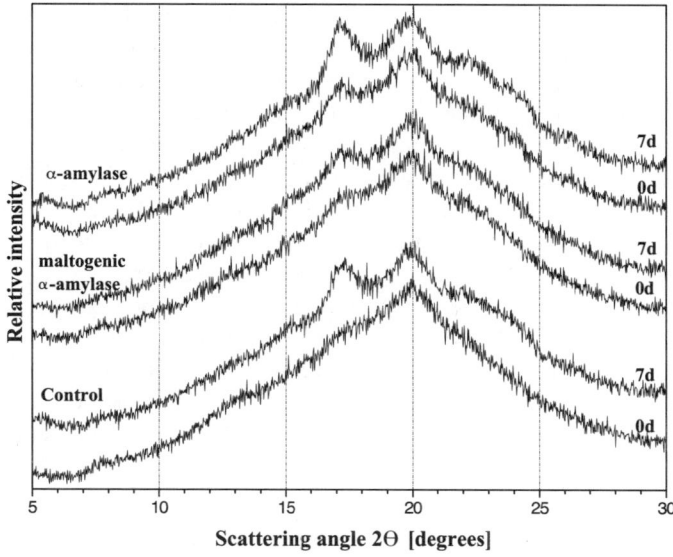

Figure 5 *X-ray diffractograms of bread crumb without enzyme addition (control) and with addition of α-amylase and maltogenic α-amylase*

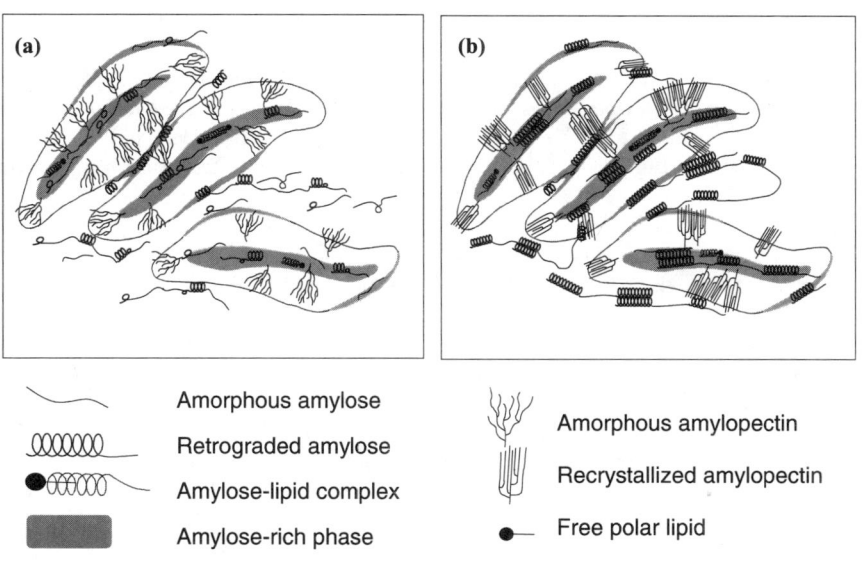

Figure 6 *Schematic presentation of a model for staling of bread as based on changes in the starch fraction. (a) fresh bread; (b) stale bread*

Changes in the starch fraction, for instance by the addition of starch degrading enzymes, have a strong impact on the mechanical properties of the bulk. The maltogenic α-amylase used in the present investigation induces large changes both in the amylose and the amylopectin fraction and by this gives an insight into how the two starch polymers contribute to the structure and texture of bread. The enzyme induces a fast gelation of amylose which is responsible for slightly increasing the initial firmness of starch in bread and model systems. By analogy with the complexation induced gelation[13] it is hypothesized that a fast formation of a partly crystalline amylose network contributes to an increase of the initial firmness. At the same time, the gelation of amylose hinders polymer diffusion and, thus, further rearrangements during aging. The addition of α-amylase clearly shows that the formation of ordered structures may be paired with a weakening of the structure to the extent that collapse occurs upon deformation. It is likely that the latter enzyme induced an excessive degradation of starch promoting the formation of dispersed crystalline regions with little interconnectivity. The challenge in the application of starch degrading enzymes as antistaling enzymes lays in finding a balance between a weakening of the intergranular starch network to prevent firming and the formation of interconnected ordered structures to increase rigidity and, thus, to prevent a collapse of the porous bread crumb structure.

References

1. J. R. Katz, *Z. Elekrochemie*, 1913, **19**, 663.
2. T. J. Schoch and D. French, *Cereal Chem.*, 1947, **24**, 231.
3. H. F. Zobel and K. Kulp, in *Baked Goods Freshness: Technology, Evaluation and Inhibition of Staling*, ed. R. E. Hebeda and H. F. Zobel, Marcel Dekker, New York, 1996, p. 1.
4. M. J. Miles, V. J. Morris, P. D. Orford and S. G. Ring, *Carbohydr. Res.*, 1985, **135**, 271.
5. L. Bowles, in *Baked Goods Freshness: Technology, Evaluation and Inhibition of Staling*, ed. R. E. Hebeda and H. F. Zobel, Marcel Dekker, New York, 1996, p. 105.
6. S. Hug-Iten, S. Handschin, S. B. Conde-Petit and F. Escher, *Lebensm. Wiss. Technol.*, 1999, **32**, 255.
7. H. Krüsi and H. Neukom, *Starch/Stärke*, 1984, **36**, 40.
8. A. S. Szczesniak, *J. Food Sci.*, 1963, **28**, 385.
9. J. Nuessli, B. Sigg, B. Conde-Petit and F. Escher, *Food Hydrocoll.*, 1997, **11**, 27.
10. H. Ottrup and B. E. Norman, *Starch/Stärke*, 1984, **36**, 405.
11. Z. Dauter, M. Dauter, A. M. Brzozowski, S. Christensen, T. V. Beier, K. S. Wilson and G. J. Davies, *Biochemistry*, 1999, **38**, 8385.
12. B. Conde-Petit and F. Escher, *Food Hydrocoll.*, 1992, **6**, 223.
13. B. Conde-Petit and F. Escher, *Starch/Stärke*, 1994, **46**, 172.
14. S. Wu, *Polymer Interface and Adhesion*, Marcel Dekker, New York, 1982.
15. M. B. Dürrenberger, S. Handschin, B. Conde-Petit and F. Escher, *Lebensm. Wiss. Technol.*, 2001, **34**, 11.
16. S. Hug-Iten, PhD Thesis, ETH Zurich, 2000.
17. H. F. Zobel, *Starch/Stärke*, 1988, **40**, 1.
18. P. Le Bail, H. Bizot and A. Buléon, *Carbohydr. Polym.*, 1993, **21**, 99.
19. S. Hug-Iten, F. Escher and B. Conde-Petit, *Cereal Chem.*, 2001, **78**, 421.

Behaviour of Amylose and Amylopectin Films

Pirkko Forssell, Riitta Partanen, Alain Buléon,[1] Imad Farhat[2] and Päivi Myllärinen

VTT BIOTECHNOLOGY, PO BOX 1500, 02044 VTT, FINLAND
[1]INRA, BP 71627, 44316 NANTES CEDEX 3, FRANCE
[2]DIVISION OF FOOD SCIENCE, UNIVERSITY OF NOTTINGHAM, SUTTON BONINGTON CAMPUS, LE12 5RD, UK

1 Introduction

There is a constant interest in using biopolymers as materials for edible films and coatings and for the encapsulation of food and pharmaceutical components. Starches are versatile biopolymers which are composed of glucose monomers. In nature, starches occur as water insoluble granules and most granules are built up of branched and linear glucose polymers called amylopectin and amylose. When preparing films of starches, the granular structure is destroyed and the product properties depend on the behaviour of amylopectin and amylose. Mechanical and permeability behaviour as well as stability define the exploitation potential of the final films.

Our aim has been to elucidate the behaviour of starch films by investigating the properties of films prepared with pure amylopectin and amylose plasticised by glycerol and water. Even though amylose is known to be a good film former[1] both amylose and amylopectin were studied in order to get better understanding of the phenomenon related to plasticisers.

2 Materials and Methods

2.1 Materials

Amylose was from potato starch (Sigma A 0512) and amylopectin was waxy maize starch (National Starch, USA). Purity of glycerol was 99% (Sigma, analytical grade).

235

2.2 Methods

Starch films were prepared by a water-casting technique. Starch was dissolved in water by heating the starch suspension (1 or 2% starch) at 140°C for 30 minutes. After cooling the system to 100°C, the hot solution was dried on teflon moulds at 70°C. Before measurements, the films were stored at RH 50% and at 20°C for one week.

Acid hydrolysis was performed for film powders (100 mg in 20 ml HCl) at 35°C for one week. The extent of hydrolysis was expressed as the soluble carbohydrates divided by the quantity of the original sample.

Tensile failure stress and strain were analysed by Texture Analyzer TX (TA, XT.2, England) using standard method of ISO 1184-1983.

^{1}H NMR measurements were performed at 20°C using a Maran NMR spectrometer (Resonance Instruments, UK) operating at resonance frequency of 23 MHz. The second moment of the FIDs was calculated using Microsoft Excel Solver.

Oxygen permeability was analysed by Mocon OX-Tran2/20 (USA) using standard method (ASTM 3985). The analysis were performed in the RH range of 50 to 90% and at 20°C. Commercial synthetic films, polyamide (PA) from Biaxis Oy and ethylene vinyl alcohol (EVOH) from Åkerlund & Rausing Oy (Finland), were used as the reference materials.

3 Results and Discussion

3.1 Effect of Plasticiser on Mechanical Behaviour

The structure of the amylose film differed a lot from that of the amylopectin film, which was especially noticeable when treating the fresh films with 2.2 M HCl (Figure 1).[2] The film made of amylopectin dissolved rapidly in acidic conditions showing its amorphous nature. The amylose film was much more stable against HCl, because only 20% was easily hydrolysed and about half of the material was still insoluble after one week of treatment.

The difference in structure of the starch films was also evident in the mechanical behaviour. The effect of glycerol on tensile failure stress and strain was investigated under conditions in which the water contents of the films were about the same (RH 50% and 20°C). Water contents were examined by determining water sorption isotherms at 20°C. The stress–strain curves for the amylopectin films were more brittle-type, but the overall effect caused by glycerol was very similar for both polymers (Figure 2).[3] Under low glycerol contents, a decrease in elongation was observed which indicated a more brittle structure. This was in agreement with the actual behaviour of the films. Above 20% glycerol, increase in elongation occurred. The highly plasticised amylopectin film lost its strength wholly, while the highly plasticised amylose film was still rather strong, having tensile stress of 10 MPa.

Tensile failure analysis under low glycerol contents demonstrated strong interaction between glycerol and both linear and branched starch polymers. But

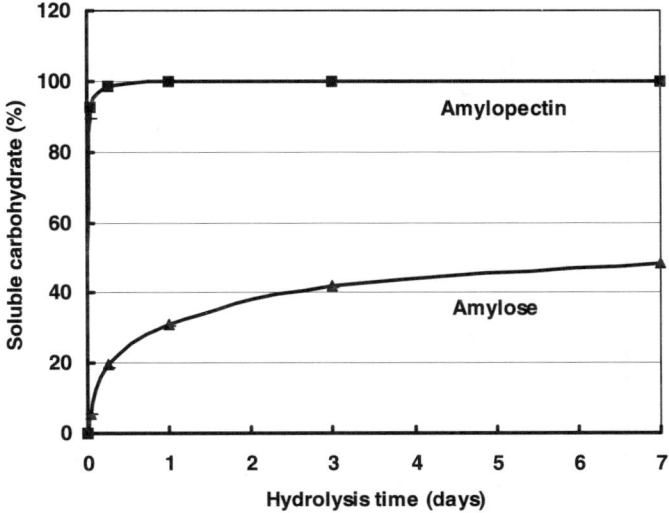

Figure 1 *The extent of acid hydrolysis of amylose and amylopectin films*

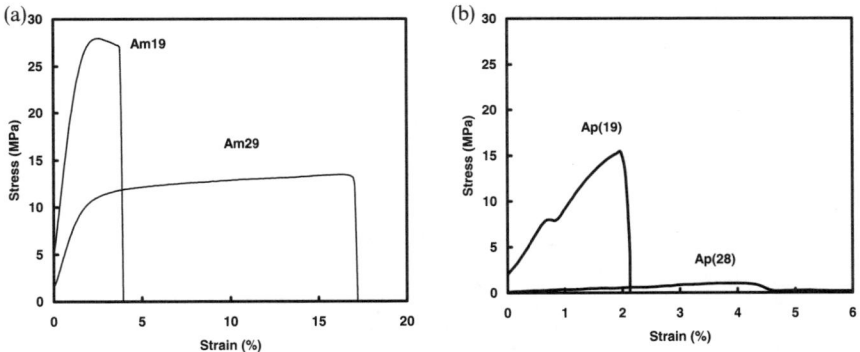

Figure 2 (*a*) *Tensile stress–strain curves of amylose films plasticised with* 19% *and* 29% *glycerol.* (*b*) *Tensile stress–strain curves of amylopectin films plasticised with* 19% *and* 28% *glycerol*

what the behaviour of the highly plasticised films meant was not clear. To get more information about plasticisation mechanism, proton NMR relaxometry experiments were conducted for the same starch films. The second moment, which is a measure of the strength of dipolar interaction, was calculated from FID. The second moment decreased with increasing water or glycerol contents, as shown for the amylose films in Figure 3, indicating an increase in the mobility of the system. Mobility was also increased in the amylopectin films when they were plasticised with water and glycerol, but the results cannot easily be analysed because the films were partially crystallised during the experiment. The highly plasticised amylose film differed from the ones with less plasticiser in being much more mobile, which is in agreement with the tensile behaviour. However, no decrease in the mobility in the film with 10% glycerol was observed.

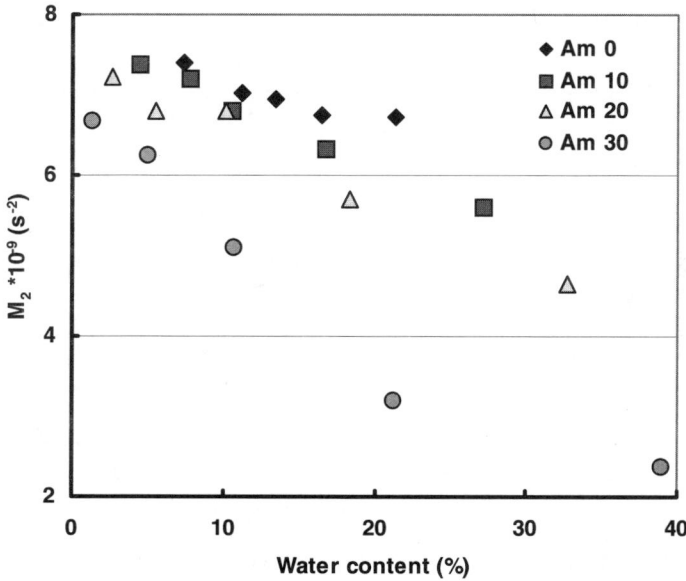

Figure 3 *Effect of water and glycerol contents on the second moment of amylose films*

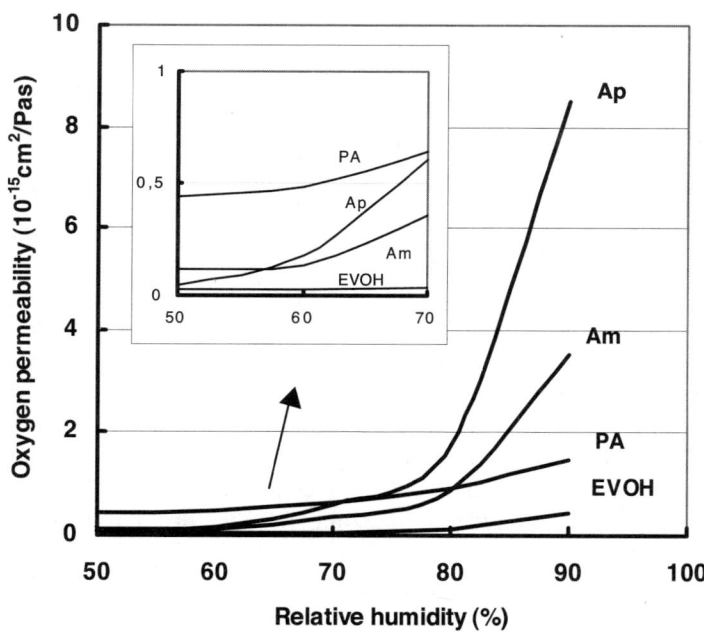

Figure 4 *Oxygen permeability of amylose and amylopectin films*

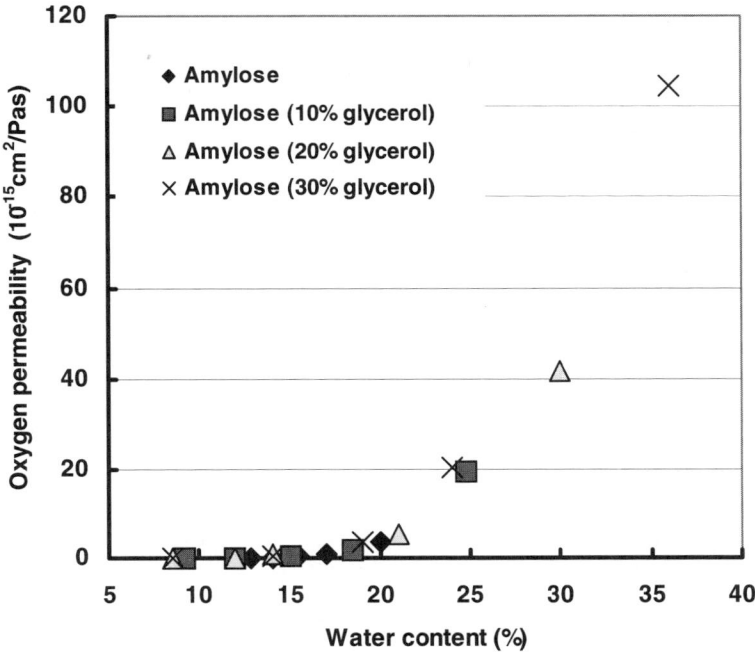

Figure 5 *Effect of water content on oxygen permeability of amylose films*

3.2 Effect of Plasticiser on Oxygen Permeability

The beneficial property of hydrophilic polymers is their ability to act as oxygen barriers. However, water and other plasticisers may greatly affect the diffusion of gases. When the starch films were only plasticised with water they were very good oxygen barriers, and the permeability observed at 20°C was as low as that of EVOH (Figure 4).[4] No difference in the permeability of the amylose and amylopectin films was detected, which is most likely due to the similar solubility of oxygen into the film materials. The presence of glycerol greatly affected the oxygen transport for both polymers, and under high humidity the barrier property was lost. The oxygen permeabilities of amylose films with different glycerol contents plotted against water content demonstrated (Figure 5) that it was water which actually facilitated the oxygen transport across the film. When comparing the oxygen permeability data with the proton NMR relaxometry it may be concluded that there are three mechanisms for oxygen transport under various water contents: very low transport under ambient humidity because of low solubility of oxygen in starch (and low mobility of the system); small flux across the films under high humidity or in the presence of glycerol, caused probably by an increase in the mobility of the system; and high flux across the highly plasticised film under high RH, which was most likely due to an increase in the mobility and an increase in the solubility of oxygen because of partial phase separation.

References

1. D. P. Langlois and J. A. Wagoner, in *Starch: Chemistry and Technology*, eds. R. L. Whistler and E. F. Paschall, Academic Press, New York, 1967, Vol. 2 pp. 451–497.
2. P. Myllärinen, A. Buléon, R. Lahtinen and P. Forssell, *Carbohydrate Polymers*, in press.
3. P. Myllärinen, R. Partanen, J. Seppälä and P. Forssell, *Carbohydrate Polymers*, submitted.
4. P. Forssell, R. Lahtinen, M. Lahelin and P. Myllärinen, *Carbohydrate Polymers*, 2001, **47**, 125.

Gel Formation by Soy Glycinin in Bulk and at Interfaces

Ton van Vliet,[1]* Anneke Martin,[1] Marianne Renkema[2] and Martin Bos[1]

[1]WAGENINGEN CENTRE FOR FOOD SCIENCES C/O WAGENINGEN UNIVERSITY
[2]CENTRE FOR PROTEIN TECHNOLOGY TNO-WAGENINGEN UNIVERSITY, PO BOX 8129, 6700EV WAGENINGEN, THE NETHERLANDS

1 Introduction

Soy proteins are applied in a wide range of food products. In this context most attention has been paid to the ability of soy proteins to form a gel upon heating. Heat denaturation and subsequent gel formation of soy proteins in bulk solutions have been extensively studied.[1-5] Besides this, studies have been performed on the interfacial tension and adsorbed amount of soy proteins and on their suitability for formation and stabilisation of emulsions and foams.[6-10] A question that, to our knowledge, has not been discussed is how far these different functional properties are mutually related.

Soy proteins consist of two major components, β-conglycinin and glycinin. Of these, glycinin is the most abundantly present and therefore this paper will be limited to aspects of the behaviour of glycinin in bulk and at interfaces. Glycinin has a compact globular form and consists of an acidic and basic polypeptide, which are linked by a disulphide bridge. At pH 7.6 and high ionic strength (0.5 M), it exists as a hexamer (an 11S globulin) with a molecular weight of about 360 kDa. At pH 3.8 and low ionic strength, glycinin is predominantly present in a trimeric form (a 7S globulin with a molecular weight of 180 kDa) and at still lower pH (~ 3) it further dissociates into a 3S form consisting of one acidic and basic unit and a molecular weight of approximately 44 kDa.[11-13]

Gel formation in bulk solutions and the resulting gel properties of pure glycinin are affected by protein concentration, heating temperature, pH and ionic strength.[1,5,14,15] Heat denaturation was found to be a prerequisite for gel formation. Gels formed at pH 7.6 exhibited a lower stiffness and could be

241

deformed to a larger extent before fracture occurs than gels formed at pH 3.8, but the fracture stress was higher at pH 3.8.[5] Furthermore, the appearance of the gels differed. Gels made at pH 3.8 were white while those at pH 7.6 were turbid. These differences indicate a difference in gel network structure between both conditions.

The suitability of glycinin for the formation and stabilisation of emulsions and foams has been studied by various authors.[7–10] The foaming and emulsifying properties of glycinin at pH 7.6 are poor, probably because of its poor adsorption behaviour. This is primarily attributed to the low hydrophobicity, the low molecular flexibility and the high molecular weight of the glycinin molecule. Dissociation of glycinin or appropriate and controlled modification of its structure by reduction, acetylation or succinylation improved foaming and emulsifying properties.[8,9] Dissociation by lowering pH and ionic strength or modification by deamidation or by reduction resulted in a reduction of the molecular weight leading to faster adsorption at interfaces (decrease of interfacial tension).[8–10,16]

Although it has been realised that the interfacial properties of glycinin are important for its foaming and emulsifying properties, only a limited number of papers has been published on this topic.[8–10,16–18] The interfacial modulus of glycinin at an air–water interface was found to depend on pH and ionic strength and to increase upon ageing. At pH 3 adsorbed glycinin forms a stiffer interfacial gel network with a higher resistance to yielding in dilation than at pH 6.7.[10,18]

This paper will focus on the effect of pH on network formation by soy glycinin systems in bulk and at interfaces. The properties of bulk gels formed at pH 3.8 and 7.6 will be compared. The same will be done for the interfacial properties and the suitability for foam formation at pH 3 and 6.7. Similarities between bulk and interfacial properties and the relation with molecular structure and foam properties will be discussed.

2 Materials and Methods

2.1 Materials

Soy glycinin was isolated from soybeans according to a modified method of Thanh and Shibasaki.[19,20] The purified glycinin was resuspended at a concentration of 135 g l^{-1} in a 10 mM phosphate buffer pH 7.8 containing 10 mM 2-mercaptoethanol and 20% glycerol and stored at $-40\,^{\circ}$C. Prior to the interfacial experiments, the protein preparations were dialysed against buffer of the desired pH and ionic strength. The protein concentration was determined by the biuret method. A phosphate buffer was used for pH 6.7 and 8.0 and a phosphate/citric acid buffer (Merck, Germany) for pH 3.0. Ionic strength of the buffers was 30 mM. For the bulk gelation experiments protein preparations were dialysed against double distilled water, freeze-dried and subsequently dispersed in 0.2 M NaCl at higher concentrations than required. The suspensions were brought to pH 7.6 with 0.5 M NaOH. For experiments performed at pH 3.8, the pH was adjusted after 1 hour using 1M HCl. Protein concentration was checked by micro-Kjeldahl analysis using 6.25 N.[21]

2.2 Bulk Rheological Measurements

2.2.1 Small Deformation Measurements. Gel formation was studied by dynamic measurements using a Bohlin CVO rheometer with a serrated concentric cylinder geometry at an angular frequency of 0.63 rad s^{-1} and a maximum strain of 0.01, which is in the linear region. Gel formation was induced by heating glycinin dispersions from 20 °C to 95 °C at a heating rate of 1 K/min, keeping them at 95 °C for 60 min and cooling down to 20 °C at a rate of 1 K/min.

2.2.2 Large Deformation Measurements. Gels for large deformation experiments were prepared by heating glycinin dispersions (8 w/w%) in cylindrical glass moulds with an inner diameter of 18 mm and a height of 115 mm. The moulds were filled three-quarter full, to enable air bubbles to escape, and placed vertically in a waterbath. Heating conditions were the same as for the dynamic measurements. Large deformation and fracture properties were determined immediately after preparation in uniaxial compression between two flat plates at 20 °C using a Zwick material-testing machine type 1425, fitted with a 50 N load cell. Gels were cut into test pieces of 20 mm height. The initial strain rate was 0.05 s^{-1}. Measurements were repeated 3–8 times and mean values with their standard deviations were calculated for fracture stress and strain.

The relative deformation at a certain stage is expressed as a Hencky strain ε_H, defined as:

$$\varepsilon_H = \ln\frac{h(t)}{h_0} \tag{1}$$

where h_0 is the original height of the test piece, and $h(t)$ the height after a deformation time t. The average stress $\sigma(t)$ in the test piece at time t is given by:

$$\sigma(t) = \frac{F(t)}{A(t)} \tag{2}$$

where $F(t)$ is the measured force and $A(t)$ the cross section of the test piece at time t. Assuming that the volume of the test piece does not change during compression and the shape stays cylindrical, $A(t) = (h_0/h(t))A_0$.

2.3 Interfacial Rheological Measurements

2.3.1 Automated Droplet Tensiometer. Surface tension and surface dilational moduli were measured by an automated droplet tensiometer (ADT) (IT concept, France) as a function of ageing time of the droplet.[22] Surface tension was determined by drop shape analysis of a gas bubble formed in a cuvette containing the protein solution. The bubble was illuminated by a uniform light source and its profile imaged and digitised by a CCD camera and a computer. The profile was used to calculate the surface tension using Laplace's equation. The dilational modulus was obtained by applying a dynamic oscillation of the bubble area. The bubble volume was 4 μl and the maximum area change during oscillation was about 8%, which was within the linear region. Temperature was 22 °C.

2.3.2 Overflowing Cylinder Measurements.

The overflowing cylinder technique used has been described by Bergink-Martens *et al.*[23] The overflowing cylinder consists of an inner cylinder (diam. 6 cm) through which the liquid is pumped at a constant flow rate (flux 31.4 cm^{3}s^{-1}). Liquid flows over the upper rim of the inner cylinder along the inner cylinder wall into the circular container formed by the inner and outer cylinder. From this container the liquid is pumped into the inner cylinder. The technique was used to measure the relative surface expansion rate d$\ln A$/dt of the top surface as a function of the length L of the falling film at the outside of the inner cylinder; d$\ln A$/dt was determined using laser doppler anemometry.[23] For water d$\ln A$/dt is independent of L but for protein solution it is not.[10,18] For low L the surface does not expand at all. Apparently an adsorbed protein layer is formed that is able to resist the shear forces exerted by the radially flowing liquid on top of the inner cylinder and the liquid flowing along the inner cylinder into the circular container. Visually, it was seen that the protein film builds up starting from the circular container and moves up over the falling film. The falling film length for which d$\ln A$/d$t \to 0$ is called L_{still}. Above L_{still} the shear forces acting on the adsorbed protein layer are able to break it. So L_{still} can be considered as a measure of the fracture or yield stress of the adsorbed protein layer.[10]

2.3.3 Steady State Expansion Measurements.

The dynamic surface tension in steady state expansion was determined using a modified Langmuir trough equipped with six barriers fixed to an endless belt which were moved caterpillarwise one after another over the liquid surface.[10,24] Surface tension was determined using the Wilhelmy plate technique. Measurements were performed going from the highest to the lowest expansion rate.

2.3.4 Surface Shear Measurements.

Surface shear properties were determined using a two dimensional Couette-type viscometer as described by Dickinson and co-workers.[25,26] A biconial disc was suspended by a torsion wire such that the disc edge was in the plane of the air/water interface. The protein solution was contained in a thermostated circular glass dish that can be rotated with a set angular speed (in this study 1.3×10^{-3} rad s^{-1}). The ensuing rotation of the inner disc was determined *via* a light beam reflected by a mirror connected to the disc onto a circular scale with a radius of 60 cm. The stress σ^s on the inner disc can be calculated from the angle of rotation of the disc θ_i and the torsion constant of the wire K by:[27]

$$\sigma^s = \frac{K\theta_i}{2\pi R_i^2} \qquad (3)$$

where R_i is the radius of the disc. For a homogeneous deformation of the adsorbed protein layer the shear strain in the protein layer near the disc γ^s is given by:[27]

$$\gamma^s = \frac{2R_0^2}{(R_0^2 - R_i^2)}(\theta_0 - \theta_i) \tag{4}$$

where R_0 is the radius and θ_0 the angle of rotation of the dish. Finally the stress is plotted as a function of strain for the adsorbed protein layers.[28]

After shearing for a longer time, a steady state interfacial stress was obtained and from this value and the average shear strain rate over the interface an apparent surface shear viscosity was calculated by:

$$\eta_{app}^s = \frac{\sigma^s}{2\Omega R_0^2}\left(R_0^2 - R_i^2\right) \tag{5}$$

where Ω is the angular velocity of the dish.

2.4 Ellipsometry

Ellipsometry measurements were performed at the Colloid and Interface Group at the Max Planck Institute (Golm, Germany) using a null-ellipsometer (Optrel). Calculations were done with the program 'Ellipsometry' designed by Plamen Petrov (version 1.31, 1997). The angle of incidence used was 50° and the wave length was set at 532 nm. Layer thickness and refractive index were calculated from the change in the ellipsometric angles Δ and Ψ assuming that the protein density profile could be described by a stratified layer model. For the refractive index increment of the protein solution 0.18 cm^3/g was taken.[29] Before measurements were started the surface was cleaned by suction of the air–water interface.

2.5 Foam Tests

Foams were made using a Ledoux whipping apparatus (Ledoux b.v., Dodewaard, the Netherlands; cylindrical sample container of 14 cm height and 9.2 cm diameter and double beaters, diameter of each beater 4 cm). Protein solutions were mixed for 2 minutes at 450 rpm. Foam drainage was monitored for 20 minutes. The evaluation of the mean bubble size, d_{21}, was followed by laser light transmission as described by Martin et al.[18] Light transmission for glycinin foams was recalculated to an equivalent mean d_{21} by using a calibration curve made for Tween 20 foam.

3 Results and Discussion

3.1 Bulk Rheological Measurements

Dynamic moduli were determined as a function of concentration at pH 3.8 and 7.6 (Figure 1). Both the storage modulus and the loss modulus (data not shown)

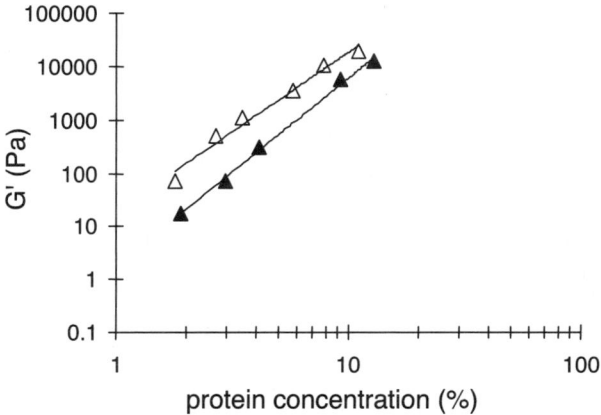

Figure 1 *Storage modulus G′ as a function of glycinin concentration for gel formed at pH 3.8*
(△) and 7.6 (▲). Added salt concentration 0.2 M, temperature 20°C

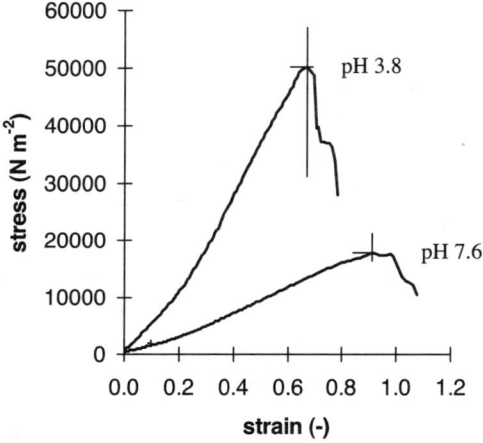

Figure 2 *Representative stress–strain curves of glycinin gels at pH 3.8 and 7.6 (indicated).*
Added salt concentration 0.2 M, temperature 20°C

were higher at the lower pH over the whole concentration range tested although
the concentration dependence varied somewhat with pH. The ratio of the loss
modulus over the storage modulus was similar. Also the large deformation and
fracture properties were clearly dependent on pH (Figure 2). Gels formed at pH
7.6 had a lower fracture stress and a larger fracture strain. The lower stiffness
(lower moduli) and fracture stress at pH 7.6 will partly be caused by the fact that
part of the protein remains in solution at pH 7.6 during heating.[20,30,31] Experiments performed at protein concentration of 1% showed that after heating all
protein is present in relatively large aggregates at pH 3.8 while at pH 7.6 part
stays dissolved. This protein will be located in the pores of the network and,
therefore, not contribute to the stiffness of the gel.

The larger fracture strain at pH 7.6 indicates that at this pH the protein
strands are likely to be somewhat more curved (tortuous) than at pH 3.8.[32,33] It

points to a difference in structure of the gels. Also the visual appearance of the gels varied with pH. At pH 3.8 they were more white and granular while at pH 7.6 they were turbid and smooth.[30,31] This implies that at pH 3.8 coarser gels are formed with larger aggregates (1 μm or larger) and probably with thicker strands and larger pores. These conclusions based on visual appearance were confirmed by Confocal Scanning Laser Microscopy.[31]

These observations indicate that besides a difference in the amount of protein actually incorporated in the gel network, there is also a difference in the spatial structure of the gel network between low and high pH. The difference in network structure may have a clear effect on mechanical properties of the gels.[32,33] Moreover, electrostatic interaction forces in and between the protein molecules will vary with pH.

3.2 Interfacial Rheological Measurements

Surface tension γ was determined at pH 3 and 6.7 as a function of ageing time using the ADT (Figure 3). At pH 3 surface tension decreases much faster indicating a faster adsorption of glycinin monomers compared with glycinin hexamers (pH 6.7). The latter have a lower surface hydrophobicity and diffusion coefficient. These results are in line with those of Wagner *et al.*[34]

Adsorbed amount Γ as a function of ageing time as determined by ellipsometry is plotted in Figure 4. Especially at pH 3 glycinin adsorbed very fast. After 1 minute, Γ was about 2.2 and 1.2 mg m^{-2} at pH 3 and 7.6, respectively. The adsorbed amount became about the same for both pHs after 1–2 hours. The ellipsometric thickness after 2 hours ageing was somewhat larger at pH 6.7 than at pH 3, *i.e.* about 12 and 8 nm, respectively.[18]

The faster adsorption rate of glycinin at pH 3 is also reflected in the ability to lower the surface tension of a continuously expanding surface (Figure 5). At expansion rates of 0.01 s^{-1} and higher, the hexameric form of glycinin was not

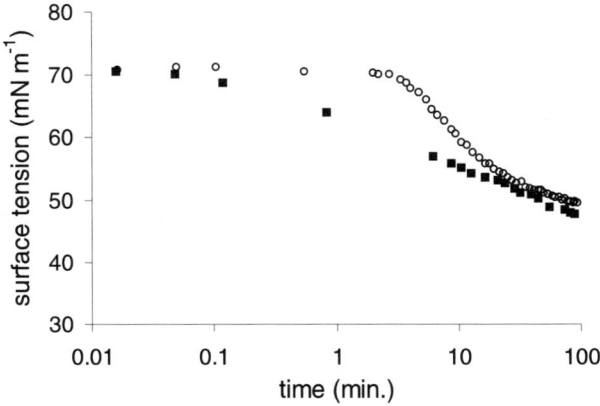

Figure 3 *Surface tension as a function of ageing time for glycinin solutions. Concentration 0.1 g l^{-1}, ($\blacksquare$) pH 3, ($\bigcirc$) pH 6.7*

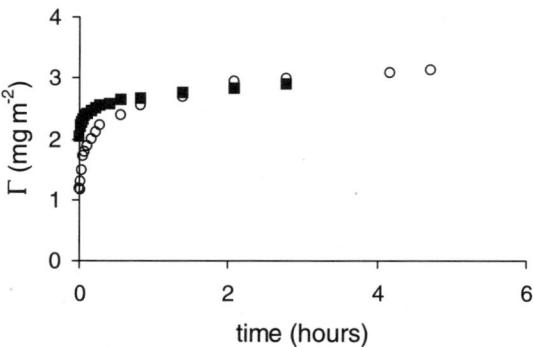

Figure 4 *Adsorbed amount* Γ *of glycinin as a function of ageing time at pH 3* (■) *and 6.7* (○). *Protein concentration 0.1 g l^{-1}*

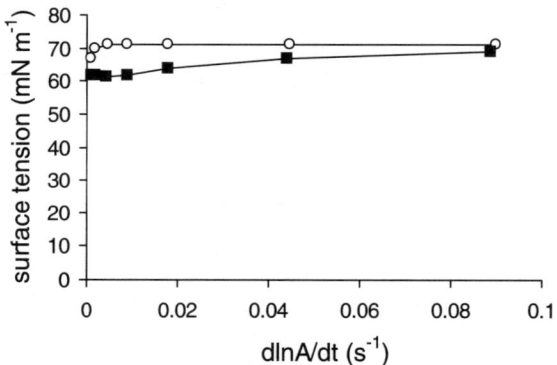

Figure 5 *Surface tension of a glycinin solution of 0.1 g l^{-1} as a function of the relative expansion rate dlnA/dt,* (■) *pH 3,* (○) *pH 6.7*

able to lower surface tension to a measurable extent while the 3S/7S form at pH 3 gave a clear lowering of the dynamic surface tension.

The observed difference in ellipsometric thickness between pH 3 and 6.7 indicates that not only the adsorption rate differs but that probably also the structure of the resulting adsorbed protein layer is different. A second indication for a difference in structure is the higher surface dilational modulus E observed at low pH after ageing for 0.03 and 1 hour (Table 1). At short ageing time, E was much higher at pH 3 than at pH 6.7. This difference is clearly related to the faster adsorption of glycinin at low pH. However, also after longer ageing times when the adsorbed amounts are about the same there was still a clear difference in stiffness (E) of the glycinin surfaces.

The overflowing cylinder data indicate also a difference in structure of the adsorbed protein layers at pH 3 and 6.7. Glycinin was able to stop radial expansion of the surface at low length of the falling film L at both pH's. However L_{still}, above which the adsorbed protein film breaks, and a radial expansion of the surface was observed, was clearly higher at pH 3 than at pH 6.7, 2.3 and 1.9 cm, respectively. This indicates that the 3S/7S form of glycinin is able to form a

Table 1 *Summary of surface rheological data for several ageing times for glycinin solutions of 0.1 g l⁻¹ at pH 3 and 6.7. For symbols used see text*

Time (h)	pH 3			pH 6.7		
	0.03	1	~ 8	0.03	1	~ 8
γ (mN m⁻¹)	61	49		70.5	50.5	
Γ (mg m⁻²)	2.2	2.7	3.0	1.3	2.6	3.1
E (mN m⁻¹)	47	60		3	40	
η^s_{app} (Ns m⁻²)		0.7	1.1		0.1	0.9
σ_f (mN m⁻¹)		2	4.2		0.2	3
ε_f (−)		0.2	0.4		0.5	0.6

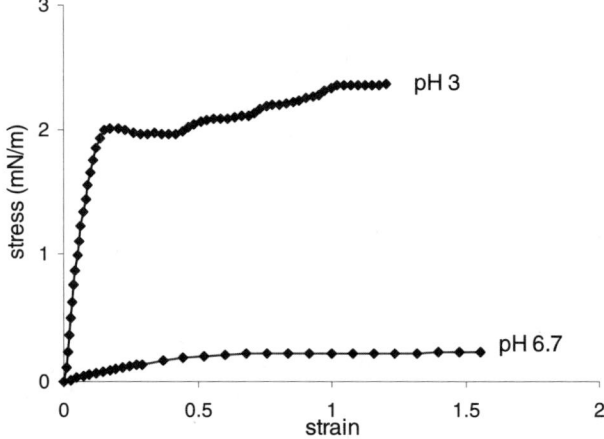

Figure 6 *Shear stress as a function of the shear strain for adsorbed glycinin protein layers after an ageing time of about 1 h. Proteins concentration 0.1 g l⁻¹. pH indicated*

stronger network than the 11S form. More likely at pH 3 a glycinin molecule can interact stronger with other glycinin molecules than at pH 6.7 after adsorption. Probably this is related to a lower stability of the molecule against unfolding after adsorption. Also the stability against heat denaturation is much lower for the 3S form present at pH 3 than for the 7S form.[33,35] At pH 3 the enthalpy for denaturation as measured by DSC is very low.[33]

The apparent surface shear viscosity η^s_{app} and the fracture/yielding stress σ_f of the adsorbed glycinin layers as determined by interfacial shear measurements also showed a strong dependency on pH (Figure 6). At low pH σ_f was much higher and the fracture/yielding strain ε_f clearly lower after an ageing time of 1 h. The pH dependence of both η^s_{app} and σ_f became smaller upon ageing but did not disappear completely during ageing for 8 h (Table 1). After ageing for 24 h η^s_{app} was the same for both pHs (1.2 Ns m⁻¹). The fracture/yielding strain did not change much upon ageing. Again, these data point to a much faster formation of a stiff protein layer at the air–protein solution interface at pH 3. It is only a matter of time for glycinin at pH 6.7 to adsorb, unfold and rearrange in the surface in such a way that also a stiff protein layer is formed (Table 1). However, the remaining difference in ε_f shows that the structure of the adsorbed layers does not become the same.

3.3 Foaming Properties

At pH 6.7 no good foam could be formed with the used equipment at a glycinin concentration up to $1 \, \mathrm{g} \, \mathrm{l}^{-1}$. Similar results have been obtained before using other methods for foam formation.[10,18] A foam could be formed at pH 8, but this was very unstable.[10] At pH 3, a very fine, stiff foam is readily formed with a uni-modal size distribution (surface average diameter d_{21} 100 μm for a glycinin concentration of 0.3–$1 \, \mathrm{g} \, \mathrm{l}^{-1}$). The height of the foam layer formed was independent of protein concentration for glycinin concentrations of 0.3–$1 \, \mathrm{g} \, \mathrm{l}^{-1}$ but lower for lower concentrations. The height decreased by about 15% over an ageing time of 20 minutes for concentrations above $0.3 \, \mathrm{g} \, \mathrm{l}^{-1}$. Due to drainage of the liquid, the foam became very stiff and brittle. No coalescence was observed during the first 30 minutes, d_{21} increased by about 15% which was likely mainly due to disproportionation.

4 General Discussion

Both bulk and interfacial rheological measurements and foaming properties show a large effect of pH on these properties of glycinin dispersions. This is undoubtedly related to the change in conformation of the glycinin molecule with pH as expressed in the presence of different amounts of the 3S, 7S and 11S form at different pH.[11–13]

An elastic network can be formed both in bulk, induced by a heating cycle causing denaturation of the protein, and at an interface, induced by adsorption, which likely caused changes in the conformation. The network formed at the air–protein solution interface can be considered as a kind of 'two' dimensional gel. At low pH, a gel is more readily formed in bulk (lower heating temperature required)[30,31] as well as at the air–protein solution surface. Likely, both phenomena are related to a lower stability of the molecule against conformational changes caused by external factors.

Considering the stress–strain curves of bulk gels and interfacial gels, some similarities are striking. Both bulk and interfacial gels formed at a low pH have a higher modulus (Figures 1 and 6) and fracture at a lower fracture strain, while the fracture stress is higher for bulk gels and for not too old interfacial gels (≤ 8 h) (Table 1). For bulk gels, these differences are caused by a difference in gel structure, as well as that at higher pH part of the protein does not contribute to network formation.[30,31,33] Data presented above for layer thickness and ε_f also indicate that there is also a difference in gel structure for the adsorbed protein gels. If the molecular basis for the effect of pH is the same for bulk and interfacial gels, is hard to answer. It is even unknown if the glycinin molecule adsorbs as a

Table 2 *Ratio of surface dilation modulus E over surface tension for glycinin solution of 0.1 g l^{-1} for pH and ageing time indicated*

	E/γ	
Ageing time	0.03 h	1 h
pH 3	0.8	1.2
pH 6.7	0.04	0.8

whole at high pH. However, we have no indications that this is not the case. Adsorbed amounts of glycinin are the same at low and high pH after an ageing time of a few hours while part of the glycinin does not contribute to gel stiffness for bulk gels at pH 7.6. Therefore, the expected difference in stiffness and σ_f for interfacial gels is smaller than for bulk gels even if the structure effect is about the same. However, although it is possible to establish some phenomenological similarities between bulk and interfacial gelling behaviour and with changes in molecular structural properties, it is not possible to present a mechanistic model for bulk and interfacial gel properties based on molecular structural properties. Therefore much more research on conformational changes of glycinin on heating and adsorption is required.

Foaming data presented above and by others show a clear relation between pH and the suitability of glycinin as foaming agent.[10,18] At low pH both foam formation and foam stability is much better than at pH 6.7. More likely the better foam formation at pH 3 is related to the much faster adsorption of glycinin at the air–protein solution interface. A similar relation between rate of adsorption and foam formation was observed before for glycinin, dissociated glycinin and glycinin modified by deamidation or reduction.[8] Due to the fast build up of an elastic network at the air–protein solution surface a foam at low pH will better resist drainage and is probably more stable against coalescence during and after foam formation. Further, it has been shown that the ratio of the surface dilation modulus E^s over the surface tension γ is an important factor determining the rate of disproportionation.[36,37] If this ratio is above 1, disproportionation will be only relatively minor. Values for E^s/γ for adsorbed glycinin layers are given in Table 2. It is clear that, with regard to this factor, fresh glycinin foams at pH 3 are much more stable against disproportionation than foams formed at pH 6.7.

Summarising, clear phenomenological similarities were observed between bulk and interfacial gelling behaviour with changes in the molecular structure of glycinin. Besides, foaming properties could be related to interfacial behaviour of glycinin in relation to aspects of the molecular structure.

References

1. A.-M. Hermansson, *J. Sci. Food Agric.*, 1985, **36**, 822.
2. F. S. M. van Kleef, *Biopolymers*, 1986, **25**, 31.
3. T. Nakamura, S. Utsumi and T. Mori, *J. Agric. Food Chem.*, 1986, **32**, 349.

4. T. Nagano, T. Akasaka and K. Nishinari, *Biopolymers*, 1994, **34**, 1303.
5. J. M. S. Renkema, J. H. M. Knabben and T. van Vliet, *Food Hydrocolloids*, 2001, in press.
6. S. Utsumi, Y. Matsumura and T. Mori, in *Food Proteins and their Applications*, eds. S. Damodaran and A. Paraf, Marcel Dekker, New York, 1997, p. 257.
7. S. H. Kim and J. E. Kinsella, *J. Food Sci.*, 1987, **52**, 128.
8. J. R. Wagner and J. Gueguen, *J. Agric. Food Chem.*, 1999, **47**, 2173.
9. J. R. Wagner and J. Gueguen, *J. Agric. Food Chem.*, 1999, **47**, 2181.
10. M. Bos, A. Martin, J. Bikker and T. van Vliet, in *Food Colloids: Fundamentals of Formulation*, eds. E. Dickinson and R. Miller, Royal Society of Chemistry, Cambridge, 2001, p. 223.
11. W. J. Wolf and D. R. Briggs, *Arch. Biochem. Biophys.*, 1958, **76**. 377.
12. W. J. Wolf and T. C. Nelsen, *J. Agric. Food Chem.*, 1996, **44**, 785.
13. I. C. Peng, D. W. Quass, W. R. Dayton and C. E. Aken, *Cereal Chem.*, 1984, **61**, 480.
14. S. Utsumi and J. E. Kinsella, *J. Food Science*, 1985, **50**, 1278.
15. T. Wongprecha, T. Takaya, T. Nagano and K. Nishinari, in *Hydrocolloids – Part 1: Physical Chemistry and Industrial Applications of Gels, Polysaccharides and Proteins*, ed. K. Nishinari, Elsevier Science, Amsterdam, 2000, p. 435.
16. M. Liu, D.-S. Lee and S. Damodaran, *J. Agric. Food Chem.*, 1999, **47**, 4770.
17. M. Yu and S. Damodaran, *J. Agr. Food Chem.*, 1991, **39**, 1563.
18. A. H. Martin, M. A. Bos and T. van Vliet, *Food Hydrocolloids*, 2001, in press.
19. V. H. Thanh and K. Shibasaki, *J. Agric. Food Chem.*, 1976, **24**, 1117.
20. C. M. M. Lakemond, H. H. J. de Jongh, M. Hessing, H. Gruppen and A. G. J. Voragen, *J. Agric. Food Chem.*, 2000, **48**, 1985.
21. AOAC, *Official Methods of Analysis*, Association of Official Analytical Chemist, Washington, DC, 1980, 13th ed.
22. J. Benjamins, A. Cagna and E. H. Lucassen Reinders, *Colloids Surf. A*, 1996, **114**, 245.
23. D. J. M. Bergink-Martens, H. J. Bos, A. Prins and B. C. Schulte, *J. Colloid Interface Sci.*, 1990, **138**, 1.
24. J. Lyklema, *Fundamentals of Interface and Colloid Science. Volume III: Liquid–Fluid Interfaces*, Academic Press, San Diego, 2000, Chapter 3, p. 3, 190.
25. E. Dickinson, B. S. Murray and G. Stainsby, *J. Colloid Interface Sci.*, 1985, **106**, 259.
26. E. Dickinson, B. S. Murray and G. Stainsby, *Int. J. Biol. Macromol.*, 1987, **9**, 302.
27. R. W. Whorlow, *Rheological Techniques*, Ellis Horwood Ltd, New York, 1992, Chapter 3.
28. A. H. Martin. M. A. Bos, M. A. Cohen Stuart and T. van Vliet, submitted for publication.
29. J. A. de Feijter, J. Benjamins and F. A. de Veer, *Biopolymers*, 1978, **17**, 1759.
30. J. M. S. Renkema, C. M. M. Lakemond, H. H. J. de Jongh, H. Gruppen and T. van Vliet, *J. Biotechnology*, 2000, **79**, 223.
31. C. M. M. Lakemond, H. H. J. de Jongh, M. Paques, T. van Vliet, H. Gruppen and A. G. J. Voragen, submitted for publication.
32. M. Mellema, J. H. J. van Opheusden and T. van Vliet, submitted for publication.
33. J. M. S. Renkema, *Formation, Structure and Rheological Properties of Soy Protein Gels*, PhD Thesis, Wageningen University, The Netherlands, 2001.
34. J. R. Wagner and J. Gueguen, *J. Agric. Food Chem.*, 1995, **43**, 1993.
35. C. M. M. Lakemond, H. H. J. de Jongh, M. Hessing, H. Gruppen and A. G. J. Voragen, *J. Agric. Food Chem.*, 2000, **48**, 1985.
36. W. Kloek, T. van Vliet and M. Meinders, *J. Colloid Interface Sci.*, 2001, **237**, 158.
37. M. B. J. Meinders, W. Kloek and T. van Vliet, *Langmuir*, 2001, **17**, 3923.

Interactions between Cellulose and Plasticized Wheat Starch – Properties of Biodegradable Multiphase Systems

Luc Avérous

UMR INRA/URCA (FARE), CERME (ESIEC), BP 1029, 51686 REIMS CEDEX 2, FRANCE

1 Introduction

Starch is gaining much attention as a cheap and totally biodegradable material. Different approaches have been made to use starch for the production of tailored materials.[1]

Native starch can be transformed to obtain easy processed starch. The so-called 'thermoplastic starch' (TPS) or plasticized starch is obtained after disruption and plasticization with water and plasticizer (*e.g.* glycerol) of native starch. Such a material has been studied for more than one decade.[1] Unfortunately, TPS properties do not fulfil all requirements in some important applications such as packaging.[2] This is mainly due to the starch hydrophilic character and to an important post-processing ageing (evolutions of the crystallinity and the water with the time, after processing).[2] TPS mechanical properties are rather poor compared to synthetic petroleum-based polymers. Whatever the plasticizer/ starch ratio, it is impossible to obtain both a sufficient modulus and an accurate impact resistance. To overcome the TPS weaknesses, research laboratories have developed two main strategies. Some authors have tried to modify the starch structure to improve the hydrophobic character of the chains, *e.g.* by acetylation.[3] This chemical approach seems to result in inferior mechanical properties, due to chain length reduction and greater product cost.[4] The second strategy is to associate TPS with another component. To preserve the biodegradable character of the multiphase system, the second component may be biodegradable polyesters (blends or multilayers)[5–9] or cellulose fibres (composites).[9–15]

In previous publications, we have tested the association of TPS with different biodegradable polyesters: polycaprolactone (PCL),[2] polyesteramide (PEA),[6] polybutylene succinate-adipate (PBSA),[7] polybutylene adipate coterephtalate

(PBAT) and unplasticized or plasticized polylactic acid (PLA).[7-8] We have shown that the different polyesters drastically increase the hydrophobic character of the blend, even at low polyester concentration. This is mainly due to the formation of a polyester thin layer at the surface. Besides, we have tested multilayer structures.[5] TPS, forming the inner layer, is protected from moisture by skin and more hydrophobic polyester layers. Such a structure is controlled by rheological factors (viscosity, *etc.*) between both molten phases during processing and inter-compatibility. Certain multilayers are easily delaminated (low peel strengths), such as, *e.g.* PLA/TPS/PLA.[9] Through the analysis of the different blends, we have shown that the compatibility between the polyester and the TPS phase gradually decreased from polyesteramide to PLA.[9] Unfortunately, blending with polyester does not solve the TPS ageing problem. The blend modulus increases several weeks after the injection moulding. Besides, except for PLA, the polyesters present rather low modulus which can be an important limitation for the TPS-based blends or multilayer applications.[9]

Surprisingly, the association between TPS and cellulose fibres (polysaccharide-based composites) has been little analysed. In the literature, cellulose fillers used in association with a plasticized starch matrix are: commercial paper fibres,[10] potato pulp microfibrils,[11-12] tunicin whiskers and paper bleached pulp.[13-14] The literature reports that these cellulose fillers improve the tensile strength. Besides, the composite water sorption seems to be decreased. This behaviour is related to the well known lower water uptake of cellulose compared to starch.[10-14]

The aim of this paper is to present the different properties of polysaccharide composites obtained by the introduction of cellulose fibres, varying in length and content, into plasticized wheat starch. We have used LDPE-based composites as reference, because without compatibilizer, the LDPE-fibres interactions are usually considered as very poor. To highlight the interactions developed between cellulose and starch, we have compared both kinds of composites. Finally, we have evaluated the post-processing ageing of TPS compared to some other systems.

2 Experimental

2.1 Materials

Wheat starch is obtained from Chamtor (France). According to the supplier, amylose and amylopectin contents are respectively 26 and 74%, and the residual protein (gluten) content is inferior to 0.2%. Glycerol used has a purity close to 99.5%. Magnesium stereate (99% purity – Aldrich) is used as a demoulding agent at low concentration, less than 1 wt%. TPS matrix behaviour varies according to the formulation. For this study, the glycerol/starch ratio is 0.25. After equilibrium at 54% RH (Relative Humidity) and 23 °C, TPS water content is 10.7%. Glass transition determined by DSC analysis is 8 °C. TPS density is 1.37.

Polyethylene (LDPE) powder is obtained from Atochem – France (Lacqtene

Table 1 *The different PE and TPS-based composites studied*

Matrix	Fibres (weight fractions)		
	SF	MF	LF
LDPE	5, 9 and 13%	5, 9, 13 and 20%	–
TPS	10 and 18%	10 and 18%	10 and 18%

1200MN18C). MFI is 24 g/10 min (190°C, 2.16 kg). Density is 0.92 (supplier information).

Cellulose fibres from leafwood are obtained from JRS (Arbocel – Germany). Different cellulose fibres with increasing lengths are tested. For a constant average diameter of 20 microns, initial average lengths are respectively 60 (SF), 300 (MF) and 900 (LF) microns. The cellulose content is greater than 99.5%. Fibres density is 1.50.

2.2 Composites Preparation

The different processing steps and the processing conditions are fully described in a previous publication.[15] After powder mixing and extrusion, injection moulding is used to produce standard dumbbells. The injected dumbbells are usually equilibrated (54% RH, 23°C) for 6 weeks before testing. The different composites elaborated are presented in Table 1.

2.3 Mechanical Properties

Moduli, tensile strengths and elongations at break are performed on a mechanical tensile tester (Instron 4204-GB) according to French standard NFT 51–034, with injected dumbbell specimens (length × thickness: 150 × 4 mm^2) conditioned at 54% RH and 23°C. Strain–stress curves are obtained with a velocity of 50 mm/min. Each mechanical parameter is determined from ten tested specimens.

2.4 Thermo-mechanical Analysis

The glass transitions of plasticized starches are sometimes difficult to determine by DSC analysis because the drop heat capacity change is quite low at the glass transition. Then, DMTA determination is preferably used to approach this transition which is clearly demonstrated by a broad tan δ peak.[6] The dynamic thermo-mechanical analyser is a TA instrument DMA-2980 (USA). The experimental protocol has been fully described in a previous publication.[6]

3 Results and Discussion

In Figure 1 are presented the main mechanical results. We show the evolution of the ratio (composite value/matrix value) *versus* fibre volume fraction. The

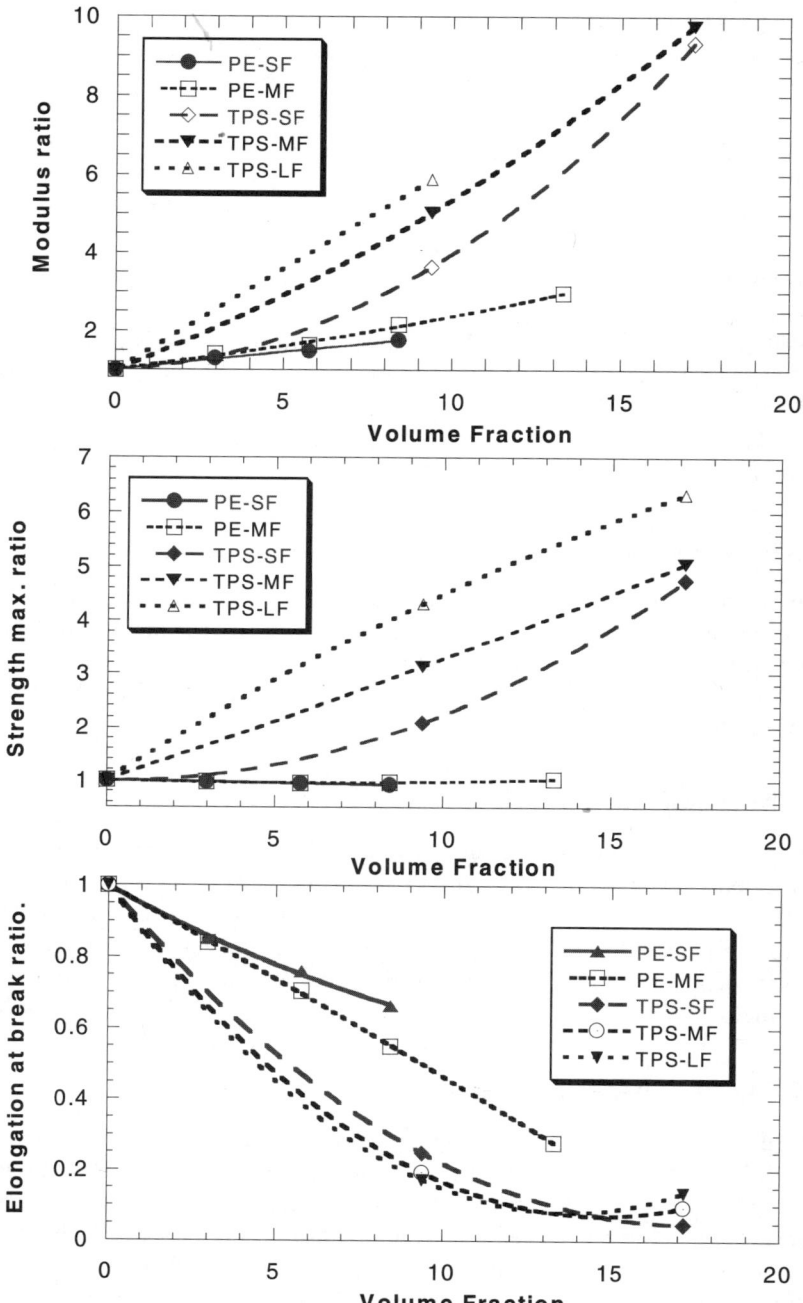

Figure 1 *PE and TPS-based composites. From top to the bottom: modulus, maximum strength and elongation at break ratios* versus *fibre volume fraction*

matrixes modulus, strength maxima and deformation at break are respectively for LDPE (111 MPa, 9.6 MPa and 102%) and for TPS (87 MPa, 3.6 MPa and 124%). The modulus ratios increase both with the fibre length and the fibre content. The evolution of PE composites modulus presents the usual linear function *versus* the filler volume fraction.[16] This evolution is much more important for TPS composites which do not follow the same trend. In Figure 1, for a volume fraction of around 9%, we can compare both TPS and PE composites results. The modulus ratio is more than twice higher than that for TPS composite, *e.g.* for MF fibres, we obtain 2.16 for a LDPE matrix and 5.03 with a TPS matrix. For PE-based composites, maximum strength ratios are quite constant whatever the fibre length. We can notice a slight decrease of this ratio with the fibre content. But, for the TPS-based composites, this strength ratio increases drastically with fibre length and fibre content. Whatever the matrix and the fibre length, elongation ratios decrease with fibre content. We can show greater diminution for TPS composites than for PE composites. These different mechanical results lead to the same conclusion: the reinforcing effect of cellulose fibres is much more efficient in a TPS matrix than in a polyolefin matrix.

DMTA results (storage modulus: E' and tan δ) are presented in Figure 2. The evolution of tan δ shows two relaxations. The main relaxation (α) associated with a large tan δ peak and an important decrease of the storage modulus can be attributed to the TPS glass relaxation. The most striking result is the strong evolution of the temperature corresponding to this relaxation ($T_{\alpha \, transition}$), by introduction of the fibres to the TPS matrix, which cannot be explained by the slight moisture contents decrease.[2] From 0 to 10 wt% fibres, the $T_{\alpha \, transition}$ shifts from 31 to 59 °C. This evolution could be linked to strong fibre–matrix interactions between the two carbohydrate products. The existence of such interactions has been assumed by Dufresne *et al.*[11-12] Such interactions decrease starch chain mobility and consequently increase the matrix glass transition. In Figure 2, the $T_{\alpha \, transition}$ presents an increase from 59 to 63 °C when fibre content ranges from 10 to 18 wt%. It seems that we have a critical value, from which the evolution of the Tg shift *versus* the fibre content decrease, which could be related to a percolation effect (fibre–fibre interactions).[12] With the introduction of fibres, the storage modulus presents an increase before and after α transition. By introducing 18 wt% fibres into TPS, the rubber plateau increases by about one decade. In a previous publication,[2] we have shown that the secondary relaxation (β) is consistent with the glycerol glass transition ($T_{\beta \, transition}$) that occurs around -50 °C. According to some authors,[6-17] we can relate the decrease of this transition with the 'free' glycerol content on TPS material. In Figure 2, we can see a $T_{\beta \, transition}$ increase with the fibre introduction. Curvelo *et al.*[14] have related this evolution with glycerol–cellulose interactions.

In Table 2 is presented the modulus evolution between 2 to 6 ageing weeks, after processing, for different multiphase systems. We can show that the introduction of fibres into the TPS matrix provokes a drastic decrease of this ageing. This is due to the fibre–matrix interactions which can create a physical 3D network (hydrogen network). These H bonds may stabilize the materials. Besides, we have tested the 'TPS–fibres–polyester' system to obtain a better water

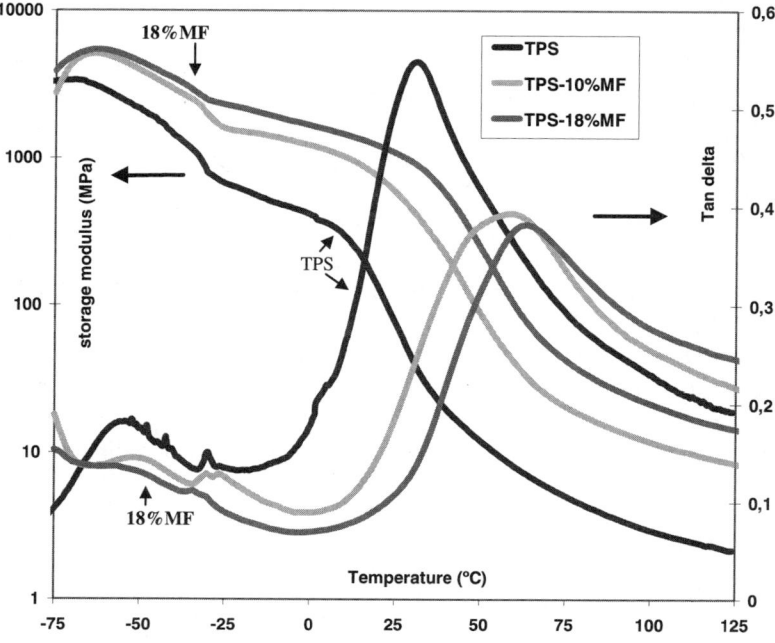

Figure 2 *Storage modulus and tan delta curves (DMTA) for TPS, TPS-10%MF and TPS-18%MF. (Dual cantilever, 1 Hz)*

Table 2 *For different systems, evolution in % of the Young's modulus from 2 to 6 ageing weeks, after injection moulding*

Systems studied	Evolution of the modulus, from 2 to 6 ageing weeks
TPS	+115%
TPS-10% MF	+23%
TPS-10% MF + 10 wt% PBAT[1]	+45%
TPS-10% MF + 25 wt% PBAT[1]	+43%
TPS-10% MF + 10 wt% PCL[2]	+35%
TPS-10% MF + 25 wt% PCL[2]	+31%
TPS-10% MF + 10 wt% PEA[3]	+35%
TPS-10% MF + 25 wt% PEA[3]	+33%

[1]PBAT: Eastar Bio Copo 146766 (Eastman-USA), d = 1.21.
[2]PCL: CAPA 680 (Solvay-Belgium), d = 1.11.
[3]PEA: BAK 1095 (Bayer-France), d = 1.05.

resistance regarding to the TPS-based composite. Compared to this latter, 'TPS–fibres–polyester' system shows a lesser stabilisation of the ageing. According to the polyester used, we obtain some variations which are mainly due to the polyester density variation (volume content variations). The values obtained for the TPS–fibres–polyester systems are intermediate between the data of the blend

(TPS–polyester) and the composite (TPS–fibres), without synergetic effect between them. In fact, the fibres embedded in the polyester phase did not participate to the cellulose–TPS network (dilution effect).

4 Conclusion

By the introduction of cellulose fibres into a plasticized wheat starch matrix, we have shown an increase of the main transition temperature (DMTA) of about 30 °C. This phenomena could be attributed to fibre–matrix interactions decreasing the starch chain mobility. These interactions have a strong influence on the mechanical behaviour. The strong evolution of mechanical behaviour of TPS composites is due to a reinforcing effect but also to the glass transition shift towards a value close to the ambient temperature. Besides, we have shown that cellulose fibres improve the ageing of the TPS system. Although the addition of polyester to a TPS–fibres system increases the water resistance, the ageing is increased compared to the TPS–fibres composites. Such cellulose fibre-based materials present enhanced performances compared to the TPS matrix. However, some investigations are currently underway to improve some properties to fulfil requirements for different applications.

References

1. D. L. Kaplan, *Biopolymers from Renewable Resources*, Springer Verlag, Berlin, 1998, Chapter 2, p. 30.
2. L. Avérous, L. Moro, P. Dole and C. Fringant, *Polymer*, 2000, **41**, 4157.
3. C. Fringant, J. Desbrières and M. Rinaudo, *Polymer*, 1996, **37**, 2663.
4. L. Avérous, L. Moro and C. Fringant, *Properties of plasticized starch acetates*, ICBT Conference, Coimbra (Portugal), 28–30 Sept. 1999.
5. P. Colonna and S. Guilbert, *Biopolymer Science: Food and Non-food Applications*, INRA Editions, Paris, 1999, p. 207.
6. L. Avérous, N. Fauconnier, L. Moro and C. Fringant, *J. Appl. Polym. Sci.*, 2000, **41**, 4157.
7. L. Avérous and C. Fringant, *Polym. Eng. Sci.*, 2001, **41**, 727.
8. O. Martin and L. Avérous, *Polymer*, 2001, **42**, 6237.
9. L. Avérous, O. Martin and L. Moro, *Plasticized Wheat Starch-based Biodegradable Blends and Composites*, 6th ISBP and 9th BEPS, Honolulu–Hawaii (USA), 12–16 December 2000.
10. U. Funke, W. Bergthaller and M.G. Lindhauer, *Polym. Degrad. Stabil.*, 1998, **59**, 293.
11. A. Dufresne and M. R. Vignon, *Macromolecules*, 1998, **31**, 2693.
12. A. Dufresne, D. Dupeyre and M. R. Vignon, *J. Appl. Polym. Sci.*, 2000, **76**, 2080.
13. M. Neus Angles and A. Dufresne, *Macromolecules*, 2001, **34**, 2921.
14. A. A. S. Curvelo, A. J. F. de Carvalho and J. A. M. Agnelli, *Carbohydr. Polym.*, 2001, **45**, 183.
15. L. Avérous, C. Fringant and L. Moro, *Polymer*, 2001, **42**, 6571.
16. R. T. Woodhams, G. Thomas and D. K. Rodgers, *Polym. Eng. Sci.*, 1984, **24**, 1166.
17. D. Lourdin, H. Bizot and P. Colonna, *J. Appl. Polym. Sci.*, 1997, **63**, 1047.

Protein Films: Microstructural Aspects and Interaction with Water

C. Mangavel, N. Rossignol, A. Gerbanowski, J. Barbot, Y. Popineau and J. Guéguen

INRA-UBTP, RUE DE LA GÉRAUDIÈRE, BP 71627, 44316 NANTES CEDEX 03, FRANCE

1 Introduction

Proteins, either from plant or animal origins, are an abundant source of raw materials and could therefore be interesting substrates for non-food uses. A lot of studies have been done through the past few years to exploit the abilities of proteins to form biomaterials, and more particularly, to form films. Such films could find some applications, for example in the packaging domain.[1]

However, one of the main drawbacks of these protein materials is their high sensitivity to water. As interaction of a protein film with water is a sum of complex phenomena, different parameters can be taken into account. The water vapor barrier properties and the plasticizing effect of water in the film (with its important influence on mechanical properties) belong to the most studied characteristics, probably because they can easily be linked to macroscopic measurements.

However, these interactions between water and protein films have never been investigated on a mesoscopic scale. We first focused our interest on film surface wettability; as protein films are often hydrophilic by nature, we tried to make their surfaces more hydrophobic. To increase protein film surface hydrophobicity, two approaches were tested. The first method was the protein hydrophobization by grafting aliphatic carbon chains on the lysine residues through an acylation reaction. The second way was to decrease the polarity of the film-forming solution by adding dioxane to the medium and thus favour the exposure of intrinsic hydrophobic zones of the protein towards the solvent.

On the other hand, we were also interested in a better understanding of film microstructure, and more particularly of network porosity.

2 Materials and Methods

2.1 Materials

Crystalline BSA, dioxane, hexanoic anhydride, diethyleneglycol and glycerol were obtained from Sigma and used without any purification.

An industrial gluten (70.6% protein, N x 5.7) was provided by Amylum (Aalst, Belgium).

2.2 Methods

2.2.1 Protein Modification. Acylation was performed at room temperature by adding dropwise the anhydride to the protein solution (2 mg/ml). To reach the maximum level of modification, three equivalents of anhydride per lysine were needed. The pH was kept at 9 during the reaction by a controlled and continuous addition of 0.5 M NaOH using an automated pH-stat titration device. The reaction was considered to be complete when the pH of the reaction medium remained constant. To remove salts and excess reagents, the protein solution was dialyzed exhaustively against water and then lyophilized.

The removal of the side-reaction products (fatty acids) noncovalently bound to the protein was performed by three washings of the freeze-dried protein sample (10 mg) in hexane (5 ml). After centrifugation (1500 g, 15 min), the supernatant was discarded.

The degree of modification was assayed by the modified *o*-phtaldialdehyde (OPA) method.[2]

The increase of BSA surface hydrophobicity after grafting was monitored by a fluorometric ANS assay.[3]

2.2.2 Preparation of Films by Casting. BSA films were obtained by casting as previously described.[4] The protein was dispersed into an alkaline medium (NH_4OH 2.5% or NH_4OH 2.5%/dioxane, 70/30) in presence of diethyleneglycol as a plasticizer (plasticizer/protein = 0.5). The film-forming solution containing 20% BSA (w/w) was cast onto a glass plate and the excess solvent was evaporated by drying at 70°C for 1 hour. The film was then peeled from the plate and stored at 20°C and 60% RH for equilibration before testing.

Gluten films were prepared according to the same procedure, except that the film-forming solution contained sodium hydroxyde (0.1 M, pH 11) and glycerol as a plasticizer (glycerol/protein = 0.45). The film-forming solution contained 12.5% gluten protein (w/w).

2.2.2 Preparation of Films by Thermomoulding. Gluten was mixed with glycerol in a mortar with a pestle (plasticizer/protein = 0.45). The blend was placed between two Teflon sheets and pressed at 120°C and 200 bars for 15 minutes in a heated press (hydraulic press 'Pinette' Shell Tellus T27).

2.2.4 Contact Angle Measurements. Contact angles were evaluated using an

coated sample with
epoxide resin

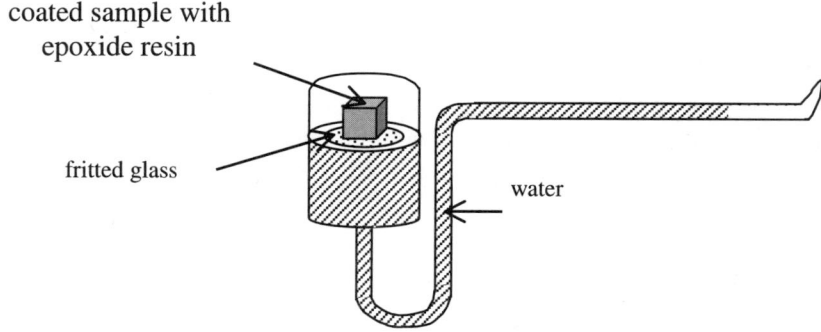

fritted glass

water

Figure 1 *Baumann device*

image analysis system Digidrop (GBX Instruments). After a drop of water was placed above the film, the angle of the tangent to the base of the drop was measured and expressed in degrees.

2.2.5 X-ray Photoelectron Spectroscopy. Protein films without plasticizer and uncoated from the glass plates were studied by XPS, using an Escalab MK II spectrometer with an Al K$_\alpha$ X-ray source (1486.6 eV of photon energy) and a three channel detector. Measurements were taken at an angle of 90°.

2.2.6 Scanning Electron Microscopy. Films were observed by scanning electron microscopy (JEOL SM 840A) with an accelerating voltage of 2.5 kV; samples were previously cut with a scalpel and gilded.

2.2.7 Water Uptake Measurements. The water absorption capability of films was determined by measurements of water rising by capillary action. The experimental device, also called Baumann apparatus is shown in Figure 1. The samples were cut in order to expose their inner structure to water rising. They were then coated on each of their faces (except the contact surface) with epoxide resin to prevent their swelling and put on fritted glass in contact with distilled water. The absorbed water volume *versus* time data is then recorded as water penetrates into the sample.

3 Results

3.1 Chemical Modification of BSA

Optimized reaction conditions previously described[5] were used in this study. Hexanoic anhydride was used to graft a six-carbon chain length onto the lysine by an acylation reaction. The use of three equivalents of hexanoic anhydride per lysine residue lead to 95% of acylation of these residues. The number of ANS

Table 1 *Contact angle values of a water drop at initial time and after 5 minutes on BSA films. Values are the mean of five replicates*

Protein/Dispersing medium	Angle at initial time (°)	Angle after five minutes (°)
Native BSA/ammonia	38 ± 4	10 (n.d.)
Native BSA/ammonia–dioxane	114 ± 4	100 ± 1
Modified BSA (hexanoic anhydride)/ammonia	114 ± 3	104 ± 1

binding sites increased from two for the native BSA to twelve for the modified BSA.

3.2 Decrease of Film Surface Wettability

Three types of films were prepared by casting: the control film was made of BSA dispersed into an ammonia solution. The effect of the chemical modification was assayed by preparing a film from hydrophobized BSA in an ammonia solution. The effect of the medium polarity was studied by preparing a film of native BSA dispersed into an ammonia/dioxane solution.

Contact angles for these films were measured at initial time and after 5 minutes (Table 1). The control film (native BSA in NH_4OH) presented a very hydrophilic surface, with a weak contact angle of 37°, falling to 10° after five minutes. Both modifications, either chemical grafting or modification of the solvent, resulted in an increase of these angles (up to 114°); moreover, these angles remained quite stable after 5 minutes, with values above 90°.

Both treatments revealed equal efficiency in decreasing surface wettability.

XPS spectra of these films were then recorded to study whether this macroscopic increase in hydrophobicity was correlated to a decrease of the atomic composition of the films surface. The spectra presented three main peaks corresponding to the elements carbon, oxygen and nitrogen. Table 2 presents the ratios of these atoms, obtained by comparing the peak areas. Both modifications induced an important decrease of the O/C and of the O/N ratios. Exposure of oxygen at the surface of the films was then much weaker compared to the control film.

The three peaks were then decomposed into single components corresponding to atoms with particular environments (Table 3). On this table, we can notice an important decrease of the protonated nitrogen exposed at the surface of the film when BSA was grafted with hexanoic anhydride. This result was expected since the acylation reaction precisely occurs on protonated nitrogen to transform it into peptidic nitrogen. Other important features on this table are the decrease of hydrophilic oxygen exposed at the surface at the expense of the increase of peptidic oxygen. In the same way, the aliphatic carbon chains have a greater exposure at the surface of the films after both modifications. These aliphatic carbon chains could come either from the C6 chain that was grafted to the protein or from intrinsic hydrophobic amino acids of the protein. Thus both

Table 2 *Quantitative analysis of ESCA measurements on BSA films*

| | Atom ratio | | |
Protein/Dispersing medium	O/C	N/C	O/N
Native BSA/ammonia	0.40	0.22	1.82
Native BSA/ammonia–dioxane	0.29	0.24	1.21
Modified BSA (hexanoic anhydride)/ammonia	0.25	0.19	1.40

Table 3 *Decomposition of C1s, O1s and N1s ESCA peaks of BSA films*

Protein/ Dispersing medium	$C_{aliphatic}$	C_α	$C_{peptidic}$	$O_{peptidic}$	$O_{hydrophilic}$	$N_{peptidic}$	$N_{protonated}$
Native BSA/ ammonia	40%	23%	37%	50%	50%	89%	11%
Native BSA/ Ammonia– dioxane	45%	20%	35%	88%	12%	85%	15%
Acylated BSA/ ammonia	50%	31%	20%	76%	24%	96%	4%

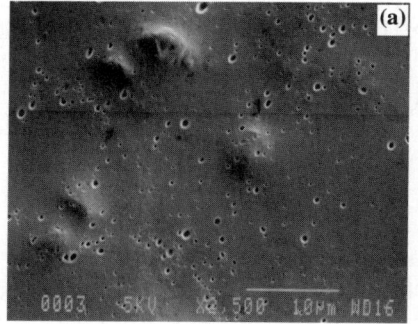 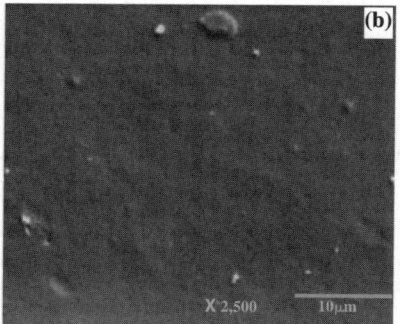

Figure 2 *SEM micrographs of gluten films surfaces at* × 2500 *magnification:* (a) *casting;* (b) *thermomoulding*

modifications lead to a burying of hydrophilic atoms towards the bulk of the film whereas more hydrophobic atoms have a better exposure at the surface.

3.3 Microstructure of the Network

Scanning Electron Microscopy (SEM) micrographs of cast and thermomoulded gluten films surfaces are presented in Figure 2, at a magnification of × 2500. The process for film-forming affects their surfaces. Indeed, the cast film surface shows some voids probably due to dissolved gases and water escaping from the film during the drying step of this process. On the contrary, the thermomoulded film has a smooth and homogeneous surface.

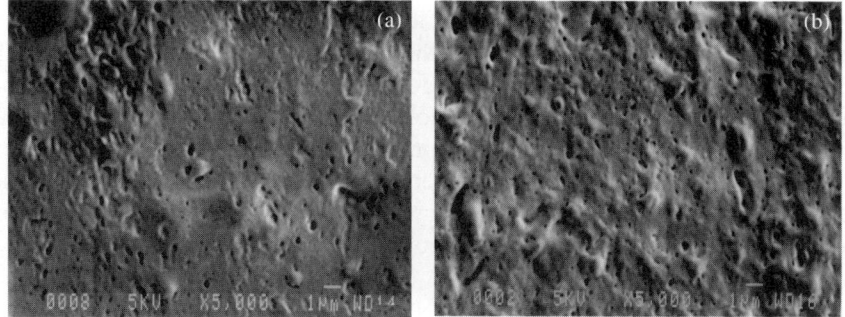

Figure 3 *SEM micrographs of gluten film cross-sections at* × 5000 *magnification:* (a) *casting* (b) *thermomoulding*

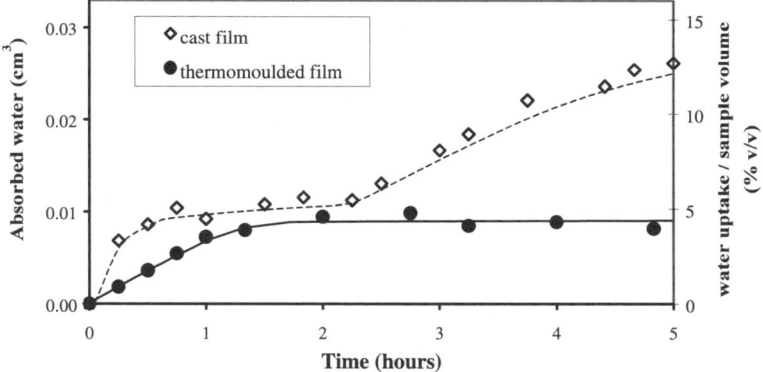

Figure 4 *Water uptake of cast and thermomoulded gluten films* ($V_{sample} = 0.2$ cm^3)

To obtain information about the inner structures, cross-sections of cut films were also examined by SEM (Figure 3). The micrographs, at × 5000 magnification, show some differences between both films, the thermomoulded sample being much rougher than the cast one. However, both films presented pores of about 200 to 500 nm diameter.

The behaviour of cast and thermomoulded gluten films brought into contact with water were compared. The kinetics of water uptake by capillarity are shown in Figure 4. Both absorption data are linear until they reach a plateau. This is the general trend for absorption data of porous medium: liquid rises into the pores until they are saturated (plateau region). Both water absorption curves presented the same plateau values, indicating that their total porosity was of the same order of magnitude (about 5% of sample volume), independent of the process. These results are in accordance with the microscopic observations of cast and thermomoulded film sections, which showed an equivalent distribution of pores. Nevertheless, the cast films presented a specific behaviour, with a second water uptake after the plateau region. This could be attributed to some rearrangement of the protein network due to hydration.

Cast films presented then a higher affinity to water than thermomoulded samples probably as a consequence of the protein solubilization step during the casting process. The films obtained by both processes presented similarities and differences: same percentages of water volume uptake at a plateau region and an equivalent porosity observed by scanning electron microscopy, but a different reaction of cast film with a second water uptake step. It is then quite clear that films behaviour in the presence of water is not only related to film porosity but eventually to a rearrangement of the network.

4 Conclusions

To have a better knowledge of the interactions between water and protein films, it is thus necessary to take various parameters into account and to simultaneously study the samples at different scales. As the surface hydrophobicity of the films could be explained by a study of their atomic composition, it is clear that the mesoscopic scale should be taken into account as the microstructure of the network also plays a role in the interaction of protein films with water.

References

1. B. Cuq, N. Gontard and S. Guilbert, *Cereal Chem.*, 1998, **75**, 1.
2. H. Frister, H. Meisel and E. Schlimme, *Fresenius Z. Anal. Chem.*, 1988, **330**, 631.
3. M. Cardamone and N. K. Puri, *Biochem. J.*, 1992, **282**, 589.
4. A. C. Sanchez, Y. Popineau, C. Mangavel, C. Larré and J. Guéguen, *J. Agric. Food Chem.*, 1998, **46**, 4539.
5. A. Gerbanowski, C. Rabiller, C. Larré and J. Guéguen, *J. Prot. Chem.*, 1999, **18**, 3, 325.

Pea: An Interesting Crop for Packaging Applications

J. J. G. van Soest,* D. Lewin, H. Dumont and F. H. J. Kappen

ATO, BOX 17, 6700 AA WAGENINGEN, THE NETHERLANDS

1 Abstract

Thermoplastic starches (TPS) were prepared by moulding pea starches with different amylose contents in the presence of plasticizers. The properties were compared with other starches such as waxy corn and potato starch. The structural properties depended on pea starch source, plasticizer content and processing conditions. The TPS materials are semicrystalline. Crystallisation of, in particular, high amylose pea starch resulting in C-type crystals was remarkable. The stiffness, strength and elongation of the materials depended on amylose content. The materials were in the glass or rubbery state depending on plasticizer, with an E-modulus of *ca.* 200 to 3500 MPa for rubbery and glassy materials, respectively.

2 Introduction

Pea is a renewable reservoir for functional macromolecules. Pea proteins or starches can be used for packaging applications, such as films, foams and controlled release systems. The functionality of the biopolymers is influenced by technological treatments and altered by physical, enzymatic or chemical modifications. This work is aimed at obtaining detailed knowledge about the structure–property relationships of pea-based biodegradable plastics.

Starch source or structure influences the properties of TPS.[1–15] Native starches differ in amylose and amylopectin ratio, degree of branching and molecular mass. During high temperature and shear processing of TPS, structural changes occur such as granule melting and polymer chain breakdown. Most studies deal with TPS based on potato, barley, wheat and maize starches. Pea starch is a less known alternative, available with a range of structural differences, such as amylose–amylopectin ratio.

In this study, TPS is prepared by compression moulding of pea starches with different amylose contents in the presence of plasticizers (water and glycerol). The influence of processing temperature is studied. The structural and mechanical properties are compared with waxy corn and potato starch based TPS.

3 Materials and Methods

3.1 Materials

High amylose (*ca.* 60–85% amylose) pea starch (HAP) was obtained from Stauderer (Germany) with a composition of starch $\sim$ 80% and protein *ca.* 1%. Normal yellow smooth pea starch (*ca.* 35% amylose; STL) was purchased from Pelmolen (The Netherlands). The composition was: starch $\sim$ 84% and protein *ca.* 5%. Water (W) and glycerol (G) were used as plasticizers. Waxy corn (Amioca, WCN) and potato starch (Farina, PN) were obtained from National Starch and Avebe, respectively.

3.2 Methods

For compression moulding, premixes were made with 100 parts of dry starch (HAP, STL, WCN or PN), glycerol (G) and water (W). The following premixes were prepared: HAP30G30W, HAP10G50W, STL30G30W, WCN30G30W and PN30G30W. Compression moulding and structural analyses were performed as described previously.[15] Stress–strain measurements were performed with a Zwick 010 Tensiometer using the following parameters: Pre-load Fv = 1 N, Speed up to pre-load = 1 mm/min, Speed for E-modulus = 1 mm/min, Test speed = 10 mm/min and Break detector dF = 50% Fmax. Mechanical properties were determined on tensile bars (3–5) after conditioning for 2 weeks at *ca.* 60% relative humidity and 20°C.

4 Results and Discussion

4.1 Structural Features

Typical diffractograms of compression moulded pea starches (HAP and STL) plasticized with water and glycerol are shown in Figure 1 and 2. Native STL starch has a C-type structure characteristic of pea starches.[16] The native HAP with the high amylose content shows more B-type structural features. Between 110 and 170 °C the native granular structure is gradually completely disrupted as observed with polarised light microscopy. The native crystallinity is molten at around 110–140°C for the STL and around 110–120°C for HAP. Above these temperatures the starch re-crystallises in B- and V-type structures. Above 180°C E-type characteristics are observed for materials containing 30% glycerol (30G). This behaviour is similar to other starches.[14] As expected from its relative high water content during the cooling down process, the amount of re-crystallised material in the HAP10G50W is significantly higher compared to the 30G30W material. Furthermore, C-type structures are formed in these TPS materials,

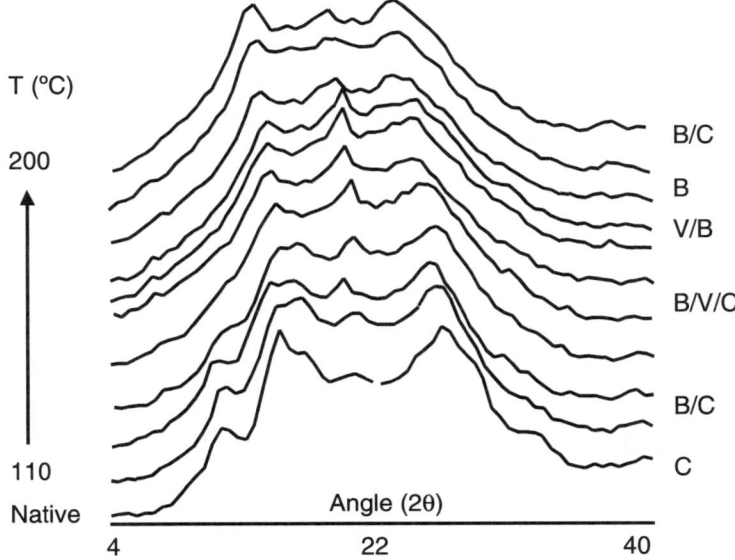

Figure 1 *Influence of moulding temperature on the X-ray diffractograms of moulded STL30G30W materials*

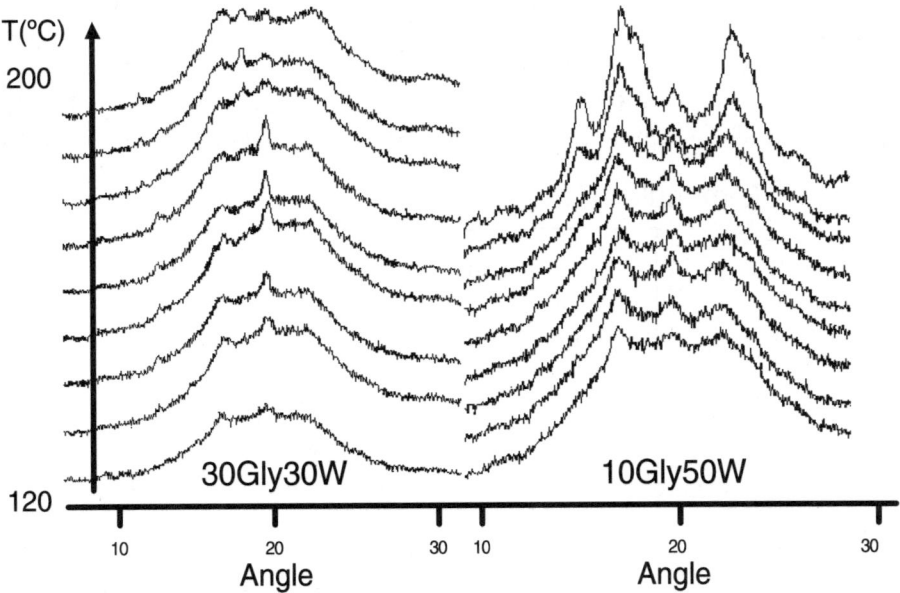

Figure 2 *Influence of moulding temperature on the X-ray diffractograms of moulded HAP30G30W and HAP10G50W materials*

especially above 180 °C. This is possibly related to a breakdown of starch molecules, which is more pronounced above 170–180 °C as previously observed in other TPS (unpublished results). Generally, B-, E- and V-type crystals are found in TPS materials and gels.[11] However, it is known that A- and C-type crystalline structures can be obtained depending on the thermal history and chain length of starch.[17]

Waxy corn (WCN30G30W) materials are amorphous after moulding at a temperature above 120°C. Potato starch materials (PN30G30W) show B-type crystallinity. The relative crystallinity (compared to native potato starch) decreases from 68% (at 110°C) to 36% (at 140°C). Above 140°C the relative crystallinity increases to a value of 70% at 190°C. These values are in agreement with values previously reported.[14] It has been shown that the total B-type crystallinity of potato starch plasticized with glycerol and water can be considered as a summation of residual amylopectin crystallinity and recrystallization of both amylose and amylopectin depending on processing conditions.

4.2 Mechanical Properties

In Figure 3, the E-moduli of HAP10G50W are shown. The water content after storage was 11–12.5% (based on total mass). Materials prepared at temperatures below 150°C and at 200°C were too brittle to be measured. At the lower temperatures, incomplete melting of the native starch granules leads to the presence of granules and a large amount of ghost structures as shown by light microscopy. At higher temperatures polymer chain breakdown takes place. This can result in inferior properties, *i.e.* lower E-modulus.[3,9] Because of the low glycerol content, the 10G50W materials are in the glassy state and exhibit E-moduli up to 3 GPa. The large differences are probably related to the low impact strength of moulded TPS materials and to inhomogeneities within the TPS materials (*i.e.* tensile bars) (unpublished results). The values of 3 GPa are

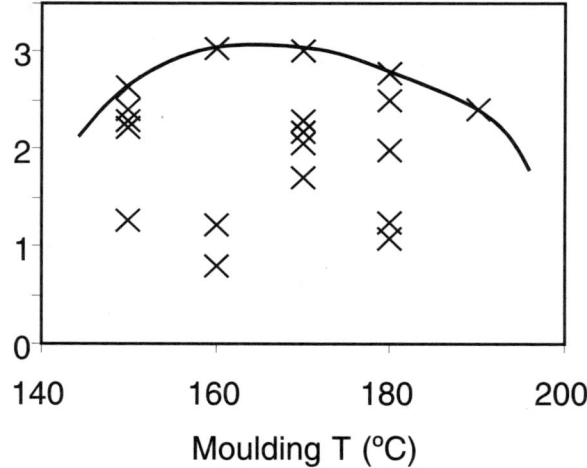

Figure 3 *Typical example of the stress–strain behaviour (E-modulus) of compression moulded glassy pea starch-based TPS materials (HAP10G50W)*

relatively high compared to values reported in the literature for other TPS materials. For instance, compression moulded potato starches, plasticized with water, glycerol and lecithin, have an E-modulus of about 1 GPa.[15] It has to be noted that the composition of the plasticizer was different after storage resulting in a water content of 6.5% (based on total mass) and a starch to glycerol ratio of 100:30.

The average stress–strain properties are shown in Figure 4. The average E-moduli are between 1.5–2.5 GPa. The strength of the glassy materials is in the range of 24–38 MPa. This value is similar to compression moulded glassy TPS based on potato starch (*i.e.* 33 MPa).[15] The elongation is low in the range of 1–3%.

Figure 5 shows the stress–strain behaviour of HAP30G30W materials. The water content after storage was 11–12% (based on total mass). The strength of the materials increases with increasing moulding temperature up to 190 °C from *ca.* 5 to 11 MPa. At 200 °C a slight decrease is observed (to about 8 MPa). The E-modulus is in the range of 200–230 MPa after processing between 120–170 °C. An increase is observed above 170 °C up to 500 MPa. The elongation increases from 7% at 120 °C to a maximum of 37% at 170 °C. Above 170 °C the elongation decreases to 7% at 200 °C.

Figure 6 shows a comparison of the properties of several native starches. Waxy corn (WCN) shows a strong increase in elongation from 5 to 340% with an increase in temperature from 110 to 200 °C. However, the E-modulus decreases from 250 to 20 MPa. PN, STL and HAP show a maximum in elongation between 150–170 °C. PN30G30W has a maximum in elongation of about 100%. STL and HAP both have a maximum elongation of approximately 35%. The

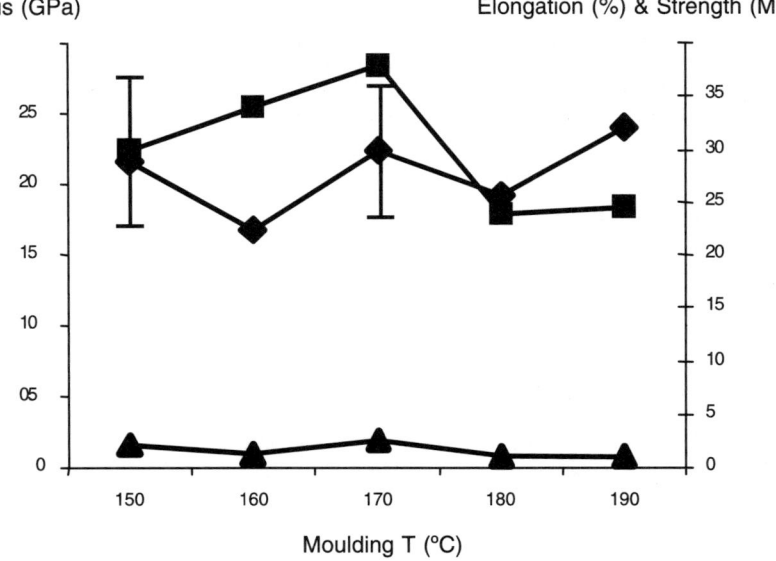

Figure 4 *Stress–strain properties of HAP10G50W;* ▲; *elongation,* ■; *strength,* ◆ *E-modulus*

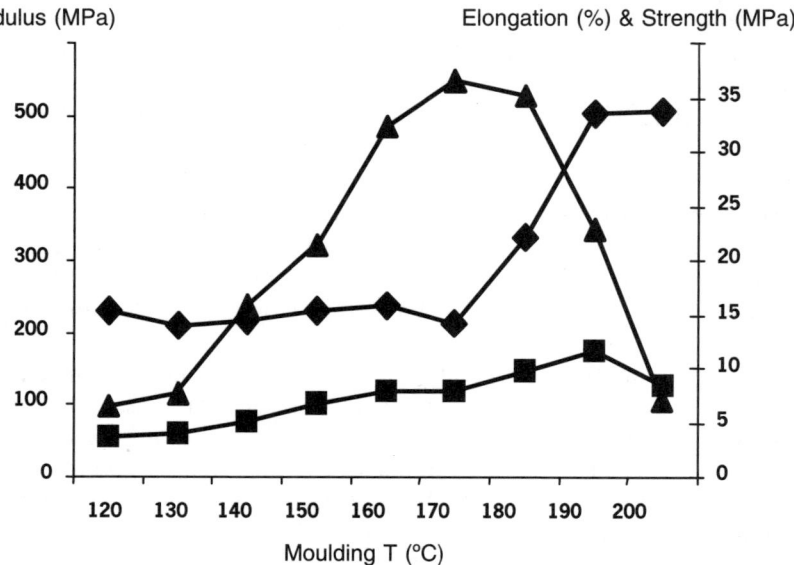

Figure 5 *Stress–strain properties of HAP30G30W;* ▲; *elongation,* ■; *strength,* ◆ *E-modulus*

E-moduli of STL and PN both show a slight minimum at 150–160 °C. The values of STL30G30W are slightly higher at temperatures above 150 °C (about 200–250 MPa) compared to 120 to 200 MPa for PN30G30W. The HAP30G30W materials have a significantly higher E-modulus at temperatures above 180 °C (> 500 MPa).

Differences in properties are explained by the amylose content. The elongation decreases and the strength and stiffness of glycerol plasticized corn-based TPS increase with increase in amylose content.[2,8] The contributions of other structural features, such as granule size, starch branching and protein content, are not clear at the moment.

5 Conclusions

The properties of pea starch-based TPS depend on amylose content and plasticizer content. The best properties are obtained at compression moulding temperatures of around 160–190 °C. High amylose pea starch gives glassy and rubbery materials with a high stiffness compared to other starches. A relatively high amount of C-type crystals after processing of high amylose pea starch at high temperatures and high water contents is observed. High amylose pea starch seems to be an interesting renewable source for the production of starch-based bioplastics.

Acknowledgements

The authors wish to thank J. S. Fijn for his initial work on thermoplastic pea

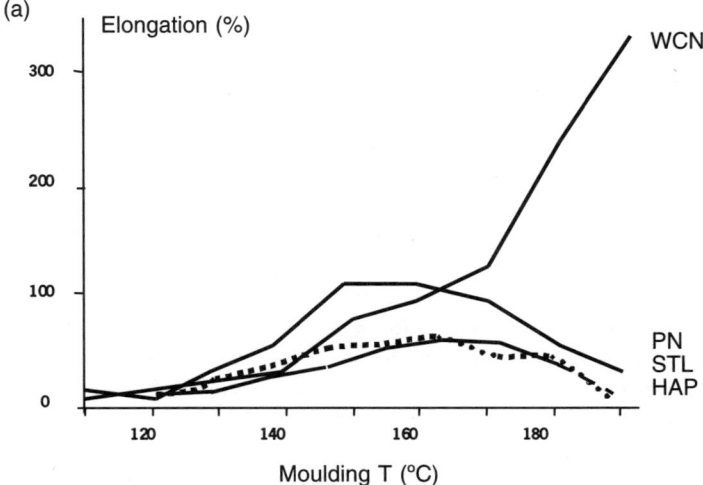

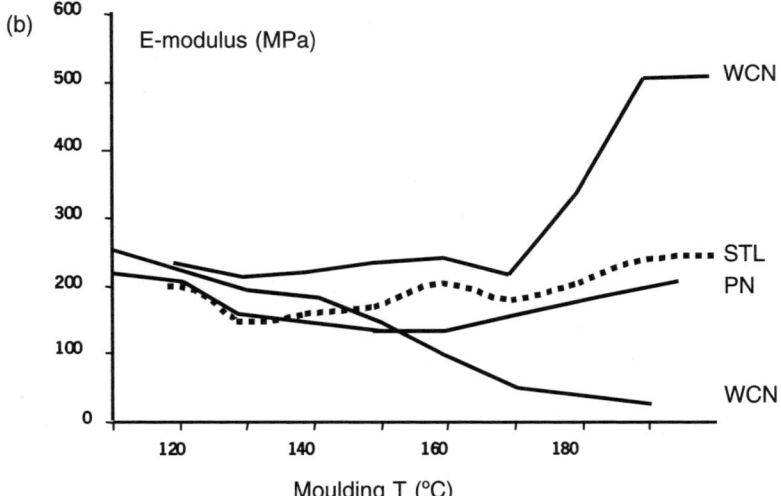

Figure 6 *Comparison of the stress–strain properties of several rubbery TPS materials based on various starches:* (a) *Elongation,* (b) *E-modulus.* —— *WCN, PN and HAP;* - - - - *STL*

starches. This work has received financial support from EC project FAIR CT98–3527 (Exploitation of the unique genetic variability of peas in the production of food and non-food ingredients).

References

1. P. Forssell, J. Mikkila, T. Suortti, J. Seppala and K. Poutanen, *J. Macromol. Sci. Pure Appl. Sci.*, 1996, **A33**, 703.

2. J. J. G. van Soest and P. Essers, *J. Macromol. Sci. Pure Appl. Sci.*, 1997, **A34**, 1665.
3. J. J. G. van Soest, K. Benes, D. de Wit and J. F. G. Vliegenthart, *Polymer*, 1996, **37**, 3543.
4. A. R. Kirby, S. A. Clark, R. Parker and A. C. Smith, *J. Mater. Sci.*, 1993, **28**, 5937.
5. J. J. G. van Soest, D. de Wit and J. F. G. Vliegenthart, *J. Appl. Polym. Sci.*, 1996, **61**, 1927.
6. J. J. G. van Soest, S. H. D. Hulleman, D. de Wit and J. F. G. Vliegenthart, *Carbohydr. Polym.*, 1996, **29**, 225.
7. S. H. D. Hulleman, F. H. P. Janssen and H. Feil, *Polymer*, 1998, **39**, 2043.
8. J. J. G. van Soest and D. B. Borger, *J. Appl. Polym. Sci.*, 1997, **64**, 631.
9. J. J. G. van Soest, K. Benes and D. de Wit, *Starch*, 1996, **47**, 429.
10. J. J. G. van Soest, *Agro Food Ind. Hi-Tech*, 1997, **8**, 17.
11. J. J. G. van Soest, *Starch Plastics: Structure–Property Relationships*, PhD Dissertation, Utrecht University, ISBN 90–393–1072–6, P&L Press, Wageningen, 1995, pp. 1.
12. J. J. G. van Soest, S. H. D. Hulleman, D. de Wit and J. F. G. Vliegenthart, *Carbohydr. Polym.*, 1996, **29**, 225.
13. P. M. Forssell, S. H. D. Hulleman, P. J. Myllärinen, G. K. Moates and R. Parker, *Carbohydr. Polym.*, 1999, **39**, 43.
14. S. H. D. Hulleman, M. G. Kalisvaart, F. H. P. Janssen, H. Feil and J. F. G. Vliegenthart, *Carbohydr. Polym.*, 1999, **39**, 351.
15. J. J. G. van Soest and P. M. Kortleve, *J. Appl. Polym. Sci.*, 1999, **74**, 2207.
16. T. L. Wang, T. Y. Bogracheva and C. L. Hedley, *J. Exp. Botany*, 1998, **49**, 481.
17. D. S. Reid (ed.), *The Properties of Water in Foods, Isopow 6*, ISBN 0751403822, Blackie Academic & Professional, London, 1998, pp. 160.

In Situ Study of the Changes in Starch and Gluten during Heating of Dough using Attenuated-total-reflectance Fourier-transform-infrared (ATR-FTIR)

Olivier Sevenou, Sandra E. Hill, Imad A. Farhat and John R. Mitchell

DIVISION OF FOOD SCIENCES, SCHOOL OF BIOLOGICAL SCIENCES, UNIVERSITY OF NOTTINGHAM, SUTTON BONINGTON, LOUGHBOROUGH, LEICESTERSHIRE LE12 5RD, UK

1 Introduction

Wheat flour can be developed to form a dough by mixing with water. Dough is a two-phase system with a gluten network and a dispersion of starch granules in an aqueous phase.[1] How water partitions between starch and gluten before and after heating has been the subject of much debate. According to Stanstedt,[2] starch gelatinisation during heating of dough leads to a transfer of water from gluten to starch with consequences on the properties of gluten. Bushuk suggested that no water is bound to gluten in bread and proposed that 77% of the total water was associated with starch and 23% with pentosans.[3] Eliasson, using Differential Scanning Calorimetry (DSC) to monitor the starch gelatinisation, calculated the amount of water associated with starch and indirectly the amount of water associated with the gluten during heating of starch/gluten mixtures.[4] In a mixture of starch/gluten/water: 10/2/9, it was calculated that 0.48 g of water was associated with 1 g of dry gluten at 100°C. Using the same method, other authors proposed a value of 0.76 g of water associated with gluten during the second DSC transition of starch gelatinisation in a dough (40% water (w. b.)).[5]

Due to the significant difference in the mid-infrared spectra of proteins and polysaccharides, it is possible to study simultaneously and independently their behaviours in mixtures.[6] In this study ATR–FTIR was used to monitor starch gelatinisation and the evolution of the water content dependent hydrogen bonding in the gluten fraction of starch/gluten mixtures as a function of heating.

2 Material and Methods

2.1 Material

Gluten and wheat starch samples were commercial products donated by Amylum Ltd, Belgium.

2.1.1 Gluten Preparation. The moisture content of samples were set using different methods: below 27% (w/w) gluten was equilibrated over saturated salt solutions, between 30 and 50% (w/w) gluten was ground with ice, stored overnight at 4°C and then left at room temperature for 4 h before use,[7] above 50% water gluten was mixed with water in a mixer. Heat-treated gluten was prepared by storing gluten (66% water (w/w)) in sealed conditions at 105°C for 30 min.

2.1.2 Starch–Gluten–Water Mixtures. Wheat starch and gluten were blended for 1 hour and then the powder was mixed with water to produce a dough.

2.2 Methods

2.2.1 FTIR Measurements. FTIR spectra were collected on heating using a heated ATR single reflectance cell with a diamond crystal (45° incidence-angle) (golden gate cell, Graseby-Specac Ltd, Orpington, UK). A Bruker IFS48 spectrometer equipped with a DTGS detector was used. At each temperature step 64 scans with a $4 \, cm^{-1}$ resolution were co-added, before Fourier transform. The spectra of water recorded in the same conditions at each temperature were subtracted from the sample spectra. The subtraction criterion was a flat baseline in the region $2500–2000 \, cm^{-1}$.[8] Before extraction of the IR absorbance values a deconvolution was applied to the spectra using Opus 3.0 (Bruker).[9] The assumed line shape was Lorentzian with a half-width of $19 \, cm^{-1}$ and a resolution enhancement factor of 1.9. The heating of the ATR cell was controlled with an Opus 3.0 macro using a heating rate of 2°C/min. The data acquisition for each spectrum occurred over a 1 min period and a temperature step of 2°C was used. The temperature control is accurate within a range of ± 1°C. For the starch gelatinisation analysis, the spectra were baseline corrected in the region $1185–800 \, cm^{-1}$ before deconvolution.[10]

2.2.2 DSC Measurements. DSC thermograms were acquired using a Perkin-Elmer DSC 7 instrument operating at 2°C/min in order to match infrared experiments. A baseline correction was performed prior to the numerical integration of the endotherm.

3 Results

3.1 Starch Gelatinisation

As expected, the infrared spectra of starch–water suspensions showed a great dependency on temperature reflecting the molecular order dimension of the gelatinisation process (spectra not shown). The IR bands with a maximum at 1047 cm^{-1} and 1022 cm^{-1} have been proposed to be sensitive to the ordered and to the amorphous fractions in native starch respectively.[10-12] The ratio of the absorbances at 1047 and 1022 cm^{-1} has been linearly correlated with the fraction of ordered starch in native–amorphous starch mixtures as quantified by wide angle X-ray diffraction (WAXS).[10] Hence, the decrease in the ratio of the IR bands at 1047/1022 cm^{-1} is considered as a loss of order in starch.

Figure 1 shows the variation in the ratio 1047/1020 cm^{-1} obtained for a wheat flour/water dough, heated at 2°C/min, plotted as a function of temperature, compared to a DSC experiment performed on the same sample with identical heating rate. The decrease in the ratio 1047/1020 cm^{-1} was closely correlated with the DSC data for flour–water mixture. This finding validated the use of FTIR to follow starch gelatinisation *in situ*.

3.2 Effect of the Moisture Content of Gluten on the IR Amide II Band

The Amide II band is mainly an in-plane C–N–H bending vibration coupled to the C–N stretching.[13] It was observed to be highly sensitive to molecular interaction *via* hydrogen bonding and consequently to hydration. As proposed in previous work there is a change in the maximum absorbance from 1512 cm^{-1} for dry gluten to 1547 cm^{-1} for dough-state gluten.[13] After deconvolution of IR spectra, the shift in the amide II band was related to changes in the maximal absorbances at 1512 and 1547 cm^{-1} (spectra not shown). The ratio of the absorbances at 1547 and 1512 cm^{-1} as a function of moisture content is shown in Figure 2.

A fully hydrated state for gluten was obtained in water conditions that were equivalent to 1.7–2.0 g of water available for 1 g of gluten. At a moisture content

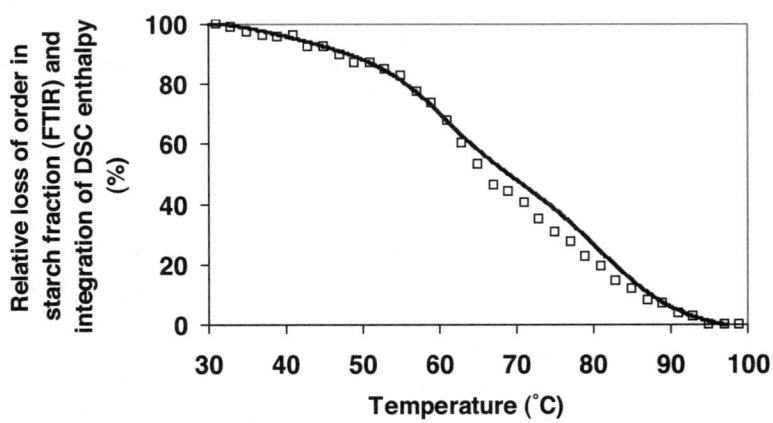

Figure 1 *Comparison of the loss of order in the starch fraction of dough during gelatinisation, detected by FTIR (□), with the integration of the DSC endotherm (solid line)*

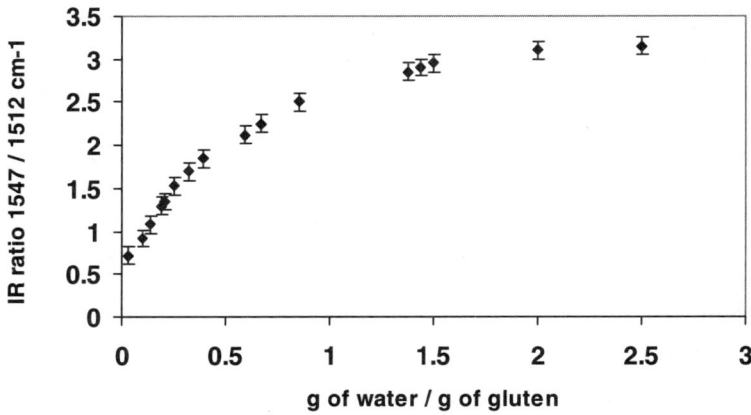

Figure 2 *Ratio of the infrared absorbances at 1547 and 1512 cm⁻¹ for commercial gluten as a function of moisture content*

lower than 66% water (w/w), the ratio of absorbances $1547/1512 \, \text{cm}^{-1}$ can therefore be used to estimate the degree of hydration of gluten.

3.3 Effect of Heating on Gluten as Detected using the Ratio $1547/1512 \, \text{cm}^{-1}$

Variation in the ratio $1547/1512 \, \text{cm}^{-1}$ was followed as a function of temperature for different initial gluten hydration levels (Figure 3). In the constant water content conditions (sealed ATR device) used for this experiment, a linear evolution of the IR ratio was observed between 25 °C and 100 °C. The fact that at a given temperature the IR ratio increased with increasing water content and the negative gradient of the dependency of this ratio on temperature is in agreement with the suggestion that additional water to native gluten caused intra-molecular hydrogen bonds within the gluten to be disrupted and to form water–gluten hydrogen bonding, whereas when heated, water–gluten hydrogen bonding would be replaced by gluten–gluten interactions.[8]

Hydrated gluten (water/gluten:2-IR ratio = 3.1) was heated from 25 °C to 45, 65, 85, 105 °C at a 2 °C heating rate with cooling at 10 °C/min to the initial temperature between each targeted temperature.

The ratio $1547/1512 \, \text{cm}^{-1}$ was followed for each heating cycle as shown in Figure 4. After heating at 65 °C and cooling, the value of the IR ratio ($\sim$3.1) was similar to that of untreated gluten. A previous study of gluten with NMR showed that after heating at 80 °C for 30 min and subsequent cooling, structures with proportions of segmental mobility comparable to native gluten were re-formed.[14] After heat treatment at 85 and 105 °C, the IR ratios after cooling were equal to 2.80 and 2.45 respectively. In order to correlate the IR ratio of heat-treated gluten with degree of hydration, the hydration of heat-treated gluten at 105 °C for 30 min was study by FTIR.

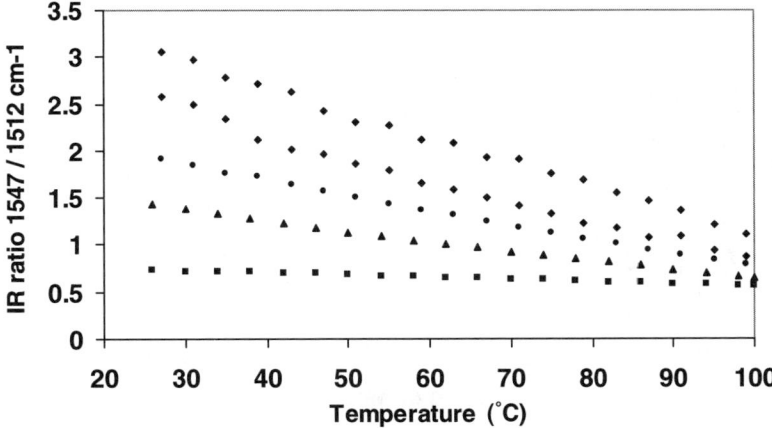

Figure 3 *IR ratio 1547/1512 cm⁻¹ as a function of temperature (heating rate 2°C/min) for gluten with different initial moisture content. From top to bottom: 2, 0.9, 0.3, 0.2, 0.05 g water/g dry commercial gluten*

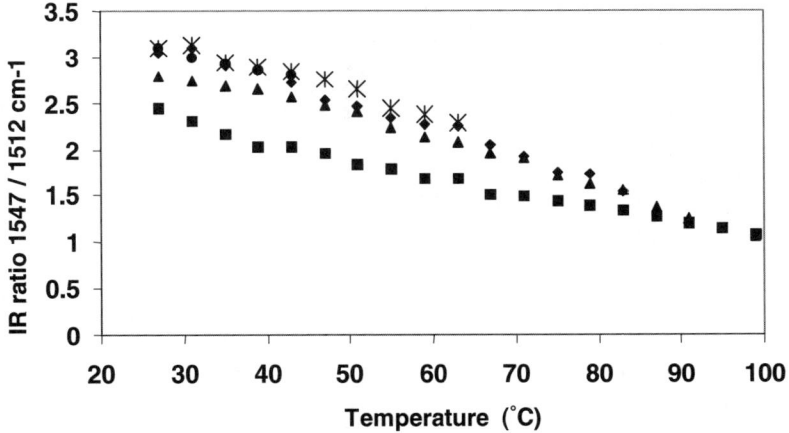

Figure 4. *IR ratio 1547/1512 cm⁻¹ for gluten (g of water/g of dry commercial gluten) during a sequential heating procedure. Heating from 27°C to 45 (●), 65 (✳), 85 (♦), 105 (▲), second run to 105°C(■)*

As shown in Figure 5, heat-treated gluten exhibited IR ratios as a function of moisture content that were close to those of non-treated gluten (*e.g.* at 46% water (w/w), IR ratio = 2.5 and 2.35 respectively before and after treatment). The water associated with heat-treated gluten at 85 and 105°C could be assigned to be 1.3 g and 0.8 g per g of gluten respectively.

3.4 Study of Dough

Similarly to the study by Eliasson, starch/gluten: 10/2 (w/w) were mixed with water on a basis of 0.9 g of water per 1 g of starch.[4]

The loss of order during the gelatinisation of starch which was followed using the IR ratio 1047/1020 cm⁻¹ is shown in Figure 6 for the mixture of

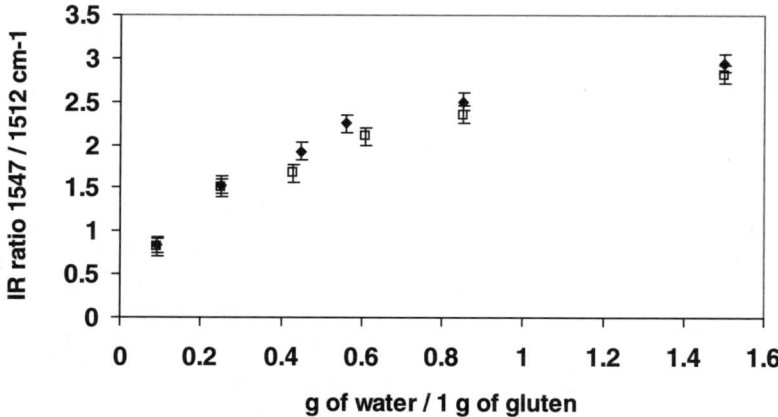

Figure 5 *Ratio of the infrared absorbances at 1547 and 1512 cm^{-1} for commercial gluten as a function of moisture content. Non heat-treated gluten ($\blacklozenge$), gluten heat-treated 30 min at 105°C in presence of 2 g of water/g of dry gluten ($\square$)*

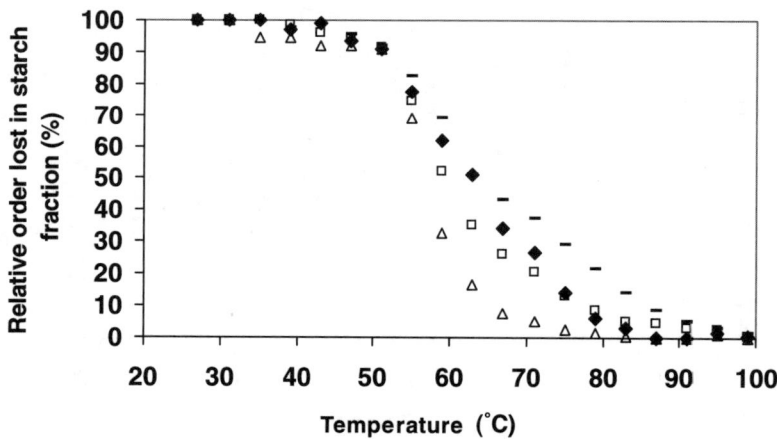

Figure 6 *Loss of order in starch during gelatinisation as detected by FTIR for starch samples (48% water ($\triangle$), 42% water ($\square$), 37% water (−) (w/w)) and starch/gluten/water: 10/2/9 mixture (47% water (w/w)) ($\blacklozenge$)*

starch/gluten/water: 10/2/9 and for starch samples with 48, 42 and 37% water (w/w). As discussed by Eliasson, the water available for starch to gelatinise depended on the quantity of gluten in the mixture.[4] Using the same heat-treatment protocol as used earlier for the gluten sample, the mixture starch/gluten/water: 10/2/9 was heat treated to 45, 67, 87 and 105°C and cooled to 25°C between each temperature step. Starch gelatinisation (IR ratio 1047/1020 cm^{-1}) (Figure 7a) and changes in gluten (IR ratio 1547/1520 cm^{-1}) (Figure 7b) were followed concurrently. Results are summarised in Table 1. At 45°C the starch gelatinisation was not initiated and the gluten IR ratio at 25°C before and after cooling was similar (IR ratio = 3.1). After 67°C, the starch gelatinisation was at an advanced stage (65% of initial order lost) and the gluten

(a)

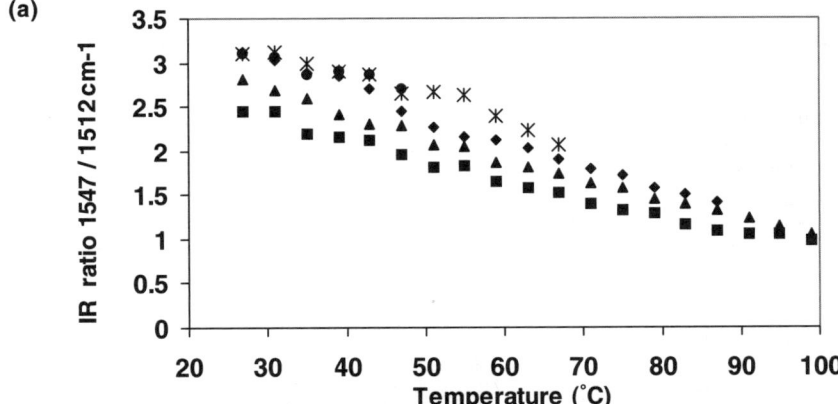

(b)

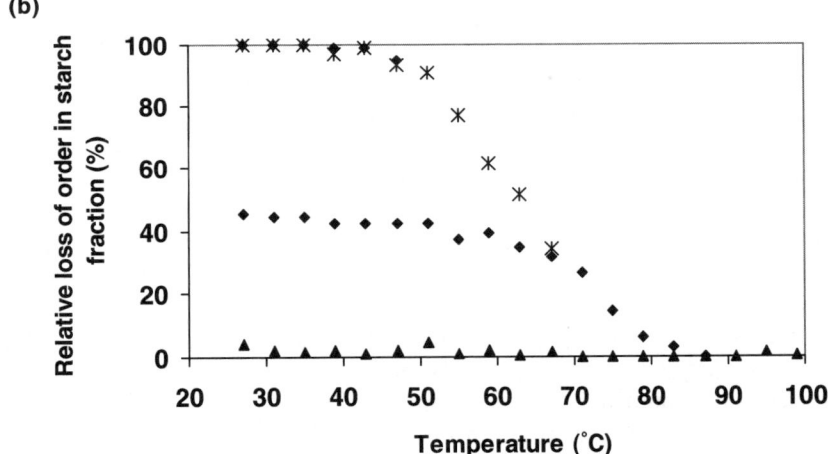

Figure 7 (a) *IR ratio 1547/1512 cm⁻¹ for starch/gluten/water: 10/2/9 during a sequential heating procedure. Heating from 27°C to 47 (●), 67 (✳), 87 (♦), 105 (▲), second run to 105°C (■). Loss of order in starch as detected by FTIR during a sequential heating procedure for starch/gluten/water: 10/2/9 (47% water (w/w)). Heating from 27°C to 47 (●), 67 (✳), 87 (♦), 105 (▲), second run to 105°C similar to the first one*

IR ratio after cooling was still similar to untreated gluten. At 87°C, the starch gelatinisation was completed. The gluten IR ratio was equal to 2.8 after cooling to 25°C which corresponded to 1.3 g of water associated with 1 g of gluten, and 0.65–0.7 g of water was calculated as being associated with 1 g of starch at the end of gelatinisation. As seen in Figure 6, starch gelatinisation behaviour in the mixture starch/gluten is intermediate between those exhibited by starch with 42 and 37% water (w/w). This is in agreement with our water partitioning calculation between starch and gluten as found using the IR gluten ratio. After heat treatment to 105°C and cooling, the IR ratio for gluten (2.45) showed that ~0.8 g of water was still associated with 1 g of gluten.

Table 1   *Sequential heating of mixture starch/gluten/water: 10/2/9*

	25°C	47°C	67°C	87°C	105°C
gluten IR ratio	3.1	3.1	3.1	2.8	2.4
starch IR ratio	0.67	0.65	0.49	0.37	0.35
g water/g gluten	> 1.7	> 1.7	> 1.7	1.3	0.8
% of loss of order in starch	Native starch	< 10	65	95–100	100

4 Conclusion

ATR-FTIR was used to study concurrently the behaviours of starch and gluten during heating. The IR ratio $1547/1512\,\mathrm{cm}^{-1}$ profiles as a function of temperature were similar for gluten, in starch/gluten mixtures, and fully hydrated gluten alone that had undergone the same heat treatment. This suggested that the dehydration of gluten does not depend on the presence of wheat starch undergoing gelatinisation. This finding is in contrast with the generally accepted belief that the gelatinisation of starch during baking does deprive gluten of its water. DSC is not sensitive to structural changes in the gluten when it is heated in the dough state. This is due to the low level of energy involved and to the occurrence of starch gelatinisation at the same range of temperatures.[15] FTIR proved to offer a molecular insight into the behaviour of starch and gluten in cereal-based products.

Results in this study are applicable for constant water levels. In baking processes, evaporation and condensation events occur in addition to changes in starch and gluten and therefore must be considered.[16] This work could be extended to monitor changes in the distribution of water between starch and gluten during bread staling.

Acknowledgement

We would like to thank the financial support of the Ministry of Agriculture, Fisheries and Food (MAFF), UK.

References

1. F. MacRitchie, *Cereal Chem.*, 1976, **53**, 318.
2. R. M. Sandstedt, *Baker's Dig.*, 1961, **35**, 36.
3. W. Bushuk, *Bakers' Dig.*, 1966, **40**, 38.
4. A. C. Eliasson, *J. Cereal Sci.*, 1983, **1**, 199.
5. S. Chevallier and P. Colonna, *Sci. Aliments*, 1999, **19**, 167.
6. I. A. Farhat, S. Orset, P. Moreau and J. M. V. Blanshard, *J. Colloid Interface Sci.*, 1998, **207**, 200.
7. P. L. Weegels, J. A. Verhoek, A. M. G., Degroot and R. J. Hamer, *J. Cereal Sci.*, 1994, **19**, 31.

8. N. Wellner, P. S. Belton and A. S. Tatham, *Biochem. J.*, 1996, **319**, 741.

9. J. K. Kauppinen, D. J. Moffatt, H. H. Mantsch and D. G. Cameron, *Appl. Spectr.*, 1981, **35**, 271.

10. J. J. G. vanSoest, H. Tournois, D. de Wit and J. F. G. Vliegenthart, *Carbohydry. Res.*, 1995, **279**, 201.

11. R. H. Wilson, M. T. Kalichevsky, S. G. Ring and P. S. Belton, *Carbohydr. Res.*, 1987, **166**, 162.

12. B. J. Goodfellow and R. H. Wilson, *Biopolymers*, 1990, **30**, 1183.

13. P. S. Belton, I. J. Colquhoun, A. Grant, N. Wellner, J. M. Field, P. R. Shewry and A. S. Tatham, *Int. J. Bio. Macromol.*, 1995, **17**, 74.

14. S. Ablett, D. J. Barnes, A. P. Davies, S. J. Ingman and D. W. Patient, *J. Cereal Sci.*, 1988, **7**, 11.

15. A. C. Eliasson and P. O. Hegg, *Cereal Chem.*, 1980, **57**, 436.

16. U. De Vries, P. Sluimer and A. H. Bloksma, in *Cereal Sciences and Technology in Sweden*, N. G. Asp (ed.), proceedings of an international symposium, Sweden, 1989, p. 174.

Effect of D$_2$O on the Rheological Behaviour of Wheat Gluten

Jacques Lefebvre, Yves Popineau and Gilbert Deshayes

INRA, CENTRE DE RECHERCHES DE NANTES, RUE DE LA
GERAUDIERE, BP 71627 – 44316 NANTES CEDEX 3, FRANCE

1 Introduction

For historical and semantic reasons (the unfortunate use of the expression
'glutenin polymers' in particular), gluten is often viewed – explicitly or implicitly
– as an entangled polymer system and its elasticity as entropy-driven as far as its
rheological behaviour is considered, whereas it is well known indeed that its
mechanical properties are strongly dependent on factors such as temperature,
addition of urea or replacement of water by D$_2$O in a way which indicates that
gluten network structure would be rather maintained through 'junction zones'
stabilised by noncovalent interactions (hydrogen bonds and/or hydrophobic
interactions between polypeptidic chains), and therefore that a strong enthalpic
contribution would be implied in its rheological properties.

The conformational stability of proteins is higher in D$_2$O than in H$_2$O. The
effect of D$_2$O has been often ascribed to a strengthening of hydrophobic interactions,[1,2] however other studies found that it reflects more likely the larger
strength of deuterium bonds and deuterium bridges.[3] The stabilising effect of
D$_2$O on intermolecular 'hydrogen' bonding of prolamins was invoked to explain
the increased strength of gluten and dough[4,5] and the enhancement of gluten
dynamic moduli[6] observed upon the substitution of D$_2$O to water at room
temperature.

In the present work the effect of temperature on the rheological behaviour of
wheat gluten in D$_2$O is compared to that in water. The viscoelastic response was
studied in shear by combining dynamical measurements and creep and recovery
tests, in order to encompass a large timescale.

2 Material and Methods

Freeze-dried gluten was prepared as previously described[7] from an Olympic × Gabo cross line, which is Glu-A1/Glu-D1 null and contains only 17 and 18 high molecular weight glutenin subunits. The gluten from this line has been extensively studied in previous work.[7]

Samples for rheological characterisation were obtained by resolution with water or D_2O containing 0.1 M N-ethylmaleimide (NEMI). In the presence of this SH-blocking agent, ageing of gluten is very slow,[8] and temperature-induced irreversible structural changes are quite limited[7] at least up to 50°C, thus allowing the study of its viscoelastic behaviour.

Rheological measurements were performed in shear using a stress controlled rheometer (Carri-Med CSL 100) operating in cone-plate geometry. Each sample is submitted successively to a first frequency sweep in range 10^{-3}–40 Hz under 3% strain, to a creep and recovery test, and finally to a second frequency sweep identical to the first one. The dynamical strain amplitude (3%) and the value of the creep stress (chosen so as to keep the maximum strain below 10%) were set in order to remain within the linear viscoelasticity domain. Creep and creep recovery were recorded during 20 h and 80 h, respectively, times which allowed the steady state to be reached in all cases. A fresh sample was used for each solvent/temperature combination.

3 Results and Discussion

Figure 1 shows examples of the mechanical spectra of gluten plotted in terms of the storage (J′) and loss (J″) compliances *vs* the angular frequency. Regardless of the solvent and temperature, all the spectra are qualitatively similar. We showed previously[7,8] that the frequency window encompasses the maximum of the loss peak which marks the onset of the transition zone at the high frequency limit of the viscoelastic plateau, and that the corresponding compliance data can be fitted with Cole–Cole functions (see Figure 1). The fit gives the plateau compliance J_N°, which is the reciprocal of the plateau modulus G_N°, the characteristic frequency ω_o of the fast (short-range) structural dissipative mechanisms underlying the peak and the spread parameter n. An overall estimate of the contribution of the slower loss mechanisms to the total viscoelastic response is given by the difference $J_e^\circ - J_N^\circ$, where J_e° is the steady state compliance which measures the total elastic deformation during steady flow. J_e°, as well as the steady state viscosity η, were obtained in the classical straightforward way from the recovery data. Examples of creep and recovery curves are given in Figure 2, showing that the viscoelastic response of gluten extends over an extremely large timescale.

Substitution of D_2O to water enhances gluten viscoelasticity. This overall effect, exemplified in Figures 1 and 2, is observed at all temperatures within the range studied. Analysis of the results indeed shows that J_N° (Figure 3), J_e° (Figure 4) and $J_e^\circ - J_N^\circ$ (Figure 4) remained higher in water than in deuterium oxide, and η lower (Figure 5). But the substitution leaves unchanged the characteristic times of the fast and slow retardation processes ($\tau_o = 1/\omega_o$ and $\tau_m = \eta J_e^\circ$, respectively), as shown in Figure 6.

One observes the same trends in the effect of temperature on gluten viscoelas-

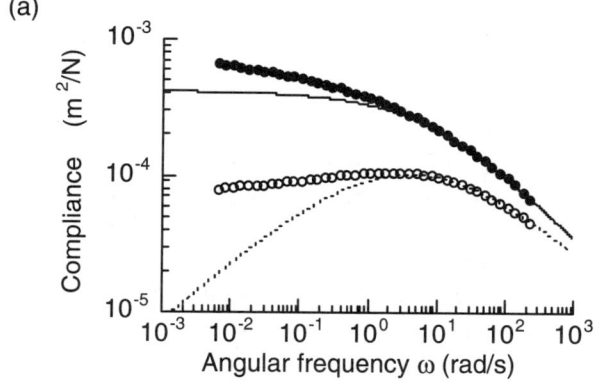

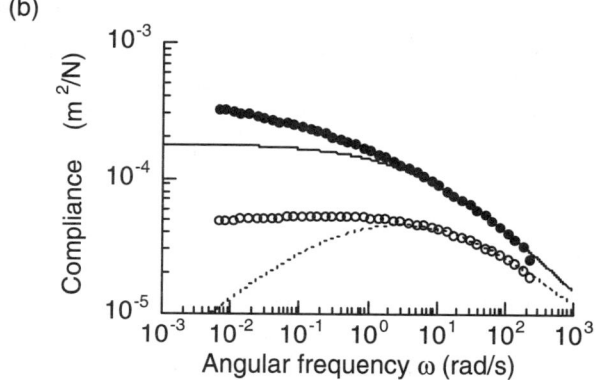

Figure 1 *Examples of mechanical spectra of gluten (Olympic × Gabo cross line -/17 + 18/-) in water (a) and in deuterium dioxide (b). Temperature: 35°C; strain amplitude: 3%. Filled symbols: storage compliance J'(ω); empty symbols: J"(ω). The lines represent the fits of the Cole–Cole functions to the high frequency data:*

$$J'(\omega) = J_N^{\circ} \left[(\omega_o/\omega)^n + \cos(\pi n/2)\right] \left[(\omega_o/\omega)^n + 2\cos(\pi n/2) + (\omega/\omega_o)^n\right]^{-1}$$

$$J''(\omega) = J_N^{\circ} \sin(\pi n/2)] \left[(\omega_o/\omega)^n + 2\cos(\pi n/2) + (\omega/\omega_o)^n\right]^{-1}$$

The procedure followed to fit the equations has been described elsewhere.[7] The symbols are explained in the text

ticity in both solvents. The plateau modulus $G_N^{\circ} = 1/J_N^{\circ}$ first decreases with temperature and then tends to increase; the blunt minimum, more conspicuous in D_2O than in H_2O, is located around 35°C in both solvents (Figure 3). An opposite variation has been observed for many thermo-reversible polymeric gels in water, with a maximum in G_N° in the 20 – 40°C region, and was interpreted as the result of the balance between an entropic contribution to the elasticity of the network, which decreases as temperature increases, and of an energetic contribution, which increases.[9,10] In the case of gluten the shape of the variations of G_N° results more probably from the balance between the respective contributions of hydrogen bonds and hydrophobic bonds, which are weakened and strengthened, respectively, as temperature increases in the range considered. As to the total recoverable compliance J_e° and to the contribution of slow retardation mechanisms ($J_e^{\circ} - J_N^{\circ}$), they vary also in a similar way in both solvents (Figure 4); they

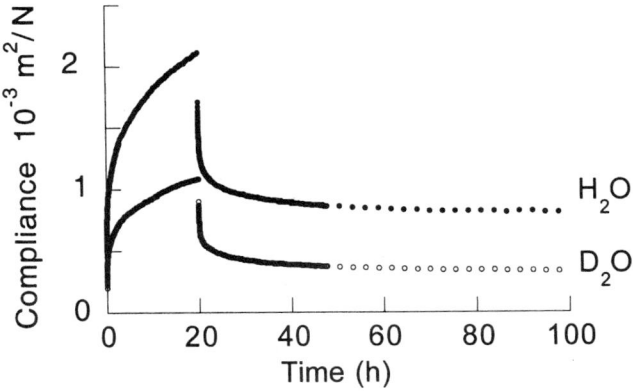

Figure 2 *Examples of creep and creep recovery curves of gluten (Olympic × Gabo cross line -/17 + 18/-) in water and in deuterium dioxide. Temperature: 35°C; creep stress was 40 Pa for the sample in water and 75 Pa for the sample in D₂O*

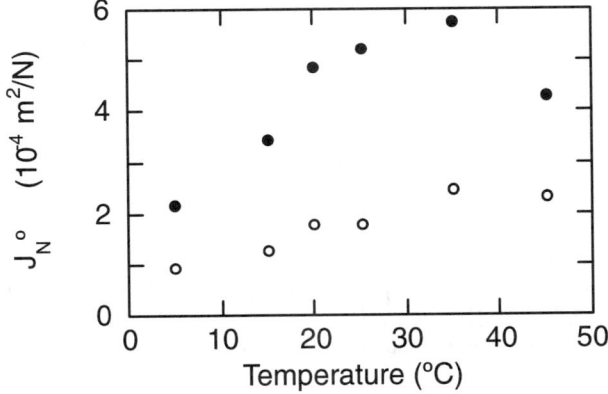

Figure 3 *Variation of the plateau compliance J_N° of gluten (Olympic × Gabo cross line -/17 + 18/-) with temperature in water (filled symbols) and in deuterium dioxide (empty symbols). J_N° was obtained from the fit of the Cole–Cole functions to the dynamic measurements data*

show a maximum, too, but located at 20°C, *i.e.* at a lower temperature than J_N°. The difference in response to temperature between slow and fast retardation processes appears also when considering their respective characteristic times (Figure 6): whereas τ_o decreases as temperature increases and can be considered to follow an Arrhenius law within the approximations involved in the treatment of the dynamic data, τ_m seems independent of temperature. The variations of the steady state viscosity with temperature are hardly significant and η can be considered as practically constant (Figure 5).

The parallel behaviours of the material in water and in deuterium dioxide suggests that replacement of H_2O with D_2O does not change its structure or the retardation mechanisms responsible for its rheological behaviour. Deuterium

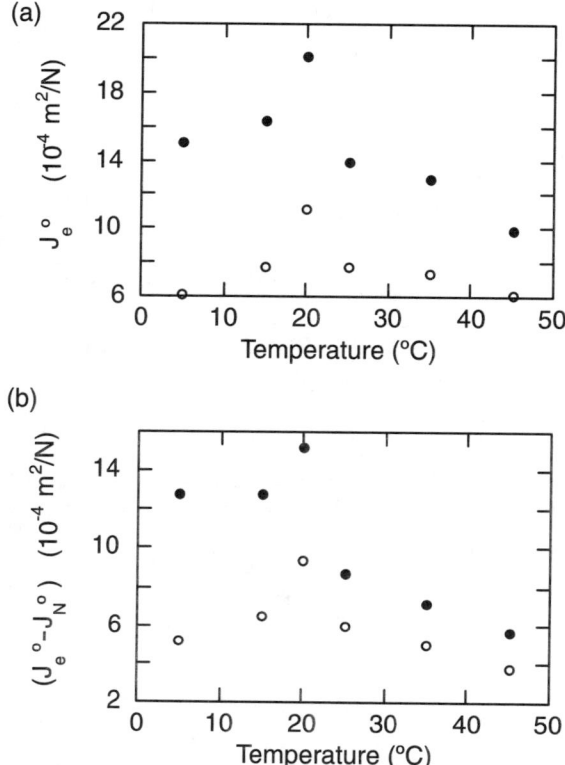

Figure 4 *Variations with temperature of the steady state recoverable compliance J_e° (a), and of the contribution $(J_e^\circ - J_N^\circ)$ of slow retardation mechanisms to J_e° (b). Gluten from Olympic × Gabo cross line -/17 + 18/-) in water (filled symbols) and in deuterium dioxide (empty symbols). J_e° was obtained from the recovery curves*

dioxide increases the cohesion of gluten network as compared to water, probably because deuterium bonds are stronger than hydrogen bonds, without affecting significantly the timescales of the retardation processes. In a somewhat similar way, it was found recently that replacement of H_2O with D_2O does not affect the mechanism of heat-induced aggregation of β-lactoglobulin.[11]

4 Conclusion

Gluten cannot be considered as an entangled system. Hydrogen bonds are largely contributing to the connectivity of the network; but hydrophobic interactions are probably also involved as pointed out by the strengthening observed above ~ 35°C. The material is rather to be viewed as belonging to the colloidal gel type. Besides, it is not thermorheologically simple. A thorough characterisation and analysis of its viscoelastic behaviour is therefore by no means an easy task.

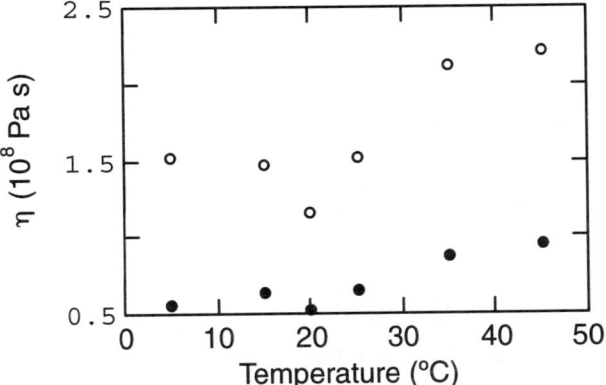

Figure 5 *Variations with temperature of the steady state viscosity η of gluten from Olympic × Gabo cross line -/17 + 18/-) in water (filled symbols) and in deuterium dioxide (empty symbols). η was obtained from the recovery curves*

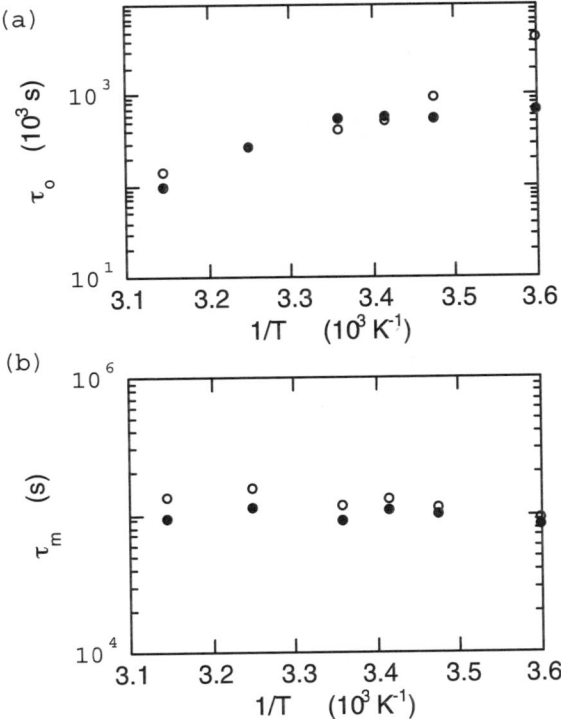

Figure 6 *Variations with temperature of the characteristic time $\tau_o = 1/\omega_o$ of the fast retardation mechanisms (a), and of the characteristic time $\tau_m = J_e^{\circ} \eta$ of the slow retardation mechanisms (b). Gluten from Olympic × Gabo cross line -/17 + 18/-) in water (filled symbols) and in deuterium dioxide (empty symbols)*

References

1. G. C. Kresheck, H. Schneider and H. A. Scheraga, *J. Phys. Chem.*, 1965, **69**, 3132.
2. J. J. Lee and D. S. Berns, *Biochem. J.*, 1968, **110**, 465.
3. R. F. Henderson and T. R. Henderson, *J. Biol. Chem.*, 1970, **245**, 3733.
4. V. L. Kretovich and A. B. Vakar, *Doklady Akad. Nauk SSR*, 1964, **155**, 465.
5. R. Tkachuk and I. Hlynka, *Cereal. Chem.*, 1968, **45**, 80.
6. A. E. Inda and C. Rha, *J. Texture Studies*, 1991, **22**, 393.
7. J. Lefebvre, Y. Popineau, G. Deshayes and L. Lavenant, *Cereal Chem.*, 2000, **77**, 193.
8. J. Lefebvre, Y. Popineau and M. Cornec, in *Gluten Proteins*, Association of Cereal Research, Detmold, 1994, pp. 180–189.
9. K. Nishinari, M. Watase and K. Ogino, *Makromol. Chem.*, 1984, **185**, 2663.
10. K. Nishinari, S. Koide and K. Ogino, *J. Physique*, 1985, **46**, 793.
11. M. Verheul, S. P. F. M. Roefs and K. G. de Kruif, *FEBS Letters*, 1998, **421**, 273.

The Influence of the Thickness on the Functional Properties of Cassava Starch Edible Films

Nívea Maria Vicentini,[1] Paulo José do Amaral Sobral[2] and Marney Pascoli Cereda[1]

[1]RESEARCH CENTER OF TROPICAL ROOTS AND STARCHES, UNESP, PO BOX 237, ZIP CODE: 18603-970, BOTUCATU, SP, BRAZIL
[2]FACULDADE DE ZOOTECNIA E ENGENHARIA DE ALIMENTOS, USP, PO BOX 23, ZIP CODE: 13630-000, PIRASSUNUNGA, SP, BRAZIL

Abstract

The properties of edible films are influenced by several factors, including thickness. The purpose of this paper was to study the influence of thickness on the viscoelasticity properties, water vapor permeability, color and opacity of cassava starch edible films. These films were prepared by a casting technique, the film-forming solutions were 1, 2, 3 and 4% (w/v) of starch, heated to 70°C. Different thicknesses were obtained by putting 15 to 70 g of each solution on plexiglass plates. After drying at 30°C and ambient relative humidity, these samples were placed for 6 days at RH of 75%, at 22°C. The sample thicknesses were determined by a digital micrometer ($\pm$ 0.001 mm), as the average of nine different points. The viscoelasticity properties were determined by stress relaxation tests with a texture analyser TA.XT2i (SMS), being applied the Burgers model of four parameters. The water vapor permeability was determined with a gravimetric method, and color and opacity were determined using a Miniscan XE colorimeter, operated according to the Hunterlab method. All the tests were carried out in duplicate at 22°C. Practically, the four visco-elasticity properties calculated by the Burgers model had the same behavior, increasing with the thickness of all films, according to a power law model. The water vapor permeability and the color difference increased linearly with the thickness (0.013–0.144 mm) of all films prepared with solution of 1 to 4% of starch. On the other hand, the effect of the variation of the thickness over the opacity, was more important in the films with 1 and 2% of starch. It can be concluded that the control of the thickness in the elaboration of starch films by the casting technique is of extreme importance.

1 Introduction

Edible films and coatings are thin materials made from biological macromolecules (biopolymers).[1] The main biopolymers used in preparing biofilms are polysaccharides[2] and proteins.[3,4] Among the most studied polysaccharides are pectin, cellulose and derivatives, alginates, carrageenan, chitosan and starch.[1,5]

Starch has been tested as coating for fruits and vegetables.[6] Vicentini *et al.*[7,8] and Oliveira and Cereda[9] studied the effect of cassava starch coating on postharvest conservation of green peppers and guavas, respectively, with positive effect only for guava. Garcia et al.[10] applied corn and potato starch coating to strawberries, finding a considerable increase in preservation time. Also, a considerable number of studies have been performed on starch film development and characterization, including structure–property relationships.[11–16]

The use of edible films in food products packaging will depend on their functional properties, viscoelasticity, optical properties (color and opacity) and water vapor permeability, that depend on the structural cohesion of the polymer, considering the effect of the formulation on the structure of the macromolecules.[17,18] A macroscopic parameter, sometimes ignored, that could influence these films' properties is their thickness.[17,19]

The viscoelasticity properties are also important, because they can supply information directly related to the form of the macromolecules. The models of the linear viscoelasticity are developed from two elements: a spring and a dashpot. Two of those elements in line constitute the Maxwell model and in parallel the Kelvin model (or Vogt).[20] Normally, those models don't represent the behavior of complex materials satisfactorily. Other models such as the Burgers model, where the Maxwell and Kelvin models are connected in line, are used to determine the modulus of elasticity (Y_1 and Y_2) and the coefficients of viscosity (η_1 and η_2).[21]

According to the Fick and Henry's laws the water vapor permeability (WVP) should not be function of thickness.[18,22] However, WVP of hydrophilic films can be function of the thickness.[19] Tomasula *et al.*[23] studied films made from CO_2-precipitated casein (CO_2-casein), with thickness varying from 0.110 to 0.150 mm, and observed that the WVP increased significantly from 2.22 to 3.80 g mm/kPa h m². The influence of the thickness on the WVP was also observed by Cuq *et al.*[19] and Sobral[24,25] that worked with films based on myofibrillar proteins from Atlantic sardines and gelatin and myofibrillar proteins from Tilapia, respectively.

Transparency (low opacity) and color are important in situations where the product should be seen, but these properties has been little studied.[19,26,27]

Starches are the most abundant polysaccharides found in nature. Due to the scarcity of published studies regarding the use of starches from tropical plants in edible films technology, this paper was aimed at studying the effect of thickness on the viscoelasticity properties, water vapor permeability, color and opacity of a film based on cassava starch, for different concentrations of film-forming solutions.

2 Materials and Methods

2.1 Cassava Starch

Commercial cassava starch (*Manihot esculenta* C.) donated by a São Paulo industry, Flor de Lotus Co., was used in this study. The composition was determined, in duplicate.[28] The sample presented density of 1.47 g/cm^3, 14.86% humidity, and 93.8% starch, of which 16.02% was amylose and 0.20% of soluble total sugars. Starch contaminants were 14% ash; 0.39% fiber; 0.12% total nitrogen; 0.11% lipids, totaling 0.76%. The starch was used without prior preparation.

2.2 Edible Film Preparation

The films were prepared by casting. The film-forming solutions were prepared with 10 (1%), 20 (2%), 30 (3%) or 40 (4%) g of starch in 1 L of distilled water heated to 70°C, in water bath with digital control ($\pm$ 0.5°C) of temperature (Tecnal, TE184) under constant stirring. These solutions remained at this temperature for 1 min for sufficient gelatinization.

The filmogenic solutions (15 to 70 g) were applied on plexiglass plates of 139.2 cm^2, previously prepared, to obtain film samples of different thicknesses. The weights were controlled on a semi-analytical scale (Marte, AS2000, $\pm$ 0.01 g). The samples were dehydrated in an oven with air circulation and renewal (Marconi, MA037), with PI control ($\pm$ 0.5°C) of temperature, at 30°C, and relative humidity of 55–65%, for 12 to 24 h. The films obtained were conditioned at 22°C and 75% RH, in desiccators containing a saturated solution of NaCl, for 6 days prior to testing. All the characterizations were carried out in a controlled environment (T = 22°C and RH = 55–65%).

Before testing, the thickness of the films was measured by a digital micrometer ($\pm$ 0.001 mm) with sensor lens of 6.4 mm diameter, at nine different points on each sample, and taking the mean of the nine measurements as the thickness.

2.3 Viscoelastic Properties

The viscoelastic properties of films were determined by stress relaxation tests with a texture analyzer TA.XT2i (SMS). The films were cut into strips of 15 mm width and 100 mm lengths and affixed to the instrument. The initial grip separation and crosshead speed were 80 mm and 0.9 mm/sec, respectively. The instrument was set for a deformation of 1%, which was held constant for 70 sec. The force required to maintain this deformation was monitored by a microcomputer in real time. The viscoelastic properties were calculated according to the Burgers model (Equation 1):

$$\sigma(t) = \frac{\varepsilon_0}{A}[(q_1 - q_2 r_1)e^{-r_1 t} - (q_1 - q_2 r_2)e^{-r_2 t}] \tag{1}$$

where q_i and r_i are parameters calculated by non-linear regression. Thus the modulus of elasticity (Y_1, Y_2) and the coefficients of viscosity (η_1, η_2) were calculated with these parameters.[21]

2.4 Optical Properties

Color values of the films were measured with a Miniscan XE colorimeter (Hunterlab). The films were placed on a white standard plate and the HunterLab color scale was used to measure color: $L = 0$ (black) to $L = 100$ (white); $-a$ (greenness) to $+a$ (redness); $-b$ (blueness) to $+b$ (yellowness). Total color difference (ΔE) was calculated by Equation 2:[29]

$$\Delta E = \sqrt{(L - L_s)^2 + (a - a_s)^2 + (b - b)_s^2} \tag{2}$$

The opacity of the film was determined according to the Hunterlab method,[24,25] using the same colorimeter. The opacity (Y) of the sample was calculated as the relation between the opacity of the sample placed over the black standard (Y_b) and that over the white standard (Y_w). The calculation of opacity was made automatically in the computer, using the program Universal Software 3.2 (Hunterlab). The opacity scale measured in this manner (from 0 to 100%) presents an arbitrary scale.

2.5 Water Vapor Permeability

Water vapor permeability (WVP) of films was determined gravimetrically at $25\,^\circ\text{C}$.[27] The test film was sealed to a glass cup containing silicagel (0% RH; 0 Pa water vapor pressure) and the cup was placed in a desiccator maintained at 100% RH with distilled water. The water vapor transferred through the film and absorbed by the desiccant was determined from the weight gain of the silicagel. The cups were weighed at 24 hr interval for 8 days. All tests were conducted in duplicate. The water vapor permeability of the film was calculated by Equation 3:[27]

$$WVP = \frac{w}{tA} \frac{x}{\Delta P} \tag{3}$$

where x is the film thickness, A is the area of exposed film (12.29 cm^2), ΔP is the vapor pressure differential across the film (3,167.46 Pa at $25\,^\circ\text{C}$), and w/t was calculated by linear regression between the weight gain of the cup and the time.

2.6 Statistical Analysis

All linear and non-linear regressions were done with the software Estat[30] and the software Excel 2000, respectively.

3 Results and Discussion

The films made from cassava starch, in their various concentrations, were transparent and homogenous; however, the thinnest (x < 0.020) or thickest (x > 0.100 mm) were more difficult to handle. Thickness of the films in this test varied from 0.013 to 0.144 mm in all cases.

3.1 Viscoelastic Properties

The coefficients of viscosity (η_i) and the modulus of elasticity (Y_i) for the Maxwell (i = 1) and Kelvin units (i = 2) according to the Burgers Model, had the same behavior (Figures 1–4), increasing with thickness of all the films, according to a power law model, with important variation in the thin films, excepting the Y_1 that present a low value of regression coefficient ($R^2 = 0.33$).

The coefficients of viscosity and the modulus of elasticity varied respectively, from 2885.65 to 9971.86 MPa s ($R^2 = 0.88$) and from 16.83 to 33.89 MPa in the first group of elements (Figures 1 and 2), from 138.18 to 388.64 MPa s ($R^2 = 0.77$) and from 35.04 to 100.86 MPa ($R^2 = 0.72$), in the second group of elements (Figures 3 and 4).

Cuq *et al.*[20] did not find influence of thickness on viscoelasticity properties of films made of myofibrillar proteins from North Atlantic sardines (0.010–0.060 mm).

3.2 Optical Properties

Figure 5 shows the results of color measurements, represented by color difference (ΔE), or the loss of color of the standard plate. The color difference increased linearly ($R^2 = 0.93$) from 1.87 to 4.33 with the thickness and the parameters of

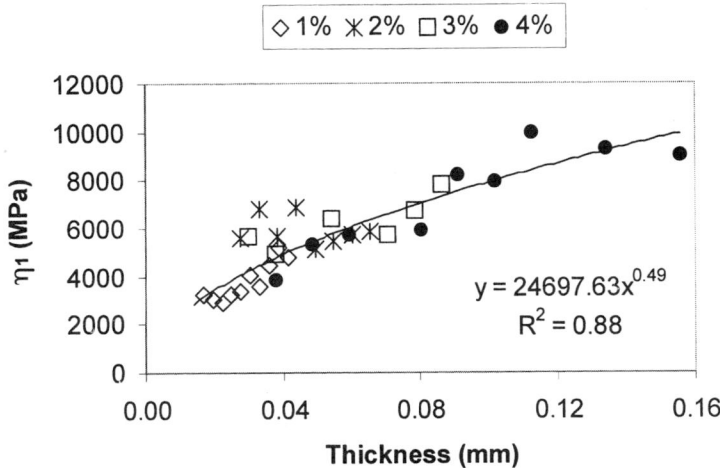

Figure 1 *Coefficient of viscosity (η_1) as a function of thickness of cassava starch films at different concentrations*

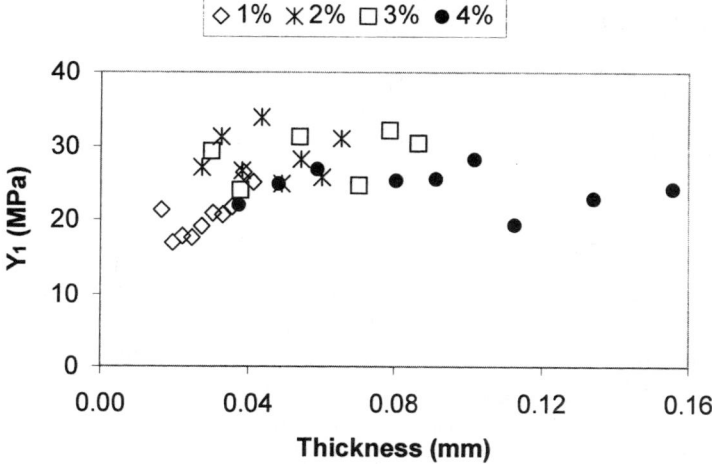

Figure 2 *Modulus of elasticity (Y_1) as a function of thickness of cassava starch films at different concentrations*

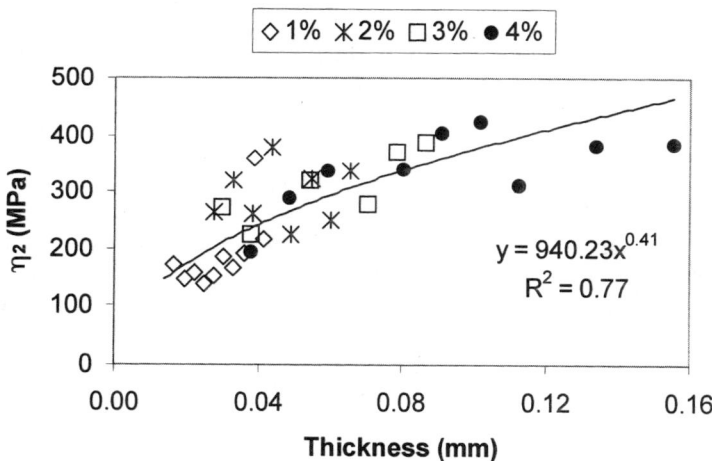

Figure 3 *Coefficient of viscosity (η_2) as a function of thickness of cassava starch films at different concentrations*

the linear equation were significant ($P < 0.05$). Sobral[24] studied the effect of thickness on color difference on films of bovine hide gelatin (0.110–0.163 mm), and observed that the color difference also increased linearly from 1.5 to 6.5 with the thickness. Similar behavior can be observed in the work of Sobral.[25]

Opacity of the films was influenced by the thickness as can be observed in Figure 6. However, this effect was more important in the films with 1 and 2% starch, which increased linearly from 1.77 to 3.67% ($R^2 = 0.88$), than in the films of 3 and 4% that increased from 1.11 to 2.06% ($R^2 = 0.84$) with slope of 27.41 and 10.17, respectively. The parameters of the equations were significant

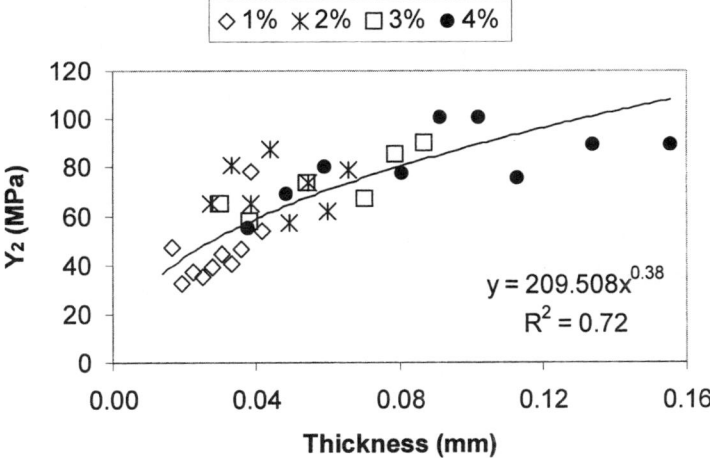

Figure 4 *Modulus of elasticity* (Y_2) *as a function of thickness of cassava starch films at different concentrations*

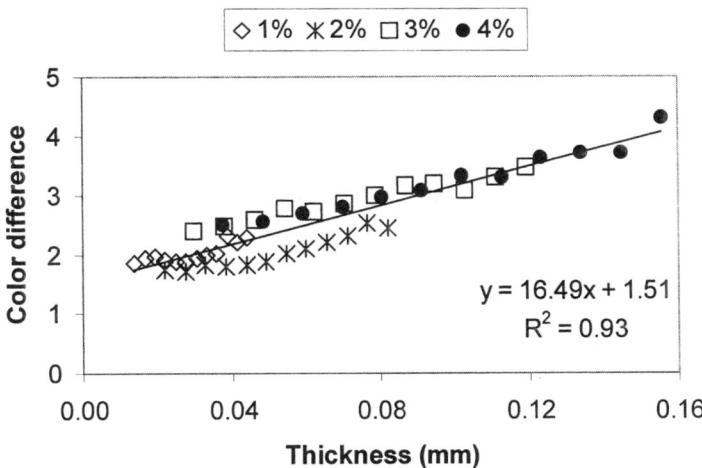

Figure 5 *Color difference as a function of thickness of cassava starch films at different concentrations*

($P < 0.05$). The cassava starch films are less transparent than the gelatin films, where the opacity tended to zero and did not vary with the films thickness,[24] but more transparent than the myofibrillar protein from Tilapia based films where opacity varied from 2 to 8 with the film thickness (0.017–0.088 mm).[25]

3.3 Water Vapor Permeability

Influence of the film thickness on the WVP can be observed in the Figure 7. The WVP increased linearly ($R^2 = 0.87$) with the thickness (0.013–0.140 mm) from 0.14 to 2.04 g mm/m² h Pa, and the parameters of the equation were significant

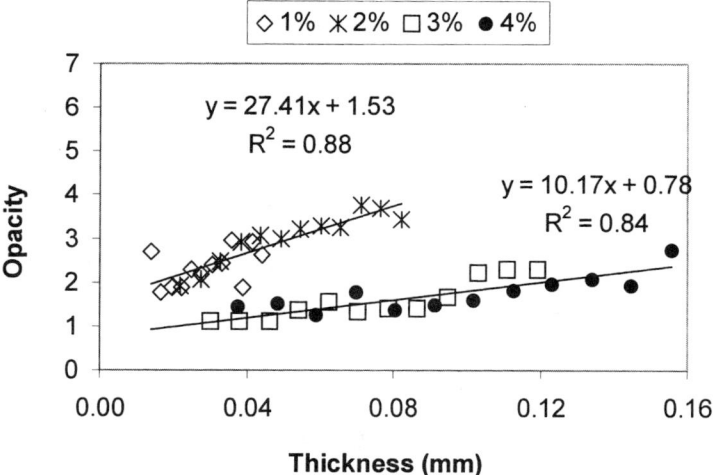

Figure 6 *Opacity as a function of thickness of cassava starch films at different concentrations*

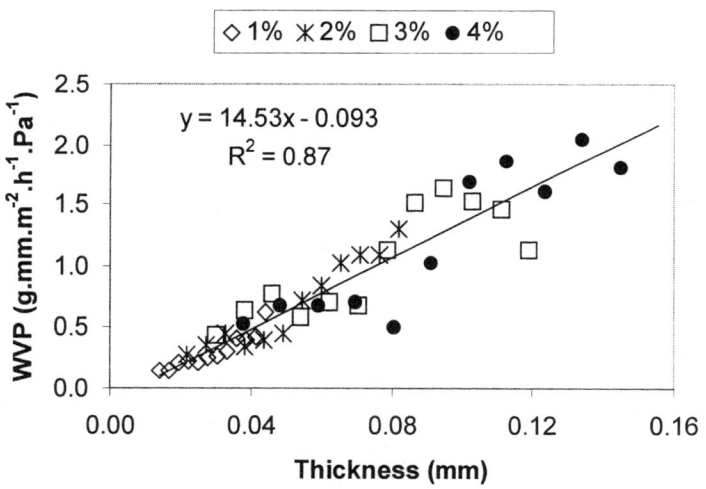

Figure 7 *Water vapor permeability as a function of thickness of cassava starch films at different concentrations*

($P < 0.05$). These results agrees with those obtained by Sobral.[24,25]

McHugh *et al.*[22] worked with sodium caseinate films and observed that WVP increase from 0.25 to 1.50 g mm/kPa h m^2 with film thickness varying from 0.01 to 0.10 mm. However, Park *et al.*[18] working with methylcellulose and hydroxy-propyl cellulose films observed that WVP was relatively constant for thickness in the range 0.023 to 0.140 mm.

4 Conclusions

The control of the thickness in the starch films elaboration for the casting technical is of extreme importance. The WVP, the color and the opacity are influenced strongly by the thickness. The variations of functional properties with thickness may be explained by sample heterogeneity which increase with thickness.

Acknowledgments

The authors would like to acknowledgments FAPESP (Proc. 98/16179-8) for the financial support.

References

1. J. J. Kester and O. Fennema, *Food Technol.*, 1986, **40**(12), 47.
2. M. O. Nisperos-Carriedo, 'Edible Coatings and Films Based on Polysaccharides', in *Edible Coatings and Films to Improve Food Quality*, J. M. Krochta, E. A. Baldwin and M. O. Nisperos-Carriedo, (eds.), Technomic Publishing Company Inc., Basel, 1994, pp. 305–336.
3. A. Gennadios, T. H. McHugh and C. L. Weller, in 'Edible Coatings and Films Based on Proteins', in *Edible Coatings and Films to Improve Food Quality*, J. M. Krochta, E. A. Baldwin and M. O. Nisperos-Carriedo, (eds.), Technomic Publishing Company Inc., Basel, 1994, pp. 210–278.
4. J. A. Torres, 'Edible Films and Coatings from Proteins', in *Protein Functionality in Food Systems*, N. S. Hettiarachchy and G. R. V. Ziegler (eds.), Marcel Dekker, Inc., New York, 1994, p. 467–507.
5. F. Debeaufort, J. A. Quezada-Gallo and A. Voilley, *Crit. Rev. Food Sci.*, 1998, **38**(4), 299.
6. N. M. Vicentini and M. P. Cereda, *Braz. J. Food Technol.*, 1999, **2**(1,2), 87.
7. N. M. Vicentini, T. M. R. Castro and M. P. Cereda, *Ciênc. Tecnol. Aliment.*, 1999, **19**(1), 127.
8. N. M. Vicentini, M. P. Cereda and F. L. d. A. Câmara, *Sci. Agric.*, 1999, **56**(3), 713.
9. M. A. Oliveira and M. P. Cereda, *Braz. J. Food Technol.*, 1999, **2**(1,2), 97.
10. M. A. García, M. N. Martino and N. E. Zaritzky, *J. Agric. Food Chem.*, 1998, **46**(9), 3758.
11. M. Stading, Å. Rindlav-Westling and P. Gatenholm, *Carbohydr. Polym.*, 2001, **45**, 209.
12. M. P. Cereda, C. M. Henrique, M. A. Oliveira, M. V. Ferraz and N. M.Vicentini, *Braz. J. Food Technol.*, 2000, **3**, 91.
13. I. Arvanitoyannis, A. Nakayama and S. Aiba, *Carbohydr. Polym.*, 1998, **36**, 105.
14. D. Lourdin, L. Coignard, H. Bizot and P. Colonna, *Polymer*, 1997, **38**(21), 5401.
15. H. G. Bader and D. Göritz, *Starch/Stärke*, 1994, **46**(11), 435.
16. W. B. Roth and C. L. Mehltretter, *Food Technol.*, 1967, **21**(1), 72.
17. R. Mahmoud and P. A. Savello, *J. Dairy Sci.*, 1992, **75**(4), 942.
18. H. J. Park, C. L. Weller, P. J. Vergano and R. F. Testin, *J. Food Sci.*, 1993, **58**(6), 1361.
19. B. Cuq, N. Gontard, J. L. Cuq and S. Guilbert, *J. Food Sci.*, 1996, **61**, 580.
20. B. Cuq, N. Gontard, J. L. Cuq and S. Guilbert, *J. Agric. Food Chem.*, 1996, **44**(4), 1116.

21. P. K. Chandra and P. J. A. Sobral, *Ciênc. Tecnol. Aliment.*, 2000, **20**(2), 250.
22. T. H. McHugh, R. J. Avena-Bustillos and J. M. Krochta, *J. Food Sci.*, 1993, **58**(4), 899.
23. P. M. Tomasula, N. Parris, W. Yee and D. Coffin, *J. Agric. Food Chem.*, 1998, **46**(11), 4470.
24. P. J. A. Sobral, *Ciência & Engenharia*, 1999, **8**(1), 60.
25. P. J. A. Sobral, *Peq. Agrop. Bras.*, 2000, **32**(6), 1251.
26. N. Gontard, C. Duchez, J. L. Cuq and S. Guilbert, *Int. J. Food Sci. Technol.*, 1994, **29**, 39.
27. N. Gontard, S. Guilbert and J. L. Cuq, *J. Food Sci.*, 1992, **57**(1) 190.
28. Association of Official Analytical Chemists, *Official Methods of Analysis of AOAC*, 15th ed., 1995.
29. A. Gennadios, C. L. Weller, M. A. Hanna and G. W. Froning, *J. Food Sci.*, 1996, **61**(3), 585.
30. D. A. Banzatto and S. N. Kronka, *Experimentação agrícola.* Jaboticabal: FUNEP/UNESP, 1989, p. 247.

Subject Index

Acacia gum
  complex coacervation of, 111–8
Acetobacter xylinus, 41
Acylation, *see* protein modification
Adsorption, 145–52
Adsorption
  kinetics of, 153–65
Aggregation, 91, 203
Air/liquid interface, 145–52
Air/water interface, 127–44, 173–8
Allergies, 16, 21
Amino acid, 63–72
α-Amylase, 226, 228, 230, 233
Amylopectin, 226, 231, 235–40
Amylose, 226, 231, 235–40
Analogs, 67
Arabic gum, 120
Arabinoxylans, 48
Arabinoxylan–protein linkages, 49
Associative phase separation, 111
Auxotrophic, 65

Baking, 282
Base wines
  aggregates of, 222–3
  composition of, 213
  films of, 217–8
  surface pressure of, 221
  surface tension of, 217
Beads, 123
Binding, 73
Biotechnology, 19

Block-copolymer model, 146–8
Bread, 226–34
Bulk rheology, 243

Caffeic acid, 48
Calcium binding, 59
Calcium speciation, 101
κ-Carraggeenan, 190–200
β-Casein, 145–52, 156, 158
Cassava, 291
Cell wall
  of tomato, 99
Cellulose, 39
Cellulose fibres, 254
CHAPS, 6
Clostridium, 49
Coacervated particles
  coalescence of, 114
Coacervation, 111, 120
Coarsening
  diffusion induced, 115
  hydrodynamics induced, 115
Coil–helix transition, 205
Complex(es)
  β-lactoglobulin–acacia gum, 112
Complex coacervation, *see* coacervation
  kinetics of, 114
Composites, 253–9
Confocal scanning laser microscopy, 112
Coniferyl alcohol, 173
Controlled release, 119–24
Correlation length, *see* light scattering

Creaming, 184–6
Creep, 285
Creep recovery, *see* creep
Critical aggregation concentration, 128
Crystal structure, 270
CSLM, *see* confocal scanning laser
   microscopy

De-esterification, 55
Dehydrogenation polymer, 173
Demethylation, 49
Depolymerised citrus pectin, 181
Deuterium oxide
   effect on gluten rheology, 284–90
DHFR, *see* dihydrofolate reductase
Diastereomer, 65
Differential scanning calorimetry (DSC),
   205
Dihydrofolate reductase, 65
Dilational surface rheology, 157
Disjoining pressure, 136
Dispersion, 93
Dough, 275, 279–82
Droplet coalescence, 114
Droplet size, 183, 185–6
Dynamic drop tensiometer
   adsorption kinetics measurements by,
      168–9
   for sinusoidal dilatation/compression,
      168
   deformation, 168
interfacial rheology by, 169
Dynamic measurements, 285
Dynamic moduli, 245–6
Dynamic scaling, 117
Dynamic surface tension, 169

E. coli, 63
Edible films, 291–300
Editing, 69
Electrostatic interactions
   of protein–polysaccharide, 111
Ellipsometry, 133–4, 245
Encapsulation, 119–24
Enterobacter, 49
Extensin, 104

Family 13 amylase, 79–87
Fatty acids, 74
Ferulic acid, 48

Ferulic acid esterase, 31–5
Films, 260–6
Films
   drainage kinetics of, 218, 222
   glycoproteins hydroalcoholic solutions
      of, 217–9
   Plateau border of, 213
Flocculation
   of coacervated particles, 114
Fluorescence, 74
Foam
   Champagne, 212
   effect of global parameters on, 215–6
   expansion of, 212, 221
foamability of, 212, 221
   stability of, 140–1, 212, 221
   influence of proteins on, 213
Foaming, 140–1, 250
Fracture properties, 246–7
FTIR, 275–6

Gel formation, 241–52
Gel mechanism, 201
Gelation, 190–200
Genes, 19
Gene polymorphism, 24
Gliadins, 120
β-Glucan endohydrolase, 3–12
Gluten, 260–6, 275, 277–9, 284–90
Gluten
   glutenin macro polymer, 91
   GMP, 91
Glycinin, 13–23, 241–52
Glycoproteins
   aggregates of, 218–9, 222
   effect of interaction with ethanol on,
      213, 221
Golgi apparatus, 4–5
Guanidine hydrochloride, 150–1

Harmonic relaxation, 157
Helix, 201–11
Helices, *see* helix, 201–11
HPAEC–PAD, 7
Hydration, 275–83
Hydrogels, 120

Incompatibility, 197
Interaction(s)
   β-lactoglulin–acacia gum, 112

Interfacial rheology, 243–5
Ionic calcium, 186
Ionic speciation, 101
Iota-carrageenan, 201–11
Isoleucine, 63
Isoleucyl–tRNA synthetase, 65
Isothermal titration calorimetry, 74
ISR1, *see* shear rheometer

Kinking-units, 203

β-Lactoglobulin, 121, 156, 158
β-Lactoglobulin
    complexation with acacia gum, 111,
        120
Light scattering
    from coacervate dispersions, 113
D-Limonene, 183
Lipid transfer protein, 73–8
LMW–GS, 24–30
LTP, *see* lipid transfer protein

Maize coleoptiles, 6
Maltosyl bovine serum albumin, 215
Mechanical behaviour of starch films
    effects of water on, 239
    effects of glycerol on, 237
Mechanical properties, 255–8
Methylcellulose
    dynamic surface tension of, 166,
        169–70
    surface dilatational properties of, 166,
        170–2
    at the air/water interface, 169–72
Microscopy, 112
Microstructure, 112, 264–5
Milk proteins, 156
Mixed-linkage β-glucans, 3–12
Molecular modelling, 80

Nanoparticles, 120
Neurospora crassa 31
Neutron reflectivity, 149
Noncovalent interactions, 284
Nu-carrageenan, 201–11
Nucleation and growth, 114
Nucleotide sequence, 26

Oil-in-water emulsion, 181–9
Optical rotation, 205

Osmotic stress, 108

Particle
    particle network, 96
PAT1, *see* profile analysis tensiometer
Pea, 267–74
Pectin, 39, 55–62, 98–110
Pectin films, 108
Pectin gels, 99
Pectin methyl esterase, 56
Peptide/pectin
    interactions of, 104
Permeability behaviour of starch films
    effects of water on, 239
    effects of glycerol on, 238
Persistence length, 137
Phase separation, 111
Phospholipids, 74
Physical properties, 291
Pichia pastoris 32
Plant cell wall, 39–47
Plasticized starch, 253–9
Plasticizer, 267–74
Polarimetry, 206
Polyelectrolyte, 59, 127–44
Polyelectrolyte network, 106
Polysaccharide–protein
    complexes/interactions, 111
Porod law, 117
Precursor, 203
Profile analysis tensiometer,
        156–7
Protein, 13–23, 63–72, 145–52
Protein bodies, 96
Protein layer
    structure of, 153–65
Protein modification, 261
Proteinase K, 6

Recombinant protein, 32
Rhamnogalacturonan, 99
Rheology, 205

Seaweed, 201
Seed, 14
Sequence analysis, 81
Shear rheometer ISR1, 158
Small angle light scattering, 112
Soy protein, 241
Soy protein isolate, 190–200

Spinodal decomposition, 114
Starch, 226, 275–7, 291–300
Starch films, 235–40
Stereoselectivity, 65
Storage modulus, 208
Storage proteins, 14, 24
Structure/function residue identification, 83
Sugar beet, 181
Surface concentration, 146
Surface dilational modulus, 146
Surface layers
  rheology of, 134–6, 158–65
Surface moduli, 243
Surface pressure, 146
Surface shear measurements, 244
Surface tension, 129–33, 243, 247
Surfactant, 127–44
Swelling, 98, 123

Temperature
  effect on gluten rheology, 287–90

Tensile properties, 271–2
Thermodynamic model, 154–5
Thermo-mechanical analysis, 255
Thermoplastic starch, 267–74
Thickness, 291–300
Thin film balance, 136–40
Topographical analysis, 79
Transient relaxation, 157
Triton X-100, 6

Viscoelasticity, 285–8

Water/air interface, 153
Wheat, 24–30
Whey proteins, 111, 119

Xyloglucan, 40, 98
X-ray photoelectron spectroscopy, 262
XPS, *see* X-ray photoelectron spectroscopy

Yeast glycoproteins, 214